Preface

Gas chromatography (GC) has been one of the most used and successful chromatographic techniques since its inception over 40 years ago and is employed by almost every type of chemical and related industry in research and development, quality control and specifications, environmental monitoring—personnel, and process chemistry, biochemical and biological research establishments, clinical and toxicological departments of hospitals and related institutions, analytical and consultant/public analyst laboratories, water companies and regulatory bodies (National Rivers Authority), and higher education teaching/research institutions at all levels.

The instrumental technique has reached a high degree of sophistication when used alone or in combination with advanced detectors, such as mass spectrometers. The applications of GC embrace many fields and although it is not possible to cover every aspect of the techniques and utilization in one text book, it is hoped that this practical approach gives a wide enough range of methods, techniques and applications within to be informative to the new student or analyst in the majority of areas.

Topics include a brief theoretical introduction, instrumentation-operation and experimental considerations, capillary column development, technology and performance, selected applications of low and high performance column modes, chemical derivatization methods for analysis by GC, applications in analytical toxicology and clinical chemistry, chiral separations by GC, environmental monitoring and analysis, the role of GC in petroleum exploration, and GC involvement in combined techniques for biological and environmental work.

The editor is aware that there is an overlap between chapters in the techniques employed, such as clean-up methods and sample preparation generally in environmental, biological assay, biochemical, clinical and toxicological fields, and the interested analyst should be able to make use of methods described by cross-reference and comparison with areas not of direct interest.

There is also an inevitable overlap in the consideration and use of detectors, columns, and stationary phases. Chapter 2 deals with instrumentation in general. However, other areas of the book, such as Chapter 6 by Flanagan, include the use of particular detectors for toxicological work. Similarly, Dawes in Chapter 3 highlights capillary columns, stationary phases, their technology and performance. Various aspects of the use of columns and stationary phases, etc. are also included in other chapters as appropriate. The techniques for monitoring of pesticides are concentrated on in Chapter 9 by Best and Dawson but are also considered in some detail in analytical toxicology by Flanagan in Chapter 6.

Preface

Derivatization methods are discussed exhaustively in Chapter 5 and should be treated as a guide to the appropriate method for modification of the target analyte rather than from the standpoint of the profiling of analytes and conditions for the chromatography.

Chakraborty in Chapter 7 describes the techniques of GC and GC–MS applied to a range of analytical problems encountered in clinical chemistry concerned with body fluids and other biological materials. Here, the use of derivatization focuses on specific important biological targets.

Taylor in Chapter 8 provides a full account of recent advances in the development/preparation of chiral phases and separations in GC with a discussion of the background principles of chirality.

Chapters 10 and 11 by Harriman and Evershed, respectively (the latter in particular), provide an insight into the role of GC in combined instrumental techniques, viz. GC–MS, which demonstrates clearly the need for the selectivity, accurate mass and scan capabilities of MS in the profiling and analysis of targets in petroleum exploration, biological and environmental areas.

The original intention with Chapter 11 was to consider the role of GC in several combined instrumental techniques, notably, GC–MS and GC–FTIR. Although the former receives attention in this chapter, because of author difficulties it was not possible to include the latter. At an extremely late stage it has been possible to include an appendix on GC–FTIR because Peter Jackson from Zeneca Specialties (with some minor assistance from the editor) has been able to step into the breach.

The chapters vary in the extent to which protocols are used and are, in fact, appropriate, but the theme of the practical approach series is largely adhered to and should be of benefit to first-time users of GC in specialized fields of application.

Finally, I wish to thank the authors for their patience in conforming as far as possible to the format of the series and for their participation and cooperation which has made my position as editor both interesting and challenging.

Salford   P.J.B.
July 1993

Contents

Contents
Contents

3. Development, technology, and utilization of capillary columns for gas chromatography   71

Peter A. Dawes

Contents

6. Gas chromatography in analytical toxicology: principles and practice   171

Robert J. Flanagan

Contents

Contents

10. The role of gas chromatography in petroleum exploration 331

Gareth E. Harriman

Contents

11. Combined gas chromatography–mass spectrometry   359
Richard P. Evershed

Appendices
A1 Combined gas chromatography–Fourier transform infrared spectroscopy   393
Peter Jackson

Contents

Contributors

KEITH D. BARTLE
School of Chemistry, University of Leeds, Leeds LS2 9JT, UK.

PETER J. BAUGH
Department of Chemistry and Applied Chemistry, University of Salford, The Crescent, Salford M5 4WT, UK.

GERRY A. BEST
Clyde River Purification Board, Rivers House, Murray Road, East Kilbride, Glasgow G75 0LA, Scotland, UK.

JAGADISH CHAKRABORTY
Department of Biological Sciences, University of Surrey, Guildford GU2 5XH, UK.

PETER A. DAWES
SGE (Australia) Pty Ltd., PO Box 437, Ringwood, Australia 3134.

J. PAUL DAWSON
Clyde River Purification Board, Rivers House, Murray Road, East Kilbride, Glasgow G75 0LA, Scotland, UK.

RICHARD P. EVERSHED
School of Chemistry, University of Bristol, Cantock's Close, Bristol BS8 1TS, UK.

ROBERT J. FLANAGAN
Poisons Unit, Avonley Road, New Cross, London SE14 5ER, UK.

GARETH E. HARRIMAN
GH Geochemical Services, 24 Higher Bebington Road, Bebington, Wirral L63 2PP, UK.

PETER JACKSON
Zeneca Specialties, PO Box 42, Hexagon House, Blackley, Manchester M9 3DA, UK.

DAVID R. TAYLOR
Department of Chemistry, University of Manchester Institute of Science and Technology, PO Box 88, Manchester M60 1QD, UK.

ANDREW TIPLER
Perkin Elmer plc, Post Office Lane, Beaconsfield HP9 1QA, Buckinghamshire, UK.

Contributors

DAVID G. WATSON
Department of Pharmaceutical Sciences, University of Strathclyde, Royal College, 204 George Street, Glasgow G1 1XW, Scotland, UK.

Abbreviations

AES	atomic emission spectroscopy
AFID	alkali flame ionization detector
AQC	analytical quality control
AUC	area under curve
BCF	bromochlorofluoromethane
BOD	biological oxygen demand
BOC	BOC-(S)-valine-a-phenylethylamide
BSA	N,O-bis(trimethylsilyl)acetamide
BSTFA	N,O-bis(trimethylsilyl)trifluoroacetamide
CD	cyclodextrin
CE	coating efficiency
CF	chloroformate
CFC	chlorofluorocarbon
CHC	chlorohydrocarbons
CHDMS	cyclohexanedimethanolsuccinate
CNS	central nervous system
COC	cold on-column
COD	chemical oxygen demand
CPI	carbon preference index
CSF	cerebrospinal fluid
CSM	cerebrospinal medium
CSP	chiral stationary phase
CV	coefficient of variation
CTS	computerized tomographic scan
CZE	capillary zone electrophoresis
DCM	dichloromethane
DCBP	decachlorobiphenyl
DCE	dichloroethylene
DCFTA	dichlorotetrafluoroacetone
pp-DDD	2,2-bis-(4-chlorophenyl)-1,1-dichloroethane
pp-DDE	2,2-bis(4-chlorophenyl)-1,1-dichloroethane
pp-DDT	2,2-bis-(4-chlorophenyl)-1,1,1-trichloroethane dichlorophenyl trichloroethane
DMDCS	dimethyldichlorosilane
DME	dimethyl ether
DMES	dimethylethylsilyl ether
DMSO	dimethylsulphoxide
'DRINS	aldrin, dieldrin, and endrin
DST	drill stem test

Abbreviations

ECD	electron capture detector/detection
EDTA	ethylenediaminetetraacetic acid
EI	electron impact/ionization
ETOAC	ethyl acetate
FAME	fatty acid methyl ester
FFA	free fatty acid
FID	flame ionization detector
FPD	flame photometric detector
f.s.d.	full scale deflection
FSOT	fused silica open tubular
FTIR	Fourier transform infrared spectroscopy
GABA	γ-aminobutyric acid
GC–MS	gas chromatography–mass spectrometry
GCOPS	gas chromatography method optimization software
GDG	glass distilled grade
GLC	gas–liquid chromatography
GLT	glass-lined tubing
GSC	gas–solid chromatography
HCB	hexachlorobenzene
HCBD	hexachlorobutadiene
HCH	hexachlorocyclohexane
HFBzA	heptafluorobenzoyl anhydride
HFBzCl	heptafluorobenzoyl chloride
HFBCl	heptafluorobutyryl chloride
HFBPACl	L(−)-N-heptafluorobutyrylphenylalanyl chloride
HFA	hexafluoroacetone
HFIP	hexafluoroisopropyl
HMDS	hexamethyldisilazane
HMIP	Her Majesty's Inspectorate of Pollution
HOC	hot on-column
HRMS	high resolution mass spectrometry
HT	5-hydroxytryptamine
HTGC	high temperature gas chromatography
HVOs	highly volatile organics
i.d.	internal diameter
IS	internal standard
LPG	liquefied petroleum gas
MBOT/C	medium bore open tubular/column
MDGC	multidimensional gas chromatography
MEOH	methanol
MNNG	1-methyl-3-nitro-1-nitrosoguanidine
MS–MS	Mass spectrometry–mass spectrometry
MSTFA	N-Trimethylsilyl-N-methyltrifluoroacetamide
MTBE	methyl t-butyl ether

Abbreviations

MTBSTFA	*N*-methyl-*N*-*t*-butyldimethylsilyltrifluoroacetamide
NBOT/C	narrow bore open tubular/column
NCI	negative ion chemical ionization
NIST	National Institute of Standards and Technology
NPD	nitrogen-phosphorus detector
OC	organochlorine
OCN	octachloronaphthalene
ON	organonitrogen
OP	organophosphorus
PAD	polychloroamino diphenylether
PAH	polyaromatic hydrocarbon
PCBs	polychlorobiphenyls
PCI	positive ion chemical ionization
PCP	pentachlorophenol
PCSD	polychlorosulphonamido diphenylether
PEG	polyethylene glycol
PER	perchloroethylene
PFBA	pentafluorobenzyl aldehyde
PFBBr	pentafluorobenzylbromide
PFBCF	pentafluorobenzyl chloroformate
PFBO.HCl	pentafluorobenzylhydroxylamine hydrochloride
PFP	pentafluoropropionoyl
PFPA	pentafluoropropionic anhydride
PHPA	p-hydroxyphenylacetic acid
PID	photoionization detector
PLOT	porous layer open tubular
PPBCl	pentafluorobenzoylchloride
PVC	polyvinyl chloride
QA	quality assurance
QC	quality control
RD	relative density
RPB	Scottish River Purification Board
RSD	relative standard deviation
SCOT	support coated open tubular
SFC	supercritical fluid chromatography
SIM/R	selected ion monitoring/recording
SPE/SE	solid phase extraction/sorbent extraction
TBAS	tetrabutylammonium sulphate
TBDMS	*t*-butyldimethylsilyl
TBDMSCl	*t*-butyldimethylsilyl chloride
TCB	trichlorobenzene
TCD	thermal conductivity detector
TDM	therapeutic drug monitoring
TFA	*N*-trifluoroacetyl

Abbreviations

TFE	trifluoroethanol
TFPA	trifluoropropionic anhydride
THF	tetrahydrofuran
TMAH	tetramethylammonium hydroxide
TMCS	trimethylchlorosilane
TMPAH	trimethylanilinium hydroxide
TMS	trimethylsilyl
TMSIM	trimethylsilylsimidazole
TPCL	N-trifluoroacetyl-L-proline chloride
TRI	trichloroethylene
TZ	trennzahl
VOA	volatile organic analysis
VOCs	volatile organic chemicals
VSA	volatile substance abuse
WBOT/C	wide bore open tubular/column
WCOT	wall-coated open tubular

1

Introduction to the theory of chromatographic separations with reference to gas chromatography

KEITH D. BARTLE

1. Introduction and history of GC

Gas chromatography (GC) is pre-eminent among analytical separation methods. It offers rapid and very high resolution separations of a very wide range of compounds, with the only restriction that analytes should have sufficient volatility. It originated as a result of experiments carried out by A. T. James and A. J. P. Martin at the National Institute for Medical Research in London; the results were published in 1952. They demonstrated the separation of carboxylic acids C_1–C_{12} by continuous partition between a liquid film on an inert support (the 'stationary' phase) and a gas (the 'mobile' phase) moving through the column of packing. These seminal results built on earlier work carried out (in Leeds), in which chromatographic separation of amino acids between two liquid phases had been demonstrated. The separation of plant pigments by simple column chromatography on columns of adsorptive solids had been known since 1906, but the technique was not commonly applied until the 1930s.

The GC instrument constructed by James and Martin 50 years ago contained most of the features of a modern gas chromatograph (*Figure 1*): a means of controlling the flow of mobile-phase carrier gas, stabilization of the temperature of the column, and a sensitive detector to determine and record the concentrations of separated constituents at the end of the column. These pioneers also introduced the concept of separation efficiency, and discussed the influence of parameters such as gas flow rate and diffusion of the sample in the mobile phase.

The influence of the uniformity of size of column-packing particles was noted by James and Martin, and led Golay in 1957 to propose the capillary column in a remarkable piece of inductive reasoning: the differing paths taken by different solute molecules as they passed through an (inevitably) non-uniform packing could be replaced by a single channel—a continuous open tube in which the stationary liquid was coated on the inner wall.

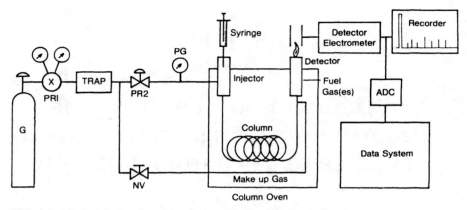

Figure 1. Modern instrumentation for gas liquid chromatography.

Capillary columns are also much more permeable than packed columns and hence can be much longer, with greater efficiency, and therefore better resolution of separated mixture components for the same stationary phase can be achieved (*Figure 2*). For many years, the full potential of capillary columns was not realized because of difficulties with the unfavourable durability of column materials and difficulties in coating with a uniform film of stationary phase. Metal columns, particularly stainless steel, were difficult to coat, while glass columns were fragile and contained catalytically active sites.

These problems were solved by the invention of the fused-silica open tubular column (FSOT) by Dandeneau and Zerenner in 1980. Such columns, manufactured by a process originally based on fibre optic technology, are both highly flexible and chemically inert. Extensive research was carried out to determine the physical chemistry principles underlying coating, and to develop means of cross-linking the stationary phase (by a free-radical mechanism) to immobilize the film. The concept of 'designer' stationary phases, synthesized specifically to produce a desired separation, also emerged in the 1980s; bonding of different groups to a polysiloxane chain has proven especially fruitful. The growth of capillary column GC has hence been rapid and now dominates over the use of packed columns. None the less, certain specific separations are still best carried out on packed columns, for example the analysis of mixtures of permanent gases.

The earliest universal detector employed in GC was the katharometer, or thermal conductivity detector, but this was quickly replaced by the much more sensitive flame-ionization detector invented in 1958. Since then other flame-based selective detectors for nitrogen and phosphorus compounds (thermionic) and sulphur compounds (flame photometric) have emerged. Electron capture detectors, specific for halogen-containing compounds, are the descendants of the early argon ionization detector. The most dramatic

2

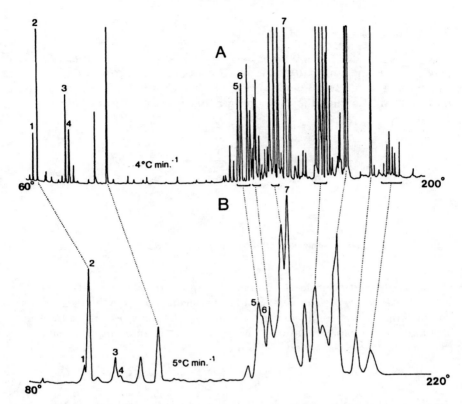

Figure 2. GC chromatograms of Calmus oil on (A) a 50 m long capillary column and (B) a 4 m long packed column.

developments in GC detection have resulted, however, from the coupling of GC with spectroscopic detection. If atomic emission spectroscopy (AES) is employed, the presence of specific elements can be verified, and several element-selective chromatograms can be recorded simultaneously. Infrared (IR) and mass spectrometers (MS) may also be linked to the end of the column (GC–IR and GC–MS) to allow identification of separated mixture constituents from the respective spectra. The recent availability of bench-top mass spectrometers has made GC–MS a routine analytical procedure.

GC has greatly profited from developments in electronics and computer technology. With autosamplers and data acquisition systems, round-the-clock operation of equipment is possible, so that very many samples per day may be analysed. GC is at the forefront of modern analysis in fields as diverse as environment, fossil fuels, foods and cosmetics, organic geochemistry, certain biological materials, and forensic science. Background to the theory and practice of GC can be found in refs 1–5.

2. The GC chromatogram

A graph of detector response versus volume of gas, or more usually, time (*Figure 3*), is the chromatogram, and it contains peaks corresponding to the elution of the solute after passage through the column. Some of the parameters characterizing the chromatographic process are also illustrated in *Figure 3*.

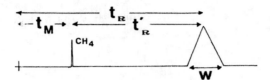

Figure 3. The GC chromatogram with retention parameters.

The solute partitions itself between the stationary phase and the gas phase with a *partition coefficient, K*, the ratio of concentrations in the two phases. K depends on the intermolecular forces and the vapour pressure of the solute. The partition or capacity ratio, k', is the ratio of the amounts in the stationary (S) and gas (G) phases, and is given by

$$k' = K \frac{V_S}{V_G} \tag{1}$$

where V_S and V_G are the volumes of stationary and mobile phases, respectively. The *phase ratio*, β is given by

$$\beta = \frac{V_G}{V_S} \tag{2}$$

so that

$$k' = \frac{K}{\beta}. \tag{3}$$

The fraction of the total time the solute spends in the stationary phase is

$$\frac{k'}{1 + k'}$$

and the fraction spent in the gas phase is

$$\frac{1}{1 + k'}.$$

If the average linear velocity of gas through the column is $\bar{u}$, the average velocity, $\bar{v}$, of the solute is given by

$$\bar{v} = \frac{\bar{u}}{1 + k'}$$

and the time it spends in the column, the *retention time*, t_R is given by

$$t_R = \frac{L}{\bar{u}} (1 + k') \tag{4}$$

where L is the column length. The *retention volume*, V_R, is the volume of gas required to elute the solute. If there is no chromatographic retention (i.e. only retention equal to the volume of gas in the column), t_R becomes t_M, the retention time of an unretained peak:

$$t_M = \frac{L}{\bar{u}}. \tag{5}$$

The part of the retention time during which the solute is held back by the stationary phase is $t_R' = (t_R - t_M)$, and from Equations (4) and (5):

$$k' = \frac{t_R'}{t_M}. \tag{6}$$

The concentration of solute as it emerges from the column corresponds to a Gaussian peak which can be characterized by the width at the base (w_b) or the width at the half-height ($w_{1/2}$).

3. Resolution in GC

How well a chromatographic separation of two components, 1 and 2, has succeeded is measured by the *resolution*, R, the ratio of peak separation to the average base width of the peaks (*Figure 4*):

$$R = \frac{t_{R,2} - t_{R,1}}{0.5(w_{b,1} + w_{b,2})}. \tag{7}$$

Clearly, the peak width, and hence the resolution, is related to the broadening of the solute band as it passes through the column. This is measured by the term *number of theoretical plates*, N, by analogy with distillation theory—the number of equilibrations or partitions of solute between gas and liquid in the column. N is given by

$$N = 16\left(\frac{t_R}{w_b}\right)^2 = 5.54 \left(\frac{t_R}{w_{1/2}}\right)^2. \tag{8}$$

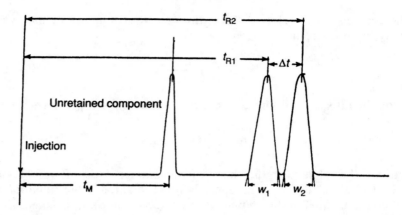

Figure 4. Resolution in GC.

Since N depends on column length, we introduce H, the *height (or length) equivalent to one theoretical plate*,

$$H = \frac{L}{N}. \tag{9}$$

Resolution, R, depends on the *relative retention*, α for the two solutes

$$\alpha = \frac{k_2'}{k_1'}\left[= \frac{t_{R,2}'}{t_{R,1}'} = \frac{K_2}{K_1}\right]. \tag{10}$$

α can be changed by varying the liquid phase. R is related to N, and the average k' for the peaks in question by

$$R = \frac{\sqrt{N}}{4}\cdot\frac{(\alpha-1)}{\alpha}\cdot\frac{k'}{(1+k')} = \frac{1}{4}\sqrt{\frac{L}{H}}\cdot\frac{(\alpha-1)}{\alpha}\cdot\frac{k'}{(1+k')}. \tag{11}$$

Thus resolution depends on the square root of column length; doubling L only increases resolution by $\sqrt{2} = 1.414$. The term involving relative retention is such that more plates are required to achieve the same resolution for early peaks than for later eluting peaks.

In practical terms, resolution is effectively complete for $R \geq 1.5$ (*Figure 4*).

4. Band broadening in GC

The flow of gas through the column and the diffusion of the solute in both gas and liquid all influence the solute band width and, hence, efficiency and resolution. Early in the history of GC it was shown (by van Deemter and co-workers for packed columns, and by Golay for capillary columns) that the height equivalent to a theoretical plate depends on the average gas velocity according to

$$H = A + \frac{B}{\bar{u}} + C_G \cdot \bar{u} + C_S \cdot \bar{u}. \tag{12}$$

The effect of the multiple gas pathways through a packed column bed is given by

$$A = 2\lambda d_p$$

where A is the term, λ is the packing uniformity, and d_p is the particle diameter. For a capillary column $A = 0$, because of the absence of a solid support.

Longitudinal diffusion in the gas phase gives rise to the B term, which is proportional to D_G, the diffusion coefficient of the solute in the gas. The C terms express the effect of resistance to mass transfer in the gas (C_G) and liquid (C_S) phases.

$$C_G \propto \frac{r^2}{D_G} \text{ (capillary column radius } r) \text{ or } \frac{d_p^2}{D_G} \text{(packed column)}.$$

For rapid diffusion and small particles, or column radius, transfer through the gas phase will be rapid enough to reduce band broadening. Slow diffusion in the liquid spreads out the band by leaving solute behind, and

$$C_S \propto \frac{d_f^2}{D_S}$$

where d_f is the liquid film thickness and D_S is the diffusion coefficient for the solute in the liquid.

The relative magnitudes of the different terms in Equation 12 for GC are shown in *Figure 5*. At low $\bar{u}$, the B term is large, but quickly diminishes with increasing $\bar{u}$; and C_G and, to a lesser extent, C_S, then dominate. The overall curve is a hyperbola—the so-called *van Deemter curve*. The smallest value of H is H_{min}, at which $\bar{u}$ is optimum, $\bar{u}_{opt}$. The greater $\bar{u}_{opt}$, the faster a sample can be analysed (see Section 6) and, in general, $\bar{u}_{opt}$ will be higher for a low density gas, such as helium, but H_{min} will be a little more favourable for a denser gas such as nitrogen (*Figure 6*). The helium curve is also flatter so that higher gas velocities may be used without significant loss of efficiency.

5. GC columns

Table 1 compares the properties of packed and capillary columns for GC. Capillary columns are much more permeable to gas flow than packed columns, and hence can be longer without using extremes of gas pressure. The efficiency per unit length is also generally greater because the film

7

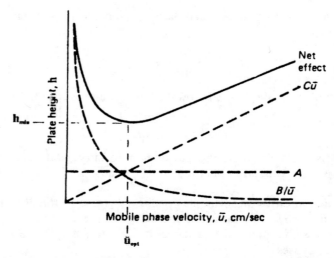

Figure 5. Relative magnitudes of the different terms in the van Deemter equation in GC.

thickness is less and the C_S term in Equation (12) is reduced. Both these effects make the separation power of capillary columns much greater than that of packed columns.

Other advantages of capillary columns over packed columns include an increase in signal-to-noise ratio in the chromatogram because the amount of column bleed from decomposing stationary phase is smaller, and also because carrier gas flow is more uniform. Moreover, peaks are sharper and less diffuse so that the detection of trace components is much easier. There is also a substantial increase in speed of analysis as measured by retention time for a given resolution and by resolution in a given time.

6. Speed of GC analysis

There are considerable advantages for the analyst in reducing the time required for a given separation. This is the time, t_p, for the solute to pass through one theoretical plate multiplied by the number of plates for the required resolution.

$$t_R = N t_p$$

since

$$t_p = \frac{H}{\dfrac{\bar{u}}{(1 + k')}}$$

8

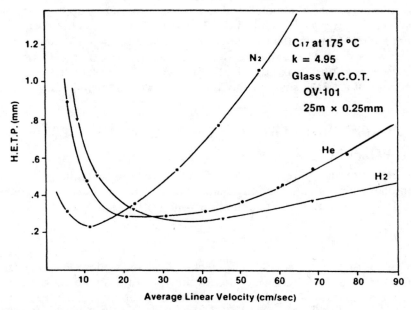

Figure 6. Van Deemter plots for different carrier gases in capillary GC.

Table 1. Comparison of capillary and packed columns

Parameter	Capillary	Packed
Length/m	5–100	1–5
Internal diameter/mm	0.1–0.8	2–4
Flow-rate/cm^3 min^{-1}	0.5–10	10–100
Pressure drop across column/kg cm^{-2}	0.1–4	1–4
Efficiency per unit length/plates m^{-1}	Up to 5000	2000–3000
Typical total efficiency/plates	150 000 (50 m column)	5000 (2 m column)
Capacity per peak/μg	<0.05	10
Liquid film thickness/μm	0.1–2	1–10

i.e. the plate height divided by the solute band velocity

$$t_R = N(1 + k') \frac{H}{\bar{u}}. \tag{13}$$

If N is eliminated between Equations (11) and (13) we have

$$t_R = 16R^2 \left(\frac{\alpha}{\alpha - 1}\right)^2 \cdot \frac{(1 + k')^3}{(k')^2} \cdot \frac{H}{\bar{u}}. \tag{14}$$

9

The term in k' has a minimum value at $k' = 2$, so that by appropriate choice of column variables it should be adjusted to this value. Otherwise, it may seem at first sight that the speed of analysis is simply inversely proportional to $\bar{u}$. However, increasing the gas velocity also increases H (see *Figure 5*) and thus the relationship is less simple. In fact $H/\bar{u}$ is obtained from the van Deemter plot as the slope of the line at values of $\bar{u}$ beyond $\bar{u}_{opt}$.

7. Retention in GC

7.1 Effect of temperature

Temperature markedly influences GC retention because it changes the partition coefficient. The retention volume per unit mass of stationary phase at 0°C is known as the specific retention volume, V_g

$$V_g = \frac{V_R}{W_S} = \frac{273R}{\gamma p M} \qquad (15)$$

where γ is the activity coefficient (a factor which multiplies the actual concentration to give the 'effective' concentration), p is the solute vapour pressure and W_S is the mass of stationary liquid with molecular weight, M. p is, in fact, a function of temperature, as described by the Clausius–Clapeyron equation

$$\ln p = \frac{-\Delta H_v}{RT} + \text{const.} \qquad (16)$$

where ΔH_v is the heat of vaporization per mole of solute and T the column temperature.

Combining Equations (15) and (16) yields

$$\ln V_g = \frac{\Delta H_v}{RT} + \ln \frac{273R}{\gamma M} + \text{const.}$$

so that, if interactions between solute and liquid do not depend on temperature, and hence γ is constant, then

$$\ln V_g = \frac{\Delta H_v}{RT} + \text{const.} \qquad (17)$$

and graphs of both $\ln V_g$, and $\ln t'_R$ (since V_g is proportional to t'_R) against reciprocal temperature are straight lines; retention decreases with column temperature.

7.2 Temperature programming

The dependence of GC retention on vapour pressure means that mixtures containing components with a wide range of boiling points cannot be separated satisfactorily in an isothermal run. The more volatile components may be well enough resolved, but the higher boiling materials will only be eluted

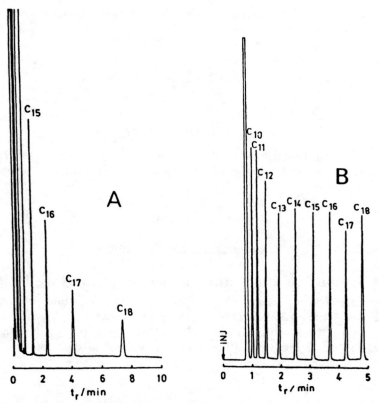

Figure 7. Comparison of (A) isothermal and (B) temperature-programmed GC separation of alkanes. (Reproduced with kind permission from Dr Alfred Huethig Publishers.)

with long retention times and very broad peaks. If the column temperature is high enough to give satisfactory peaks for the less volatile compounds, the low-boiling constituents will be less well-resolved (*Figure 7A*).

The solution is to raise the column temperature during a chromatographic run, so that for a homologous series peaks emerge at regular intervals (*Figure 7B*). The effect of this procedure on retention may be predicted by writing an expression similar to Equation (16) but in terms of the partition coefficient, K.

$$\ln K = \ln \frac{RT}{V_L} + \frac{\Delta H_v}{RT} + \text{const.}$$

where V_L is the molar volume of the stationary liquid. Since $k' = K/\beta$, then, to a good approximation,

$$\ln k' = \frac{\Delta H_v}{RT} + \text{const.} \tag{18}$$

11

We may use Equation (18) to determine the average temperature rise from T_1 to T_2 ($= \Delta T$), required to halve k'.

The k' ratio is

$$\ln 2 = \frac{\Delta H_v}{RT} \cdot \frac{\Delta T}{T}$$

whence

$$\Delta T = 0.693 \, RT^2/\Delta H_v$$

where T is the geometric mean of T_1 and T_2. If the operating temperature is nearer the solute boiling point, T_b, and, using the Trouton rule which states that

$$\frac{\Delta H_v}{T_b} = \text{const.} \tag{19}$$

the required ΔT is 20 °C at 70 °C and 22 °C at 120 °C.

7.3 Dependence of retention on solute properties in GC

Retention time may be correlated with a number of solute properties by making use of the relationships discussed in section 7.1. A graph of $\ln t'_R$ against $\ln p$ is linear with a gradient of -1. Moreover, plots of boiling point, T_b, against the number of carbon atoms contained in the molecule, n, are linear for a wide range of homologous series, i.e.

$$T_b = a + bn \tag{20}$$

and the Trouton rule (Equation 19) also holds for such series, so that

$$\Delta H_v = c + dn \tag{21}$$

where a, b, c, and d are constants. Now $\ln p$ is linearly related to n from Equations (16) and (21) and to T_b from Equations (16) and (19). Consequently, linear relations between $\ln t'_R$ and n, and $\ln t'_R$ and T_b are expected. *Figure 8* illustrates such a plot for n-alkanes.

These relationships hold only if the activity coefficients, γ, are constant. This is so if molecular interactions between solute and stationary liquid are the same, as might be expected for the same compound types. Values of γ differ between homologous series, so that different linear plots between $\ln t'_R$ and, e.g. carbon number, are observed. It is noteworthy that the origin of selective separations in GC lies in the same variations in γ between compounds which contain different functional groups.

7.4 Retention indices

The above relations lead to a scheme for compound identification from retention data. The value of $\log t'_R$ relative to $\log t'_R$ for the series of n-alkanes

12

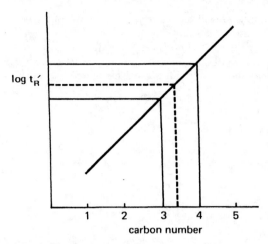

Figure 8. Basis of the Kóvats' retention index system, a graph of log t'_R versus carbon number. The index is obtained by interpolation (dotted line).

is used, since relative values are easier to measure than absolute values. For isothermal chromatography, the Kóvats' retention index I for a given compound is a number indicating its retention relative to the n-alkanes which are assigned retention indices equal to 100 times the carbon number (i.e. 800 for n-octane, etc.). Linear interpolation on log t'_R scale between values for adjacent n-alkanes gives I. Thus for a compound with retention time $t'_R(x)$ eluted between n-octane and n-nonane, retention times $t'_R(8)$ and $t'_R(9)$, will have retention index $I(x)$ given by

$$I(x) = 800 + 100 \frac{\log t'_R(x) - \log t'_R(8)}{\log t'_R(9) - \log t'_R(8)}. \tag{22}$$

A similar expression applies to the case of temperature-programmed GC, except that log t'_R is replaced by the retention temperature, T_R, the temperature at which the compound is eluted. The temperature-programmed retention index, $I_p(x)$ for a compound eluted between two n-alkanes with carbon numbers z and $z + 1$ is given by

$$I_p(x) = 100z + \frac{T_R(x) - T_R(z)}{T_R(z + 1) - T_R(z)}. \tag{23}$$

References

1. Ettre, L. S. (1971). The development of chromatography. *Anal. Chem.*, **43**, 20A–31A.
2. Ettre, L. S. and Zlatkis, A. (ed.) (1979). *75 years of chromatography*. In *J. Chromatogr. Library*, Vol. 17. Elsevier, Amsterdam.

13

3. Purnell, H. (1962). *Gas chromatography*. John Wiley, New York.
4. Lee, M. L., Yang, F. J., and Bartle, K. D. (1984). *Open tubular column gas chromatography: theory and practice*. Wiley-Interscience, New York.
5. Jönsson, J. A. (ed.) (1987). *Chromatographic theory and basic principles. Chromatographic science series*, Vol. 38. Marcel Dekker, New York.

Gas chromatography instrumentation, operation, and experimental considerations

ANDREW TIPLER

1. Introduction

Of all the analytical techniques, gas chromatography (GC) offers the largest choice of instrumental components. While this feature contributes to the enormous power and flexibility of the technique, it does present the analyst with the often daunting task of selecting and operating the optimum hardware configuration to suit a particular analysis. A good understanding of the choice of components and their operation is a prerequisite for a successful chromatographic analysis.

This chapter is divided into three sections. Section 2 discusses the functional components of a modern gas chromatograph and reviews some of the more popular designs. Section 3, which forms the bulk of the chapter, considers the practical aspects of GC. Finally, Section 4 describes the principles involved in the processing of chromatographic signals to generate meaningful results.

2. Instrumental components and function

This section discusses the functional components of a gas chromatograph and reviews the options available and their operating principles.

2.1 Fundamental components

A gas chromatograph (see *Figure 1*) is essentially a device which enables a small amount of sample to be introduced into an inlet system where it is vaporized and passed into a chromatographic column. To provide suitable conditions for chromatography, the column is held within an oven and a flow of inert carrier gas passes through it. A detector is fitted at the column exit to monitor the separated components as these elute from the column. The detector provides an electrical signal which is amplified and fed to a

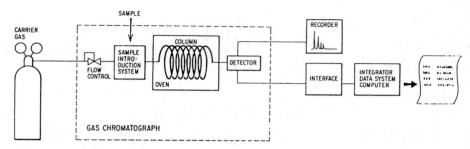

Figure 1. Basic components of a GC system.

recording or data-processing device from which meaningful results can be obtained.

The performance of the chromatograph, and hence the quality of the results generated, depends not only on the design of the components but also on how carefully they are controlled—particularly with respect to temperatures and gas flow rates. Most modern chromatographs, therefore, use a microprocessor at the heart of the control system. The use of such technology has the added benefits of an improved user interface, better programmability, method storage, external control, and intelligent system diagnostics.

2.2 The chromatographic oven

The partition of solutes between the carrier gas and the stationary phase is highly dependent on the temperature of the chromatographic system. While this is a desirable attribute in that it allows conditions to be tailored to suit a particular analysis, it also places extreme demands on the temperature controlling system if a repeatable separation is to be achieved. The chromatographic column must be heated uniformly and should match the set temperature at all times. These requirements apply, not only under isothermal conditions, but during temperature programmes where the rate may be in excess of 30°C/min. As a result, the most effective way of heating columns is to use a temperature-controlled air bath oven.

Chromatographic ovens are essentially boxes, containing an electric heating element, in which the chromatographic column is mounted. The heat from the element is distributed throughout the oven by means of a powerful circulatory fan. Good air circulation is necessary to maintain a uniform temperature along the entire length of the column. A temperature sensor is located at a carefully selected position in the oven. The output from the sensor is fed into the oven temperature control system where it is compared with the expected temperature and adjustments are made automatically (normally using a microprocessor) to the heating element temperature as appropriate.

Table 1. Cryogenic media for subambient oven control

Medium	Cooling principle	Minimum temperature	Comments
Liquid nitrogen	Latent heat of vaporization	$-160\,°C$	Needs a Dewar vessel pressurized with helium
Liquid carbon dioxide	Joule–Thompson effect	$-77\,°C$	Faster at cooling than liquid N_2

The problems with such a 'closed box' system are that it is slow to cool down to the initial set temperature after a temperature programme, and control is poor at near-ambient temperatures. To address these limitations, many manufacturers use a proportional cooling mechanism or 'smart door' which draws ambient air into the oven under these conditions.

For some applications, in particular gas analyses, subambient temperatures are required. In such instances, liquid nitrogen or liquid carbon dioxide is normally introduced into the oven. *Table 1* lists some of the attributes of these cryogenic media. Although the cooling principle is different for the two media, the means of introduction is essentially the same—the pressurized medium is fed through a rugged solenoid value which is switched on and off (again under microprocessor control) to allow pulses of the cooling medium to enter the oven. By adjusting the period and/or frequency of these pulses, the degree of cooling can be controlled.

Ovens are available in a range of sizes. Large ovens provide better access and are able to accept more injectors, detectors, etc. They do, however, have a greater thermal capacity and may be slower to heat or cool, and a uniform temperature may be more difficult to maintain.

2.3 Pneumatics

A gas chromatograph uses gases for various functions. The most important of these is the carrier gas which acts as the chromatographic mobile phase. Many detectors require support gases in order to function. A flow of 'make-up' gas is sometimes required to reduce the effects of 'dead' volumes in some system components. This section considers the supply and application of the various gases used in GC.

2.3.1 Gas supplies

Table 2 lists the gases normally used in GC. Gases are normally supplied from pressurized cylinders, although commercial devices are available to provide air or nitrogen filtered from the atmosphere and hydrogen by chemical or electrolytic means. It is extremely important that all gases are as clean as possible as any contamination will interfere with the chromatography or cause

Table 2. Gases used in gas chromatography

Gas	Application	Comments
Helium	General carrier or make-up gas	Excellent but can be expensive Cannot be used with ECD except with another make-up gas
Nitrogen	General make-up gas or packed column carrier gas	Cheap Not good for capillary carrier gas as gives long run times
Hydrogen	Carrier gas for capillary columns Combustion gas for FID, NPD and FPD	Cheap Highly explosive—needs oven trip sensor and careful venting Best carrier gas for capillary
Air	Combustion gas for FID, NPD and FPD Pneumatically driven hardware mechanisms	Cheap and readily available
Oxygen	Combustion gas on some FPDs	Not normally required
Argon	Carrier gas for TCD	Determination of helium
Argon/methane	Make-up and packed column carrier gas for ECD	Better linearity and selectivity than nitrogen but poorer detection limits
Helium/hydrogen	Carrier gas for TCD	Determination of hydrogen

noisy detector signals. This is especially true for carrier gases where high purity (> 99.999%) cylinder gases should be obtained.

Gas supplies should be connected to chromatographs using solvent-rinsed and oven-baked copper tubing. Polymer tubing should be avoided as oxygen from the atmosphere is able to permeate the tubing walls and enter the gas stream within. Oxygen will degrade most column stationary phases at elevated operating temperatures.

As an additional precaution, it is usual to fit in-line filters to the supply lines just before they enter the chromatograph as shown in *Figures 2a* and *2b*. Some chromatographs have filters fitted internally—in such cases, external filters should still be used as a precaution.

2.3.2 Pneumatic components

There are several types of pneumatic component used in gas chromatographs:

i. Pressure regulators

These devices maintain an adjustable constant pressure, downstream of it, over a range of flow rates (typically 1 to 500 ml/min). Pressure regulators are available in a variety of operating ranges (e.g. 0 to 30, 60, or 100 lb/in^2). Lower range devices give finer regulation, so choose the lowest range regulator suitable for the chromatographic method.

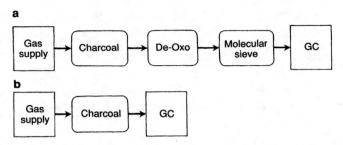

Figure 2. Typical gas filter configurations (a) for carrier and make-up gas supplies and (b) for detector combustion gas supplies.

ii. Back-pressure regulators
These are similar to the above but maintain a constant pressure *upstream* of the device.

iii. Flow controllers
These controllers maintain a constant gas flow even though the downstream impedance may vary. Flow controllers are available in a variety of operating ranges (e.g. 0 to 10, 20, 100, or 200 ml/min). Lower ranges give a finer degree of control.

iv. Needle valves
Needle valves are used to present adjustable impedances to flows of gas passing through them.

v. Gas snubbers
Gas snubbers provide fixed impedance restrictions to gas flow. Many impedance values are available.

vi. Pressure gauges
Pressure gauges are mechanical devices with dials indicating gas pressures. They are available with various operating ranges (e.g. 0 to 30, 60, or 100 lb/in^2).

vii. Pressure transducers
Transducers are electronic devices which can be used to generate a voltage proportional to gas pressure. Additional hardware and software is required in the chromatograph to utilize such a signal. They have the advantage of being able to suspend chromatograph operation if the pressure drifts or the gas supply runs out.

viii. Remote controllers
These controllers are electromechanical devices which perform the same function as pressure regulators or flow controllers but are controlled by electronics and software. They tend to be expensive but have the great

advantage in that they can be set up automatically and can be programmed during chromatography for specialized applications.

2.3.3 Pneumatic configurations

The components described in Section 2.3.2 may be configured in several ways to provide the necessary control for the various gases required in a gas chromatograph. *Figure 3a–f* show typical configurations and *Table 3* lists their typical applications.

Table 3. Recommended applications of pneumatic configurations shown in *Figure 3*

Application	Configuration					
	a	b	c	d	e	f
Packed column carrier gas	Yes[a]	Yes	No	No	No[b]	No[b]
Capillary column carrier gas	Yes	No	Yes	Yes	No[b]	No[b]
Make-up gas	No	Yes	No	No	Yes	Yes
Detector combustion gas	No	Yes	No	No	Yes	Yes

[a] Isothermal chromatography only.
[b] Control not precise enough for carrier gas.

2.4 Sample introduction

There are many ways of introducing a sample into a chromatographic column. The inlet system is chosen by taking into account the sample matrix, the components in the sample which are to be determined, and the type of chromatographic column to be employed. The role of the inlet system is to introduce all, or a representative portion of the sample, into the chromatographic column.

2.4.1 Packed column injector

Figure 4 shows a schematic diagram of a typical packed column injector. A liquid sample (typically 0.1 to 10 μl) is injected by syringe, through a silicone rubber septum into an inert heated zone within the injector. This injection zone comprises either a glass liner (flash vaporization injection) or the extended inlet of a glass-packed column (on-column injection). Adapters are available which allow a 0.53 mm internal diameter (i.d.) capillary column to be fitted to a packed column injector to allow either flash vaporization or (hot) on-column injection. After injection, the vaporized sample is swept by carrier gas into the chromatographic column. This injector can be used with a gas syringe for gaseous samples or with a pelletizing syringe for solid samples but quantitative performance may be poor.

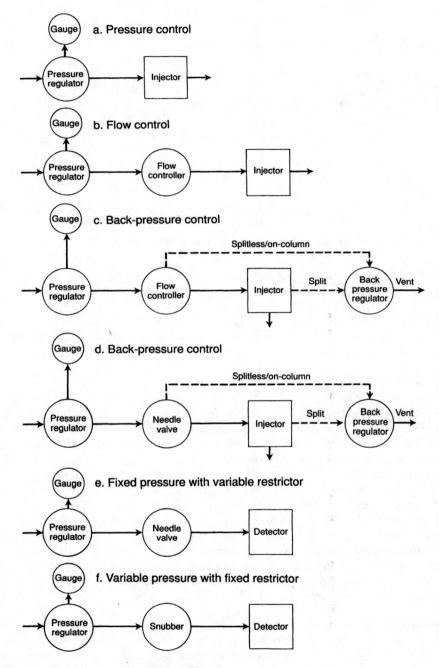

Figure 3. Typical pneumatic configurations.

21

Andrew Tipler

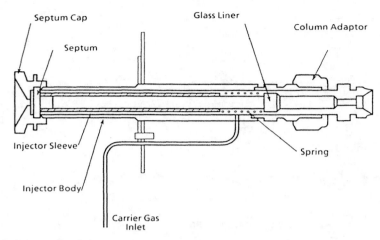

Figure 4. Schematic diagram of a typical packed column injector.

2.4.2 Split/splitless injector

Figure 5 shows a typical design. Sample introduction is performed, in the same way as for a packed column injector, into a heated glass liner. This injector is used with a capillary column, the inlet of which is inserted into the heated liner. A switchable vent is provided at the liner outlet to enable either split or splitless modes of injection.

i. Split injection

In this mode, the vent remains open during injection (and, usually, during the analysis). This has the effect of accelerating the passage of sample vapour through the liner and across the column inlet and thus reducing the 'band-width' of the sample entering the column, therefore improving peak resolution. Split injection can be used for gas (gas syringe) or solid (pelletizer) samples but is usually used for liquid samples where an injection volume of 0.1 to 1.0 μl is normally used. The ratio of the vent flow rate to the column flow rate is termed the *split ratio* and is typically in the range 200–300:1— lower split ratios give peak broadening and higher split ratios make carrier gas control difficult and reduce the available time for complete sample vaporization, thus affecting the performance of the splitting process. Because most of the sample is vented, the amount of sample entering the column is greatly reduced, making this technique suitable for undiluted or concentrated sample mixtures but unsuitable for trace level determinations.

ii. Splitless injection

Splitless injection is suitable for trace level determinations in dilute liquid samples. It cannot be used for gas injections (no focusing) or solid injections (too much sample). The vent remains closed during splitless injection to

22

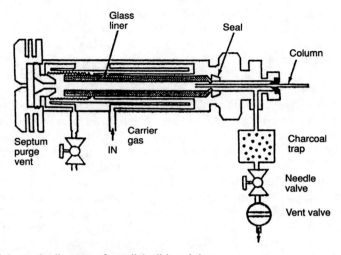

Figure 5. Schematic diagram of a split/splitless injector.

effect total sample transfer to the capillary column. The vent is opened about 30 sec after injection to purge residual traces of solvent from the liner, which would otherwise cause a tailing solvent peak which would obscure early eluting component peaks. The bandwidth of the sample entering the column will depend on the type of solvent, how much is injected, carrier gas pressure, and injector temperature. This bandwidth is much too large for capillary chromatography and so some form of secondary focusing must be employed:

Cold trapping

Cold trapping uses an initial oven temperature of at least 100°C below the boiling point of the components of interest. As these components enter the column, they condense as a narrow band on the column walls and remain there until a temperature programme is initiated.

Solvent effect

The solvent effect (1) uses an initial oven temperature of typically 20°C below the boiling point of the solvent. As the sample is heated within the injector, the solvent, if it is more volatile than the components of interest, is the first to vaporize and enter the column. Some of the solvent will condense and collect on the column walls. As the components of interest pass into the column, they will partition into the condensed solvent, thus slowing down and giving a focusing effect. The column oven is then temperature-programmed and the solvent vaporizes, leaving components deposited in narrow bands. The solvent effect cannot be used in instances where the components of interest are more volatile than the solvent. One problem associated with the solvent effect is that, if too much solvent is condensed—particularly if the solvent has a very different polarity to the column stationary phase and will not wet the surface—it can be physically driven along the column by the carrier gas. This *solvent flooding* effect may fractionate the sample along the length of the column and give distorted or multiple peaks for some components.

Retention gap

The *retention gap* technique (2) overcomes the solvent flooding effect and provides a further degree of focusing. A length of deactivated and normally uncoated capillary tubing (typically 2 m × 0.53 mm i.d.) is connected between the injector and the column. Conditions are set up as for the solvent effect technique. The condensed solvent and retained components are allowed to spread along the length of this retention gap tubing—it is important that all the solvent vaporizes before it reaches the column inlet or the focusing effect will be lost. The oven is then temperature-programmed and the solvent vaporizes leaving the components of interest deposited along the retention gap. As the temperature increases, each component will vaporize in turn and travel with the carrier gas to the column inlet. As the component enters the column, it will partition into the stationary phase and slow down causing a focusing effect.

As can be seen from the focusing techniques above, the choices of solvent and initial oven temperature are very critical in splitless injection. Injection volumes tend to be larger than in split injection (1 to 10 µl is typical, but larger injection volumes are possible with extended retention gaps) but care must be taken with the speed of injection to prevent sample *flashback* into the carrier gas supply lines or on to the septum (0.5 µl/s is a suitable rate). Most injectors use a *septum purge* gas to clean the septum and prevent deposited material (or volatile material from within the septum itself) from entering the column which may cause *ghost* peaks in subsequent chromatography.

2.4.3 Temperature-programmed vaporization injection

This injector (see *Figure 6*) is an enhancement to the split/splitless injector and is functionally equivalent except that the injection liner temperature is able to be rapidly heated or cooled. Sample injection is made into a cool liner, thus eliminating potential problems with sample thermolysis and mass discrimination during the injection process. After the syringe needle has been removed, the temperature of the liner is rapidly increased to effect sample vaporization. Component vaporization occurs relatively slowly (unlike the explosive effect when injecting into a hot zone), giving rise to a more controlled transfer to the capillary column. As each compound is vaporized it is swept from the liner into the column and thus will not be subjected to a temperature higher than necessary. Its advantages over classical split/splitless injection are that it should deliver a more repeatable injection, handle labile components, and exhibit less mass discrimination. The PTV can be used for

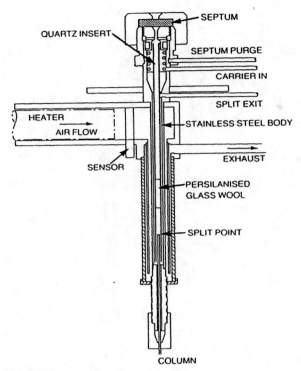

Figure 6. Schematic diagram of a PTV injector.

split or splitless injection and the techniques (split ratios, secondary focusing etc.) described in Section 2.4.2 apply here.

A similar approach to the PTV can be used for cold on-column injection (see Section 2.4.4) where a 0.53 mm i.d. retention gap or column can be introduced into a cold liner with needle guide and enable direct injection by a normal syringe into the retention gap or column. The liner is heated after injection to move all the sample into the column residing within the oven. A good summary of the practical operation of this injector is provided elsewhere (3).

2.4.4 Cold on-column injection

This injection technique (4) gives the highest performance but, unfortunately, is the most difficult to use. It can be viewed as an alternative to splitless (classical or PTV) injection. The injector (see *Figure 7*) is essentially a cooled needle guiding system to allow direct insertion of a very narrow syringe needle into the inlet of a narrow bore (down to 0.2 mm) capillary column. A critical factor is the design of the sealing system, which cannot be a traditional septum as the needle will not be strong enough to penetrate it. Various

Andrew Tipler

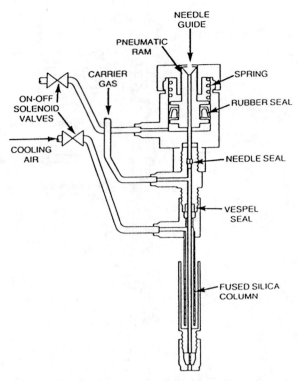

Figure 7. Schematic diagram of a typical cooled on-column injector.

designs are offered by different manufacturers including mechanical valves, compression seals, and pneumatic rings. The seal is opened and the syringe needle is pushed through the seal and into the column to a depth where injection actually occurs within part of the column held within the oven. The syringe is removed, the seal closed, and the oven is temperature-programmed to effect sample vaporization. Note that there are no intermediate vaporization steps—this is what gives this technique advantages over the others. It is important that vaporization does not occur within the syringe needle or this will affect performance, so the injector is kept cool during injection with a jacket through which air or a subambient medium is flowing. Like splitless injection (see Section 2.4.2 ii) component focusing is required so those techniques apply here. The column oven must not be operated above the boiling point of the solvent or explosive vaporization will occur, giving rise to possible pneumatic contamination, sample loss, and poor peak shape. The use of a retention gap is particularly recommended as it also acts as a *guard column* which prevents involatile material from entering the analytical column.

26

2.4.5 Gas sampling valve

A gas sampling valve (GSV) is the best way of injecting gaseous samples (see *Figure 8*). The sample is passed through a sample loop of calibrated capacity (typical values are 0.1, 0.2, 0.5, 1.0, 2.0, and 5.0 ml). The pressure inside the loop is allowed to equilibrate to either atmospheric pressure (via a vent) or to an elevated pressure (using a backpressure regulator) and then the valve is rotated (either manually or using a pneumatic or electromechanical actuator) to allow carrier gas to sweep the sample into the column. The valve may be heated to prevent condensation of the less volatile components. For capillary column injection, either a splitter is required between the valve and the column to reduce the sample bandwidth, or a purpose-designed micro-GSV should be used. Besides sample injection, GSVs which are available with 4, 6, 8, or 10 ports can be plumbed in a variety of ways to effect column switching techniques such as backflushing or multidimensional chromatography.

2.4.6 Liquid sampling valve

This is similar in operation to the GSV but uses an internal sampling 'loop'. It is only suitable for injecting liquified gases, e.g. liquid petroleum gas, aerosol propellants, etc. The sample is introduced under pressure and at ambient temperature (to prevent premature sample vaporization). The valve is rotated and the sample is pushed, by carrier gas, through a small bore transfer tube into a heated chamber, where it vaporizes, and then into the chromatographic column. For capillary columns, a splitter between the vaporization chamber and the column may be necessary.

2.4.7 Pyrolyser

This is a specialized accessory for a gas chromatograph. It is used for the qualitative examination of materials such as polymers and rubbers. A small amount of the sample is placed within a glass tube, which is then placed within a heater (curie point or ohmic heating), and raised to a temperature of typically 1000 °C. This has the effect of breaking the material down into lower molecular weight fragments which are carried by carrier gas, with splitting for capillary columns, to the chromatographic column. The nature of the fragmentation is often characteristic of the material under examination and GC is used to produce *pyrograms* which can be used for sample identification.

2.4.8 Headspace analyser

Solid or liquid samples are placed in a thermostatted and sealed vial with an inert gas at a controlled pressure occupying the headspace. The system is allowed to stand for a fixed period during which time volatile components diffuse out of the sample matrix and partition and equilibrate with the headspace. A fixed volume of the headspace is then sampled and transferred, via a splitter or cryogenic trap for capillary columns, to the chromatographic

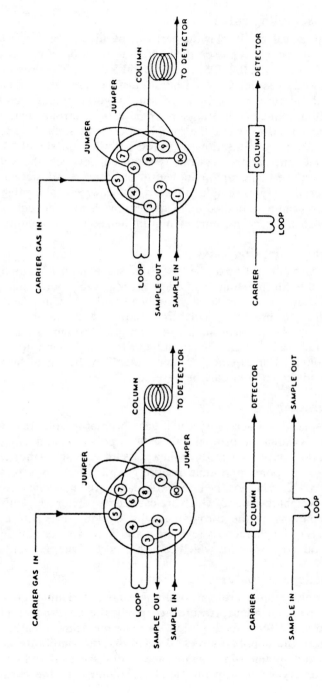

Figure 8. Schematic diagram of a 10-port GSV configured for simple injection into a single packed column.

column. A headspace analyser is very useful in many applications where the volatile content of *dirty* matrices is to be determined—for example: alcohol in blood, aroma in food stuffs, and fragrances in various formulations. A comprehensive treatment of this technique is reported in the literature (5).

2.4.9 Thermal desorption

This is similar to the headspace analyser except that, rather than using a static equilibrium system to extract volatile material, the sample is held in a heated tube through which an inert gas passes continuously. The effluent from the sample tube is cryotrapped either on a temperature-programmable adsorbent bed (two-stage thermal desorption) or in the column inlet itself. Care must be taken with the latter to prevent blockage of the column because of ice formation which will occur if the sample contains water. Thermal desorption is best suited for the analysis of solid samples. It is also very applicable to the determination of trace level vapours in atmospheric samples, where the sample has been drawn through or allowed to diffuse into a thermal desorption tube packed with a suitable adsorbent which retains the analytes of interest.

2.4.10 Purge and trap

Thermal desorption is also used in conjunction with a liquid sparging device to enable the technique known as 'purge and trap'. This device is used to determine the volatile content of liquids. It is used mainly for the determination of trace levels of organic pollutants in water. A relatively large sample of water (this can be up to 100 ml) is placed in a vessel, which may be thermostatted, and a stream of a very clean inert gas is bubbled through it. This has the effect of extracting the volatiles from the sample and after a short period of time (typically 5 to 10 min) most of these will have been removed. The eluting gas is passed through an adsorbent trap, which may be cooled, which retains and focuses the extracted analytes. By using an in-line drier and/or a hydrophilic adsorbent, water, which would cause problems with the subsequent chromatography, is not retained on the trap. The adsorbent is then heated to effect the transfer of trapped analytes to the chromatographic column in a narrow band and chromatography is initiated.

This technique has advantages over static headspace extraction in that larger samples can be handled; a greater proportion of the analytes are transferred to the chromatographic column and potential problems with water are reduced.

2.4.11 On-line supercritical fluid extraction

The sample is held in a reinforced cell through which a supercritical medium such as carbon dioxide is passed. To maintain the supercritical state, the sample cell is thermostatted, in an oven, above the medium's critical temperature and pressurized, by a purpose designed pump, above its critical pressure.

The pressurized extract is fed by a transfer line to the gas chromatograph where it is depressurized through a restrictor and deposited, normally via a splitter, into a cryotrap. After collection, the extract is thermally desorbed and carried by carrier gas to the chromatographic column. The selectivity of the extraction process can be altered by adjusting the density (by pressure and/or temperature) of the supercritical medium or by adding a modifier (such as methanol) to alter the polarity of the extraction medium.

The technique is suitable for the extraction of relatively high molecular weight compounds and, because no high temperatures are involved, for labile compounds from a variety of solid or liquid sample matrices. Liquid samples should be retained, prior to extraction, on a suitable adsorbent such as glass beads, filter paper or diatomaceous earth. Supercritical fluid extraction is particularly suitable for 'dirty' sample matrices such as soil, plant material, food stuffs etc.

2.4.12 Autosamplers

Many of the injection techniques described above can be automated using specialized hardware available from many manufacturers. Of particular interest is the liquid autosampler, which is able to inject small amounts of sample into packed or capillary column injectors. These devices use either a conventional syringe which is primed with sample or a 'flow-through' syringe through which sample under pressure is forced. The filled syringe is then driven through the injector septum and the sample expelled into the liner. Chromatographic conditions remain as for manual injection. Samples are placed in sealed vials and loaded into a carousel which can be indexed by the sampling mechanism to enable each sample in turn to be drawn into the syringe. Carousels usually hold up to 100 sample vials and often have provision for solvent and waste vials to enable the syringe and associated plumbing to be cleaned after each injection. The current vial position can be communicated to a data-handling system, via a binary coded decimal (BCD) interface, so that the chromatogram and final results are correlated with the correct sample.

A recent development has been the 'high-speed' autosampler which is able to perform the injection process within a fraction of a second. This is useful for heated injectors as the metallic syringe needle is prevented from becoming hot inside the liner and so reduces mass discrimination and catalytic decomposition effects.

Autosamplers have two main advantages over manual injection: analyses can be performed in the absence of the analyst, thus freeing time for other work or making sampling possible while the analyst is away from the laboratory; and because the injection process is highly repeatable, much more precise analytical results can be obtained. Most modern analytical laboratories rely heavily on these autosamplers to increase throughput and performance.

2.5 Gas chromatographic detectors

2.5.1 Classes of detector

There are many types of detector that are used in gas chromatography. The reason for this proliferation concerns the need for detection systems to provide selective responses to particular groups of compounds to simplify the chromatograms from complex samples. Different detectors give different types of selectivity. The degree of selectivity may be classed as follows.

(a) *Non-selective* (or universal) detectors respond to all compounds which differ from the carrier gas.

(b) *Selective* detectors respond to a range of compounds which have some common chemical or physical property.

(c) *Specific* detectors respond to a single chemical compound.

Figure 9 shows two chromatograms, produced using detectors of differing selectivity, obtained from a single chromatographic run. Much more informa-

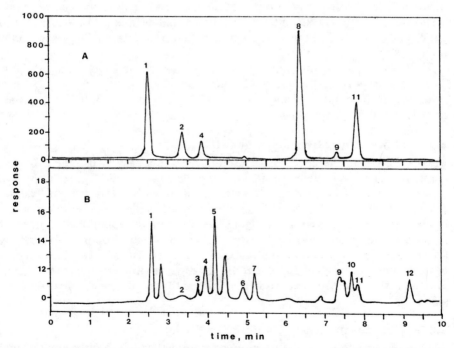

Figure 9. Chromatograms of air sample obtained from a single capillary column connected to two detectors. 1, air; 2, freon 12; 3, methyl chloride; 4, freon 114; 5, vinyl chloride; 6, methyl bromide; 7, ethyl chloride; 8, freon 11; 9, vinylidene chloride; 10, dichloromethane; 11, trichlorotrifluoroethane; 12, 1,1-dichloroethane. A, ECD; B, FID.

tion can be derived from such data than from a single trace from a non-selective detector.

Besides selectivity, detectors can be classed according to how they respond to the amount of analyte passing through.

i. Concentration-dependent detectors

The detectors produce a signal which is related to the concentration of solute in the gas stream present in the detector at a point in time. In operation, these detectors are normally non-destructive and so can be used in series with other detectors. Their response is expressed in units of signal per analyte concentration. Because of their sensitivity to analyte concentration, any dilution of the column effluent with a make-up gas will lower the response. This is unfortunate as many of these detectors will require a make-up gas when used with capillary columns in order to prevent peak-broadening effects.

ii. Mass flow dependent detectors

These detectors produce a signal that is related to the rate at which solute molecules enter the detector. These detectors normally generate a signal as a result of some destructive process occurring in the solute molecules. Their response is expressed in units of signal per analyte mass flow. Response is generally unaffected by the addition of a make-up gas although this is normally unnecessary.

Table 4 summarizes the main detection systems used in gas chromatography. Two very important detection systems: the mass spectrometer (MS) and Fourier transform infrared spectrometer (FTIR) are not included here as they are discussed in Chapter 11.

2.5.2 Flame ionization detector (FID)

This is the most popular detector used in GC. It is easy to use, gives a very stable response and is sensitive to most organic compounds. *Figure 10* shows a typical design. The column effluent is mixed with hydrogen and passes through a jet into a chamber through which air is passed. The hydrogen is ignited to produce a continuous flame. As compounds from the column enter the flame, they undergo combustion. A very small proportion (typically 0.001%) of the carbon atoms undergoes ionization during this combustion process. An electrode, which is polarized with respect to the jet, collects these ions and the resulting electrical current is amplified to provide the chromatographic signal.

2.5.3 Thermal conductivity detector (TCD)

This detector is able to detect any compound that has a different thermal conductivity to the carrier gas. It is not as popular as the FID because of its generally poorer detection limits. The column effluent enters a cell in which is located an electrically heated tungsten-rhenium filament. The temperature of

Table 4. Summary of GC detectors

Detector	Type[a]	Support gases	Selectivity	Detectability[b]	Dynamic range[b]
Flame ionization (FID)	MF	H_2 & Air	Most organic compounds	100 pg	10^7
Thermal conductivity (TCD)	C	Reference	Universal	1 ng	10^7
		Make up[c]			
Electron capture (ECD)	C	Make up[d]	Halides, nitrates, nitriles, peroxides, anhydrides, organometallics	50 fg	10^5
Nitrogen-phosphorus (NPD)	MF	H_2 + Air	Nitrogen Phosphorus	10 pg	10^6
Flame photometric (FPD)	MF	H_2, Air + possibility O_2	Sulphur, phosphorus, tin, boron, arsenic, germanium, selenium, chromium	100 pg	10^3
Photo-ionization (PID)	C	Make-up[c]	Aliphatics, aromatics, ketones, esters, aldehydes, amines, heterocyclics, organosulphurs, some organometallics, O_2, NH_3, H_2S, HI, ICl,Cl_2, I_2, PH_3	2 pg	10^7
Hall electrolytic conductivity (HECD)	MF	H_2, O_2	Halide Nitrogen Nitrosamine Sulphur		

[a] MF, mass flow; C, concentration.
[b] Typical limit for a favourable compound.
[c] With capillary columns.
[d] Must be nitrogen or methane (5 to 10%) in argon.

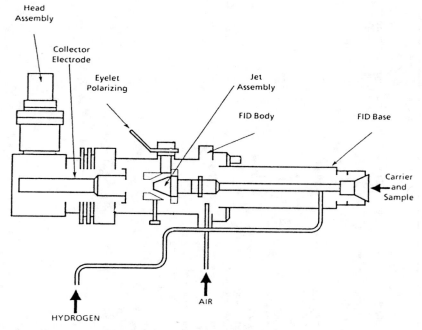

Figure 10. Schematic diagram of a typical FID.

this filament depends on the rate of heat dissipation through thermal radiation, conduction through the filament mounting connections, heat transfer by mass flow of the carrier gas, free convection and, most importantly, conduction by the carrier gas to the cell walls. The first four factors can be compensated by using additional filaments and cells in a bridge arrangement (see *Figure 11*) located within the same detector block but will require an additional reference gas to flow through one-half of the bridge (R2 and R4). Using this arrangement, the temperature, and hence resistance of the analytical filaments (R1 and R3), will vary with the thermal conductivity of the gas passing through it. With a constant current applied, this resistance is proportional to the voltage difference across the bridge and this forms the basis of the output signal.

2.5.4 Electron capture detector (ECD)

The ECD is extremely sensitive to molecules containing highly electronegative atoms, such as halides. It is therefore a popular detector for trace level determinations of pesticides and halocarbon residues in environmental samples.

The ECD (see *Figure 12*) consists of an ionization chamber containing a radioactive source (Ti^3H, Sc^3H, or preferably ^{63}Ni) which emits β-particles. During operation, there is a constant stream of nitrogen or 5–10% methane

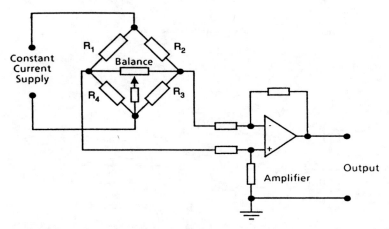

Figure 11. Schematic diagram of the filament bridge arrangement in a typical TCD.

in argon flowing through the cell which is ionized by this emission, causing the liberation of free thermal electrons. A positively charged electrode collects these electrons, thus generating a small constant current known as the *standing current*. If an electrophilic species enters the cell, it will react with the free thermal electrons, thus reducing the standing current. The output signal is derived by amplifying and inverting the standing current. The means of polarizing the collector is critical to performance and most modern detectors use a pulsed mode of operation to improve sensitivity, stability, linearity, and selectivity.

The flow rate of gas through the detector depends on the cell geometry, but is generally greater than 10 ml/min. Thus for capillary columns a make-up gas of nitrogen or argon/methane must be used—enabling the use of helium or hydrogen as the carrier gas.

2.5.5 Nitrogen-phosphorus detector (NPD)

This detector (also known as the thermionic emission detector) is similar in design to the FID but with an important difference: an electrically heated

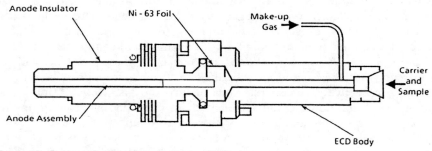

Figure 12. Schematic diagram of a typical ECD.

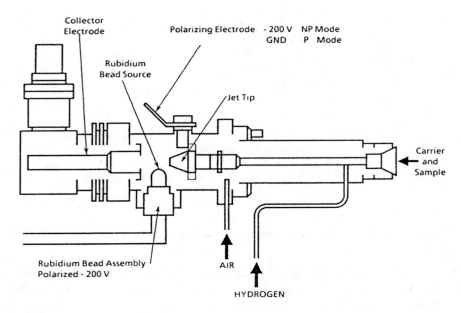

Figure 13. Schematic diagram of a typical NPD.

silicate bead doped with an alkali (such as rubidium) salt is mounted between the jet and the collector (see *Figure 13*). A very low hydrogen flow rate (typically 2 ml/min) is mixed with the carrier gas and burns as a plasma flame as it makes contact with the heated bead. The collector is maintained at a positive electrical polarity with respect to the bead and jet.

The exact mechanism for producing a response has been the subject of some controversy but Kolb's theory seems to be widely accepted: at the operating temperature, the bead substrate is electrically conductive and some of the alkali ions are able to acquire an electron and are thus converted to the atomic form. These atoms are relatively volatile and are emitted into the plasma, where they quickly react with combustion products, and are ionized again and recollected on the negatively polarized bead. This cyclic process gives rise to the background signal and explains why the bead continues to function over an extended time period. If a compound containing nitrogen or phosphorus elutes from the column into the plasma, these molecules will burn and react with the excited atomic alkali to form cyan or phosphorus oxide anions, respectively. These reactions disturb the alkali equilibrium in the plasma and additional alkali is released into the plasma, thus promoting further ion formation and increasing the signal. Cyan anions are only formed from molecules containing nitrogen bonded to a covalently bonded carbon atom, and so a poor response is produced from compounds such as nitrogen gas (thus this can be used as a carrier gas), carbamates, ureas, barbiturates, etc.

The formation of cyan anions can be suppressed by increasing the reaction temperature with an increased flow of hydrogen (electrical heating of the bead is unnecessary under these conditions) allowing a phosphorus-selective mode of operation. The jet is usually grounded under this mode.

2.5.6 Flame photometric detector (FPD)

This detector is popular because it gives a highly specific response to compounds containing sulphur or phosphorus and so is particularly important in environmental applications where pesticides or herbicides containing these elements are being determined in complex matrices. The FPD has also been used for detection of molecules containing boron, arsenic, germanium, selenium, chromium, and tin.

The column effluent is mixed with hydrogen and is burnt in a cell through which air is passed (see *Figure 14*). As compounds eluting from the column

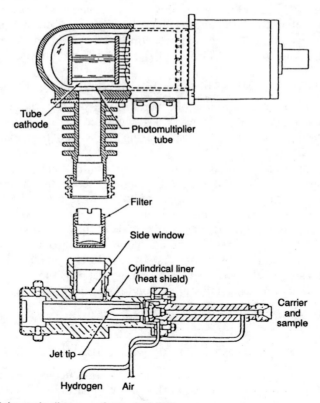

Figure 14. Schematic diagram of a typical FPD.

enter and burn in the flame. Some of these form chemiluminescent species which emit light at a wavelength characteristic of specific elements they contain. This emission is monitored by a photomultiplier tube which provides the chromatographic signal. To reduce noise and give a selective response a suitable narrow bandpass optical filter is placed between the flame and the photomultiplier tube. The selectivity of the detector is easily modified by changing the type of filter.

2.5.7 Photoionization detector (PID)

This detector is becoming popular in the oil industry and for environmental analyses because it offers a better degree of selectivity than the FID.

The carrier gas from the column enters a cell and is irradiated with ultraviolet light. Some molecules entering the cell are sufficiently excited to undergo ionization. A pair of electrodes, with a potential difference across them, monitor these ions and the resulting current is amplified to provide the detector output signal. The selectivity of the response is determined by the energy of the emitted radiation. Lamps are available with energies of 9.5, 10.0, 10.2, 10.9, and 11.7 eV. Only those molecules with an ionization potential less than the lamp energy should undergo ionization and produce a response. In practice, a diminished response will also be given by molecules with an ionization potential of up to 0.3 eV above the lamp energy.

2.5.8 Hall electrolytic conductivity detector (HECD)

This detector can be used in one of several modes of operation (see *Table 4*). It is particularly popular, when used in the halogen mode, for the determination of volatile halocarbons in water using purge and trap extraction (see Section 2.4.10).

The column effluent is mixed with hydrogen or air (depending on the mode of operation) and passes into a heated catalytic reaction tube where oxidation or reduction reactions occur. The reaction products are passed (through a scrubber in some cases) into a conductivity cell through which an electrolyte (such as propanol) is pumped. The conductivity of the electrolyte is continually monitored by two electrodes. As the analyte reaction products dissolve in the electrolyte, its conductivity changes and a response is produced.

The different modes of operation are made possible by selecting the appropriate reaction gas, catalyst, reaction temperature, in-line scrubber, and electrolyte.

3. Operation and experimental considerations

This section considers the steps necessary to produce a chromatographic method. Method development is, perhaps, the most time-consuming activity associated with GC, and so a systematic approach is recommended.

3.1 Where to start

The most efficient way to approach a GC analysis (or indeed any other analysis) is to find out if it has been successfully performed before. Possible sources of information are:

(a) *Within the laboratory*. In many cases, the analysis is not new and, in such cases, the analyst would normally be presented with a documented method containing instrumental details and operating conditions. If such a method is not available, then discussions with colleagues and a review of departmental archives may prove fruitful.

(b) *From publications*. There are several journals dedicated to the subject of chromatography and many of the papers describe gas chromatographic methods for analysing specific sample types. *Table 5* gives a list of some of the most useful journals to the gas chromatographer. Each paper will often provide references to previous papers on the same subject. In addition to journals, many books have been written on the subject which contain details of specific gas chromatographic analyses.

(c) *From regulatory authorities*. Many gas chromatographic analyses must conform to guidelines specified by a regulatory authority. In some cases, the exact chromatographic conditions are specified.

(d) *From suppliers of columns and instruments*. Many suppliers produce application notes describing practical methods for chromatographic analyses and would be pleased to supply these on request. They can also help with practical advice and may be able to consult, on your behalf, other chromatographers known to be involved in similar work.

3.2 Installation and preparation of the chromatograph

This section deals with the basic steps involved in preparing the chromatograph prior to use. Careful preparation can save much time later when attempting to improve chromatographic performance.

Table 5. Journals containing information on gas chromatography

Journal	Publisher
Journal of Chromatography	Elsevier
Journal of Chromatographic Science	Preston Publications Div.
Chromatographia	Friedr, Vieweg, & Sohn
Journal of High Resolution Chromatography	Huthig
LC/GC International	Aster Publishing Corp.
Chromatography Abstracts	Elsevier/Chromatography Society

3.2.1 Instrument configuration

Before any chromatography can be performed, suitable instrumentation must be available. Section 2 describes many different components and their designs which comprise a modern gas chromatograph. The user must, therefore, start by making decisions as to the optimum chromatographic hardware configuration (injector, detector, and pneumatic components) to suit the analysis. If a method exists, then the selection process is simply reduced to reproducing the configuration described in that method. In cases where a new method is being developed, the chromatographer must make personal judgements as to suitable columns and hardware. This will become easier with experience—*Figures 15* and *16* show flow diagrams which illustrate typical approaches to

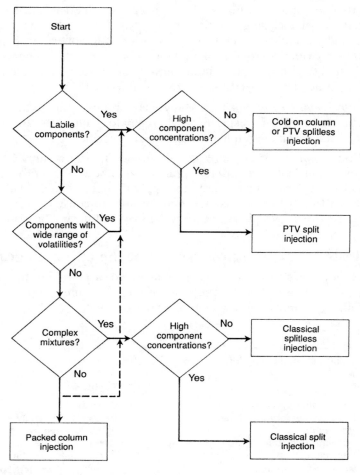

Figure 15. Choosing an injector.

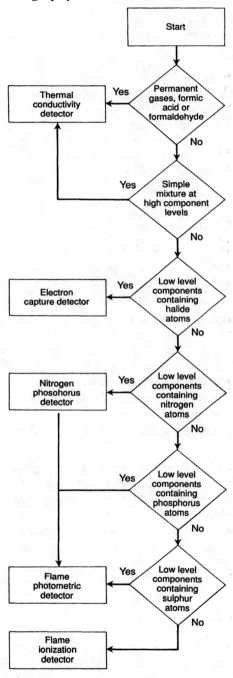

Figure 16. Choosing a detector.

Andrew Tipler

choosing injectors and detectors respectively. Selection of pneumatic components has already been discussed in Section 2.3. Column selection is discussed in Chapter 3.

3.2.2 Preparation of the injector

Many chromatographic problems can be traced to the injector, so it is worth spending a little time checking this before starting the analysis. The general approach to checking an injection system is to ensure that it is clean, undamaged, leak-tight, and that all functional components (such as heaters, solenoid valves, sealing mechanisms) operate correctly. Routine operations, such as septum replacement and liner preparation are described in *Protocols 1* and *2* respectively.

Protocol 1. Septum replacement

1. Unscrew the septum cap and remove it.
2. Inspect the old septum—this can reveal problems with the type of septum being used, with the way it is fitted, or with the injection technique.
3. Discard the old septum.
4. Select a new septum suitable for the analysis. Low temperature septa tend to be softer, last longer, but release higher levels of 'septum bleed' into the carrier gas. Some soft septa have a PTFE laminate to reduce the bleed level. High temperature septa are harder (pre-puncturing is recommended), have a shorter life, but give less bleed. A useful tip is to pre-condition a new septum by installing and leaving it in a heated but unused injector until it is needed.
5. Fit the septum into the septum cap, taking care not to leave fingerprints on its surface.
6. Replace the septum cap. The cap should be tightened to apply *slight* pressure to the septum. Overtightening will cause compaction of the septum centre and may result in excessive fragmentation in use. Undertightening will allow gas leakage.
7. Test the installation by pushing a syringe needle through the septum. A smooth slight resistance should be felt. If obstructions are apparent, check the installation.

Protocol 2. Liner preparation

1. Unscrew the septum cap and remove it.
2. Pull out the glass injection liner and inspect it. Discard the liner if damaged and replace it with a new one.

42

3. Clean the liner with a suitable solvent and ensure no particulate matter or residues are left in it. If necessary, place the liner in solvent in an ultrasonic bath or soak it in a suitable cleaning agent (e.g. concentrated nitric or chromic acid).

4. If reactive analytes are to be examined, the liner should be deactivated with a suitable agent such as dichloromethylsilane (DMCS), hexamethyldisilazane (HMDS), or orthophosphoric acid.

5. Check the interior of the injector body and remove any particulate matter. Ensure that the column port is clean.

6. Where the injection mode demands a packed liner (see Section 2.4), loosely pack the liner with deactivated glass or quartz wool according to the manufacturer's instructions.

7. Re-install the liner. All liners employ some sealing mechanism to ensure that the carrier gas flows along the correct path. This could take the form of a ferrule, an 'O'-ring, or a spring-loaded ground glass butt joint. Check the seal and replace the sealing components if necessary.

8. Replace the septum as described in *Protocol 1*.

3.2.3 Column installation

The procedure for column installation will depend upon the type of column, injector, and detector that are to be used.

i. Packed columns

Packed columns are the easiest to fit because of their robustness and the relatively high carrier gas flow rates employed. Packed injectors and detectors are normally ported with 1/4 or 1/8 inch male connectors ('Swagelok' or similar). Packed columns are then easily connected using a suitable nut and ferrule. Brass nuts and ferrules should be avoided as these will tend to anneal to the stainless steel injector and detector fittings at high temperatures. Several different types of ferrule can be employed as listed in *Table 6*. *Protocol 3* gives the procedure for packed column installation.

Protocol 3. Installation of a packed column

1. Examine the column and reject it if it is damaged at either end.

2. Check the injector and detector ports and ensure that they are compatible with the outer diameter of the column. Most manufacturers supply adapters to enable either 1/4 or 1/8 inch columns to be connected.

Protocol 3. *Continued*

3. Hold the column in position in the oven and check that the positions of the two ends align with the injector and detector ports. If this is not the case, then a metal column may be bent to fit the ports, whereas a glass column must be exchanged for one of the correct geometry.

4. Check the injector and detector ports and make sure they are clean and free from bits of glass, ferrule, packing material, etc. Use a jet of air to blow debris out of the injector.

5. Slide a suitable nut on to the column inlet followed by a ferrule. In the case of a soft ferrule such as graphite, a backing ferrule can be placed between the ferrule and the nut—this will minimize ferrule extrusion. Make sure that no particles from the ferrule enter the column.

6. Insert the column into the injector following the manufacturer's recommendations. For on-column injection, the end of the column should be pushed right through the injector until it makes contact with the septum. For flash vaporization injection, the column should be inserted until it reaches a natural stop.

7. Withdraw the column by approximately 1 mm to allow movement of the column as the ferrule compresses. Care must be taken with on-column injection so that the column does not seal itself against the septum as this would prevent carrier gas flow.

8. Hand-tighten the column nut on to the injector port. Using a suitable wrench, further tighten the nut until the ferrule just starts to grip the column. Use the wrench to rotate the nut a further quarter of a turn. Do not overtighten.

9. Connect the column to the detector in a similar manner, although this is normally left until the carrier gas has been connected and checked for leaks.

ii. Capillary columns

Capillary column installation demands more care on the part of the chromatographer because of the delicate nature of these columns, their low thermal mass, and the much lower carrier gas flows involved. Connections are normally made using a nut and ferrule in a similar manner to packed columns. Fittings of 1/16 inch are easier to use and have a lower thermal mass for capillary columns. Wide bore (0.53 mm i.d.) capillary columns can be operated at gas flows similar to packed columns and so can be used with packed column injectors. Most modern capillary columns are manufactured from polyimide-coated fused silica. *Protocol 4* describes the procedure for installing this type of column.

Table 6. Ferrules used in connecting columns

Material	Application	Comments
Stainless steel	Stainless steel packed and capillary columns	Very robust and reliable Cannot be removed once fitted Can damage column if overtightened No temperature limit
PTFE	Glass-packed columns	Easy to fit Good for delicate columns Highly inert Low temperature limit (280°C) Will creep and extrude in use needing frequent re-tightening and eventual replacement
Graphite	All packed and capillary columns	Easy to fit Suitable for high temperatures Easily removed and re-used Highly adsorptive—cannot be used in chromatographic pathway Will tend to creep and extrude in use
Vespule	All packed and capillary columns	Inert Robust Needs care when fitting delicate columns Difficult to remove/re-use Moderate temperature limit
Graphitized vespule	All packed and capillary columns	Combines the advantages of graphite and vespule—high temperature and inertness but still difficult to remove

---

Protocol 4. Installation of fused silica capillary columns

1. Examine the column and check for fractures and discolourations. Fractures may be repaired using a low dead-volume union. Discolouration may be indicative of involatile sample residue or stationary phase degradation. If present, this will normally occur in the first few centimetres of the column and can be easily removed by breaking off this section of the column (see step 5).

2. Suspend the column (normally wound on a metal mandrill) in the oven on the capillary mounting bracket supplied with chromatograph. The column should be carefully positioned so that it is concentric with the oven fan and no section of the column should be in contact with the

Protocol 4. *Continued*

oven walls or any metal bracket or other hardware. This is to minimize any variations in temperature along the length of the column.

3. Make sure that the injector and detector ports are clean and free from debris.

4. Unwind approximately 20 cm of the column from the mandrill and thread on to it a suitable column nut and ferrule. For 1/16 inch fittings the ferrule may be reverse-fitted (i.e. tapered end towards the nut) so check the manufacturer's instructions for correct orientation. For 1/8 inch fittings with a graphite ferrule, a back-ferrule should be used to reduce extrusion.

5. Using a suitable scribe or knife, score the polyimide coating of the column within 2 cm of the end. Carefully bend the column in this region and a clean break should result. Examine the new end of the column (with a magnifying glass if possible). The end should be square and free from cracks and bits of fused silica or polyimide. Re-cut the column if necessary. Cut the column after threading on to the ferrule as this will ensure that no ferrule fragments are left in the column.

6. The position of the end of the column inside the injector is extremely critical for good performance, so follow the manufacturer's instructions carefully in this respect. The column position is normally defined as a distance from the back of the column nut to the end of the column. Using a felt-tip pen or typewriting correction fluid, mark the column at the specified distance from the end of the column. Do not use an abrasive marker as this would almost certainly fracture the column. Carefully insert the end of the column into the injector and push the ferrule and nut on to the injector port. Hand-tighten the nut. Use a wrench to further tighten the nut until the ferrule just starts to grip the column. Slacken the nut slightly and adjust the position of the column until the mark is aligned with the back of the nut. Retighten the nut until the ferrule just starts to grip the column and then rotate it by a further quarter turn.

7. Connect the column to the detector following the same procedure as steps 4–6. For many detectors, the capillary column is pushed right through the detector port and into the detector jet or cell to minimize dead-volume effects. If this is not possible, then a make-up gas may be required (see Section 2.3). Refer to the manufacturer's instructions for guidance.

Note: the column is usually connected to the detector after leak testing the carrier gas (see *Protocol 5*).

3.2.4 Application of carrier gas

Connect a suitable carrier gas to the instrument (see Section 2.3) and perform a system leak test as described in *Protocol 5*.

Protocol 5. Performing a carrier gas leak test

1. Disconnect (if necessary) the column from the detector. Seal the column outlet with a suitable blanking union (packed columns) or by pushing the column into a piece of injector septum (capillary columns).

2. Connect carrier gas to the instrument via an on/off toggle valve between the filters and the chromatograph.

3. Pressurize the system to about 30 lb/in^2 (most chromatographs have a gauge or transducer to monitor column inlet pressure). In the case of a split/splitless injector ensure that the split and septum purge vents are sealed.

4. Listen for any hissing sounds which would indicate a serious leak. Eliminate any apparent leaks.

5. Using a Pasteur pipette, apply a drop of water/isopropyl alcohol (50/50, v/v) mixture to all gas joints and look for bubble formation. Eliminate any apparent leaks.

6. Note the reading on the inlet pressure gauge. Turn off the carrier gas inlet toggle valve. Monitor the gauge reading. It should not decrease over a period of 15 min. If a decrease is observed, open the toggle valve and repeat steps 5 and 6.

7. When you are satisfied that the system is leak-free, remove the column outlet seal (capillary columns should be checked for blockage by pieces of septum and re-cut as necessary), open the toggle valve and connect the column to the detector (see *Protocols 3, 4,* and *7*).

Leaks should be investigated at all the following potential sources:

- unions in plumbing external to the instrument
- internal connections between the instrument's pneumatic units
- injector septum, septum purge and split vent lines (if appropriate)
- injector-column union
- possible column fracture

Leaks can normally be cured by a *slight* tightening of the appropriate nut in a union. If this is not effective, replace the ferrule and try again. Leaks (and permanent mechanical damage) can be caused by overtightening a nut and ferrule—so take care. Do not use soapy water for leak testing as this may introduce contamination into the chromatographic system and may weaken the polyimide coating of fused silica capillary columns.

3.2.5 Conditioning the column

The stationary phases used in GC are generally mixtures of compounds; they may also contain residual solvents, reagents, and impurities used during the manufacturing process. It is necessary to remove the volatile content of the stationary phase prior to chromatography to prevent an excessive background signal ('column bleed') or potential contamination of the detection system.

Many capillary columns now use stationary phases which are chemically bonded to the internal walls of the column (see Chapter 3) and may be supplied as being 'pre-conditioned'—even so, some minor reconditioning is a wise precaution. Columns should be reconditioned after long storage or when becoming contaminated. The general procedure for column conditioning is given in *Protocol 6*.

Protocol 6. Column conditioning

1. Install the column in the chromatographic oven and connect it to the injector but not the detector (see *Protocols 3* and *4*).

2. Apply carrier gas and eliminate any leaks (see *Protocol 5*). Ensure that carrier gas filters are installed (see Section 2.3.1) and are in good condition. Set a nominal carrier gas flow or pressure (see *Tables 8–10*).

3. Allow several minutes to pass before heating the column to ensure that all traces of air are swept from the column.

4. Apply a slow temperature programme (e.g. 5°C/min) from 40°C to 10°C below the column temperature limit (refer to manufacturer's documentation). Hold the column at the upper temperature for 30 min.

5. For a new column, step 4 should be repeated several times. Most chromatographs offer synchronization signals which allow a 'ready out' output signal to be connected using short lengths of wire to a 'start in' input to enable automatic cycling of the temperature programme.

6. Cool the oven and connect the column to the detection system.

7. If, in use, excessive column bleed is apparent, further column conditioning should be applied.

3.2.6 Preparing the detector

There are many types of GC detector and many designs of each type from different manufacturers. It is difficult, within the scope of this text, to cover all the steps involved in the set-up of all these detectors. There are, however, a few universal rules and procedures that can be applied and these are covered in *Protocol 7*.

Protocol 7. General preparation of a GC detector

1. Check that the detector column port and internal components (jet, collector, etc.) are clean and free from particulate matter.
2. Check that the signal and any polarizing voltage cables and their connectors are in good condition and fitted properly.
3. Ensure that the chromatographic column is well conditioned (see *Protocol 6*) and connect it to the detector (see *Protocols 3* and *4*).
4. Apply carrier gas to the column as described in Section 3.3.2.
5. Set the detector temperature to 50 °C above the expected highest column temperature. Do not heat the column until this temperature is attained or contamination may accumulate in the colder detector.
6. If required, apply the detector support gases (hydrogen, air, make-up gas, reference gas etc.) at the manufacturer's recommended flow rates. These gases should be cleanly filtered (see Section 2.3.1) or noisy baseline signals may result.
7. Where appropriate, ignite the detector flame (FID or FPD), heat the alkali bead (NPD), or apply filament current (TCD).
8. Set the column oven to a convenient isothermal temperature and leave the system to equilibrate. In the case of a TCD, it may take several hours or, for an ECD, it may take several days to reach a stable background signal and allow low level trace analyses—when not in use, it is recommended that these detectors are left heated with carrier gas flowing through them to maintain them in an operating condition.
9. Check that the detector is functioning correctly by chromatographing a suitable analyte. Manufacturers usually test detectors prior to delivery using a simple test mix. They should be able to provide the user with such a test mix, the chromatographic conditions, and the expected results to confirm correct detector performance.

3.3 Chromatographic method development

Before any analysis can be performed, a suitable method must be established. Method development is, essentially, the process of setting up the chromatographic hardware with suitable conditions and validating the performance with samples of known composition. The quality of the analytical results obtained will be highly dependent on this exercise.

3.3.1 Sample preparation

A liquid sample may be injected directly into a GC system but many samples must undergo extraction, concentration, and/or derivatization procedures before any chromatography can be performed. The use of techniques such as headspace extraction, thermal desorption, purge and trap, and supercritical fluid extraction can minimize this effort in many instances (see Sections 2.4.8, 2.4.9, 2.4.10, and 2.4.11, respectively). Where these are not possible or inappropriate, the chromatographer is faced with the task of injecting a liquid sample, solution, or extract. Decisions must be made concerning the type of solvent and degree of dilution. The factors which affect these decisions are the type of column, injector and detector, and the nature of the analytes and the sample matrix. The solvent must also completely dissolve the analytes and must not co-elute with them during chromatography.

Table 7 lists the range of analyte concentrations suited to the various column types and injection systems when a flame ionization detector is used. With more sensitive detectors (e.g. ECD) much lower concentrations can be used, but the upper limit is imposed by the column which will exhibit peak

Table 7. Analyte concentration ranges

System	Concentration range (%, w/v)
Packed column	0.0005–100
0.53 mm capillary column with splitless injection	0.0001–0.1
Capillary column with split injection	0.005–0.5
Capillary column with splitless injection	0.00005–0.005
Capillary column with PTV split injection	0.001–0.5
Capillary column with PTV splitless injection	0.00005–0.005
Capillary column with cold on-column injection	0.00005–0.005

distortion if too much is injected. With splitless and cold on-column injection the boiling point and polarity of the solvent is critical to achieve a good solvent-focusing effect and to prevent solvent flooding effects (see Section 2.4.2 ii).

3.3.2 Carrier gas flow rate
Column efficiency, hence peak resolution, is affected by the carrier gas flow rate (or to be more precise, linear gas velocity). Each column will have an optimum flow rate for the carrier gas used. *Tables 8–10* list typical flows, linear gas velocities, and pressures, respectively.

i. Measuring gas flows
In the case of packed columns (or 0.53 mm i.d. capillary columns at high flow rates), it is convenient to measure the gas flow rate directly using a rotameter, electronic transducer, or a bubble flow meter (see *Protocol 8*).

Protocol 8. Measuring packed column carrier gas flow rates with a bubble flow meter

1. Install the column (see *Protocol 3*) and ensure the system is leak-tight (see *Protocol 5*). Set the oven temperature to suitable value (see Section 3.3.4).
2. Ensure that the bubble flow meter is clean. Rinse it if necessary. Introduce approximately 2 ml of soap solution into the rubber bulb.
3. Connect the rubber tubing to the detector vent using a suitable adapter if necessary. Some detectors are not sealed, in which case disconnect the column from the detector and attach the rubber tube directly to the end of the column.
4. Squeeze the rubber bulb until a film of soap solution is carried by the gas flow up into the glass body of the flow meter.
5. Using a stopwatch, time the passage of the film between any two of the calibration marks on the flow meter body.
6. Calculate the gas flow rate from the following equation:

$$\text{Flow rate (ml/min)} = \frac{V \times 60 \times T_c \times (P_a - P_w)}{t \times T_a \times P_a}$$

Where: V = volume of gas passed (ml); t = time taken (s); T_c = column temperature (K); T_a = ambient temperature (K); F_w = vapour pressure of water at ambient temperature; and P_a = atmospheric pressure.

7. Repeat steps 4 to 6 until a consistent result is obtained.

For capillary columns, it is more practical and meaningful to measure the carrier gas linear gas velocity (see *Protocol 9*).

Protocol 9. Measuring carrier gas linear velocity in capillary columns

1. Connect the column to the detector.
2. Set up the detector with support gases and light the flame if necessary.
3. Attach a recorder or integrator to the detector output.
4. Set the column oven to an isothermal temperature similar to the top temperature expected for the analysis.
5. Inject (see Section 3.3.3) a compound[a] known to have little retention on the column.
6. Record a chromatographic trace and measure the residence time (t_0) of the unretained solute.

Protocol 9. *Continued*

7. Calculate the mean linear gas velocity from the following equation:

$$\text{Mean linear velocity (cm/sec)} = \frac{\text{Column length (m)} \times 100}{\text{Unretained solute } t_0 \text{ (min)} \times 60}$$

[a] Methane (natural gas) or butane (lighter fluid) are convenient compounds as they are unretained on many columns, their elution will be apparent (as a slight disturbance in some instances) on most detectors.

ii. Optimizing gas flows

The flows and linear gas velocities given in *Tables 8–10* are only intended for guidance. For critical work, the carrier flow or velocity should be optimized for the chromatographic column being used. *Protocol 10* outlines the procedure for optimizing the carrier gas flow/velocity for maximum chromatographic efficiency.

Table 8. Typical optimum gas flows for packed columns

Column internal diameter (mm)	Flow rate (ml/min)
2	20
3	40
4	60

Table 9. Typical optimum carrier gas linear velocities for capillary columns

Carrier gas	Linear gas velocity (cm/sec)
Hydrogen	30–40
Helium	20–30
Nitrogen	10–20

Table 10. Typical carrier gas pressures to give optimum linear carrier gas velocities in capillary columns in 16/in^2

Column length (m)	0.15			0.22			0.32			0.53		
	H$_2$	He	N$_2$	H$_2$	He	N$_2$	H$_2$	He	N$_2$	H$_2$	He	N$_2$
10												
12	13	13	6	5	5	2.5	2.5	2.5		1	1	–
25	25	25	13	10	10	5	5	5		2	2	1
50	50	50	25	20	20	10	10	10		4	4	2

Protocol 10. Optimization of carrier gas flow/velocity

1. Prepare a solution (typically 0.1%, w/v) of an analyte that gives a good detector response and is easy to chromatograph on the column being used.

2. Set the carrier gas flow/velocity according to *Tables 8–10*.

3. Set the oven to a suitable isothermal temperature to give an analyte retention time of 5 to 10 times longer than (t_0) (see *Protocol 9*). This temperature may need adjustment by trial and error.

4. Connect a chart recorder or integrator to the detector output.

5. Inject an aliquot (typically 1 μl) of the solution (see Section 3.3.3). For capillary columns, split mode injection (see Section 2.4.2 i) must be used to give good peak shape under isothermal conditions.

6. Record the resultant chromatogram at a fast chart speed, keeping the analyte peak 'on scale'.

7. Measure the peak retention time (t_R) and width (W_h) at half the peak height using a ruler (alternatively some data handling systems can generate this information directly).

8. Calculate the height equivalent to a theoretical plate (HETP) as described in Section 3.3.5 i.

9. Repeat steps 5 to 8 at carrier gas flows/velocities covering, at 10% increments, the range −20% to +20% of the flow/velocity set in step 2.

10. Plot a graph of HETP against carrier flow or velocity and determine the flow/velocity which gives the minimum HETP. If the minimum lies outside the tested range, extend this range until the minimum is found.

11. Adjust the carrier flow/pressure to the optimum value.

3.3.3 Syringe injection techniques

To introduce a liquid sample into an injection system requires the use of a low capacity syringe. The injection technique used is critical to good chromatographic performance. A good technique should give:

- repeatable peak areas
- low mass discrimination
- no contamination from previous sample injections
- good peak shape

Injection by an autosampler will nearly always give better performance than manual injection because of its more precise control of the syringe mechanics.

There are several manual injection techniques that have been established— each depends on the nature of the sample, the type of injector, the type of syringe, the sample volume to be injected, and the personal preference of the analyst performing the work. The following techniques are provided for guidance only—the chromatographer should develop his/her own techniques and check them using suitable standard samples.

i. Cleaning the syringe

To prevent contamination of the sample, the syringe must be carefully cleaned prior to use. Often this is just a matter of rinsing the syringe at least five times with a suitable solvent, ensuring that the plunger is fully withdrawn during the rinsing process. In instances where a sequence of samples with high and low analyte concentrations are to be injected, the cleaning process must be more rigorous. The best approach in this instance is to use two syringes: one for high concentration samples and the other for low. Where a single syringe is to be used, increase the number of rinses, use an ultrasonic cleaning bath or try leaving the syringe needle inside an unused heated injector to 'bake out' any residue.

The syringe should also be cleaned immediately after making an injection to prevent dry sample residues making the plunger stick and thus causing increased wear and possible mechanical damage to the syringe.

ii. Making an injection

There are two types of syringe that are used for liquid injection in GC.

(a) *Plunger-in-needle* where the sample resides in the syringe needle prior to injection and is suitable for injection volumes up to 1 μl.

(b) *Plunger-in-barrel* where the sample is drawn into the syringe barrel and is best suited for injection volumes in excess of 1 μl.

Table 11 lists the typical application of each type.

For a non-vaporizing injector (PTV or cooled on-column), the syringe

---

Table 11. Use of syringes for liquid injection

Injection type	Plunger-in-needle	Plunger-in-barrel
Packed column	Good	Good
Classical split	Good	Limited [a]
Classical splitless	Limited [b]	Good
PTV split	Good	Good
PTV splitless	Good	Good
Cold on-column	Unusable	Good [c]

[a] Poor precision and discrimination at low volumes.
[b] Normally needs at least 1 μl injection volume.
[c] Needs very narrow needle for columns with less than 0.53 mm i.d.

---

technique is less critical and is usually a case of priming the syringe with a required volume of sample (see *Protocol 11*), introducing the needle into the injector, depressing the plunger, and finally withdrawing the needle.

For heated injectors, the technique is far more important and steps must be taken to ensure that high quantitative precision and low mass discrimination are achieved. *Protocols 12* to *15* are designed to optimize injection performance with heated injectors.

Protocol 11. Priming a syringe

1. Push the syringe plunger fully home.
2. Submerge the end of the needle in the sample.
3. Slowly withdraw the plunger to its limit and wait a few seconds for the sample to enter the syringe (for viscous samples a longer wait may be necessary).
4. Remove the syringe from the sample and discard its contents on to a piece of absorbent tissue paper.
5. Repeat steps 1 to 4 a further four times.
6. Submerge the end of the needle in the sample.
7. By repeated (at least five times) rapid depression of the plunger followed by slow withdrawal, try to eliminate air bubbles from within the liquid held in the syringe.
8. Remove the primed syringe from the sample and examine the liquid for the presence of air bubbles (plunger-in-barrel syringes only). If bubbles are present, invert the syringe and tap the barrel to dislodge the bubbles. Slowly depress the plunger to expel the air from the syringe. Repeat step 7 if insufficient sample volume remains.
9. With the needle pointing downwards, hold the end of the needle against the inner wall of the sample vial above the sample itself and slowly depress the syringe until the required sample volume is achieved.

Protocol 12. Injecting with a plunger-in-needle syringe

1. Prime the syringe with the required sample volume according to *Protocol 11*.
2. Fully insert the needle into the injection port.
3. Quickly depress the plunger.
4. Leave the needle in the port for a timed period (typically 5 sec) to allow the full vaporization of the sample from the syringe needle.
5. Withdraw the needle from the injector.

Protocol 13. Hot needle injection with a plunger-in-barrel syringe

1. Prime the syringe with the required sample volume according to *Protocol 11.*
2. Withdraw the plunger so that the sample is fully drawn into the syringe barrel.
3. Fully insert the needle into the heated injector port and leave it for about five seconds to heat up.
4. Depress the plunger and then quickly withdraw it again.
5. Withdraw the needle from the injector.

Protocol 14. Solvent plug injection with a plunger-in-barrel syringe

1. Prime the syringe with 1 µl of a suitable solvent according to *Protocol 11.*
2. Place the end of the needle in the sample and further withdraw the plunger to take up the required volume of sample.
3. Repeat steps 2 to 5 of *Protocol 13.*

Protocol 15. Air plug injection with a plunger-in-barrel syringe

1. Withdraw the plunger to allow 2 µl of air to enter the barrel.
2. Place the end of the needle in the sample and further withdraw the plunger to take the required volume of sample.
3. Repeat steps 2 to 5 of *Protocol 13.* The exact sample volume may be measured in the syringe barrel prior to injection.

3.3.4 Setting the oven temperature

The column oven may be operated in one of two modes: isothermal or temperature-programmed. The choice depends on several factors including sample composition, column type, and mode of injection. *Figure 17* indicates a typical approach. Generally, isothermal conditions are used for simple mixtures and temperature-programming is used for complex mixtures or, in the case of capillary columns, when the injection technique demands it (e.g. splitless or cooled on-column). *Protocols 16* and *17* describe the steps involved in establishing oven temperature conditions.

Optimizing a temperature programme can be an extremely time-consuming

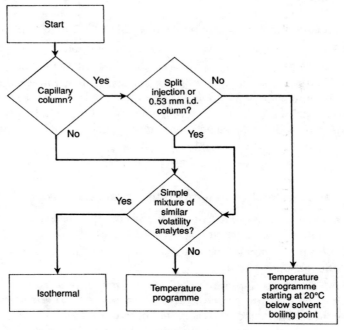

Figure 17. Selection of GC temperature conditions.

task. For highly complex mixtures, slight variations in the programme may bring about significant changes in the chromatographic separation. It is very difficult to predict the relationship between the programme and the resultant separation, especially if many of the peak identities are unknown. Development of a temperature programme is often, therefore, a matter of trial and error. Close review of the literature or use of chromatographic prediction software, such as 'Drylab GC' (6) is recommended for critical work.

Protocol 16. Establishing isothermal conditions

1. Set up the GC with the column, injector, and detector under their normal operating conditions.

2. Set a temperature programme to run at 10°C/min from a near-ambient temperature to near the maximum temperature limit of the column (see manufacturer's details).

3. Inject the sample, run the temperature programme, and record the chromatogram.

4. When the run has finished, locate the last eluting peak and calculate its elution temperature from its retention time and the temperature programme.

Protocol 16. *Continued*

5. Set an isothermal oven temperature at about 20 °C below this value.

6. Reinject the sample and examine the chromatogram.

7. All peaks of interest should be well separated from the solvent peak and from each other. If this is not the case, then reduce the oven temperature slightly and re-inject. If a satisfactory separation still cannot be achieved, then a different column (higher stationary phase loading or a different selectivity stationary phase) or a temperature programme may be required.

Protocol 17. Establishing temperature programmed conditions

1. Repeat steps 1 to 3 of *Protocol 16*. In the case of capillary columns, set the initial oven temperature as described in Section 2.4.2 ii.

2. Establish the elution temperatures of the first and last peaks of interest and the last peak to elute.

3. Set the temperature programme initial temperature to 20 °C below the elution temperature of the first peak of interest. Where a capillary column is being used, the initial temperature may be dictated by the mode of injection (see Section 2.4).

4. Set an isothermal period of at least 1 min (this will improve retention time repeatability and peak shape).

5. Set a moderate programming rate (e.g. 5 °C/min for capillary columns or 10 °C/min for packed columns) up to 10 °C below the column temperature limit.

6. If the initial temperature is set lower than the elution temperature of the first peak of interest, to support the mode of injection, set a fast initial programming rate (e.g. 20 to 30 °C/min) to the elution temperature of the first peak of interest, then set a second ramp as described in step 5.

7. Set an isothermal period, following the ramp, of 5 min.

8. If peaks elute after the last peak of interest, set a fast ramp up to 10 °C below the temperature limit of the column and hold the column under these conditions until all peaks have eluted. Failure to do this will result in the residual peaks eluting in subsequent chromatography. Better methods of removing unwanted late-eluting components would be to use pressure programming to increase the pressure, hence flow rate, of the carrier gas to speed up elution or, preferably, to use a technique called backflushing to reverse the direction of carrier gas flow through the column to quickly drive components back out through the column inlet. Refer to the manufacturer's literature for further details.

3.3.5 Chromatographic interpretation and validation

Having established a working chromatographic method, it is useful to quantify and document its performance. This enables the user to assess the quality of the subsequent analytical results and to act as a benchmark if the analysis is to be repeated at a later time. Several performance indices can be calculated and Section 3.3.5 i to ix describe some of the most common.

i. Chromatographic efficiency

Efficiency is a measure of how much a band of analyte broadens as it passes through a column. As column efficiency is increased, resolution (see Section 3.3.5 ii) will also increase enabling more complex samples to be examined. High efficiency, however, is no guarantee for high resolution as no improvement in co-eluting peaks would be apparent. Efficiency can be expressed in several ways but the two below are most widely used:

$$\text{Number of theoretical plates } (n) = 5.545 \cdot \left(\frac{t_R}{W_h}\right)^2$$

$$\text{Height equivalent to a theoretical plate (HETP), cm} = \frac{L}{n}$$

Where: t_R = retention time of peak (sec); W_h = peak width at half height (sec); and L = column length (cm).

A good column should give about 2000 plates/metre (packed) or 5000 plates/metre (capillary) and so calculating the number of theoretical plates gives a good measure of system performance.

ii. Resolution

Resolution describes the degree of separation between a pair of adjacent peaks and is calculated as shown below. A minimum resolution of 1 should be sought for all peaks of interest in the chromatogram.

$$\text{Resolution, Rs} = \frac{t_{R2} - t_{R1}}{W_{h1} + W_{h2}}$$

Where: t_{R1} = retention time of first peak; t_{R2} = retention time of second peak; W_{h1} = width at half height of first peak; and W_{h2} = width at half height of second peak. These values must have the same units.

iii. Sensitivity

Sensitivity describes the response of the detector per unit mass per unit time (mass-flow-dependent detectors) or per unit concentration (concentration-dependent detectors) for a given analyte. Sensitivity should not be confused with detectability. The latter is a more meaningful measure of analytical performance as it takes into account the level of background noise and thus has more practical value. Sensitivity, however, can be used to confirm correct detector operation and to make comparisons between different designs.

iv. Detectability
Detectability is the ability to discern a chromatographic peak from the background noise and thus defines the lowest amount of analyte that can be analysed. Detectability can be expressed in two ways:

(a) The limit of detection is usually calculated as the mass of analyte that gives a peak with an amplitude twice that of the background noise. This is highly dependent on peak shape as sharp peaks (e.g. capillary chromatography) will give lower detection limits than broad peaks.

(b) The limit of quantification is expressed as the mass of analyte that can be reliably quantified (i.e. integrated). It is usually calculated as the mass of analyte that gives a peak with an amplitude five or ten times greater than the background noise.

v. Accuracy
Accuracy is the ability to generate results that truly represent the content of the analyte in the sample. Accuracy can be assessed by analysing blank samples with a known amount of added analyte (recovery determinations). If the precision is poor, then averaging the results of multiple analyses may be necessary to achieve good accuracy.

vi. Precision
Precision expresses the ability of the method to achieve the same result from the same sample on the same instrument in succession. Qualitative precision expresses the repeatability of peak retention times and quantitative precision expresses the repeatability of peak areas (or heights). These figures are normally calculated as relative standard deviations. A good method will have a qualitative precision of less than 0.05% and a quantitative precision of less than 1%. Good precision is necessary for reliable results.

vii. Reproducibility
This is often confused with precision but it has a slightly different meaning. Reproducibility is the ability of the method to achieve the same result from the same samples but using a different instrument, column, operator, etc. Good reproducibility is necessary for a reliable method.

viii. Dynamic range
The dynamic range defines the range of analyte levels that can be analysed. It can be expressed simply as the ratio between the largest amount of analyte that can be determined against the smallest.

ix. Linearity

Linearity is similar to dynamic range but applies to the ratio of the largest analyte amount to the smallest between which the detector response is linear (typically within 5%). Good linearity simplifies quantification. If the linearity is poor then a multi-level calibration would be necessary to achieve accurate results.

4. Data handling

4.1 Introduction

So far, the instrumental components and techniques that are necessary to produce chromatography have been considered. The detector output often comprises a very small analogue current or voltage. This section considers the necessary steps involved in extracting meaningful information from this signal.

4.2 Potential information from chromatography

Before examining the mechanics of producing analytical results, we should first consider what information could be derived from the chromatographic signal:

(a) The retention time of a peak can be used to give an indication of identity of the compound responsible for that peak. Note that chromatography alone cannot provide positive proof of peak identity as it is possible for different compounds to coelute. Positive confirmation of compound identity must involve a supporting independent technique such as mass spectrometry.

(b) The amount of compound present is directly related to the area and height of a peak. For most detectors this relationship is linear.

(c) The shape of a peak and the nature of the background signal allow a measure of confidence to be placed in the results produced.

Chromatography can thus be used for both qualitative and quantitative determinations. Any general data handling system must, therefore, start with a reliable means of measuring peak retention times and peak areas and/or heights.

4.3 Detector signal processing

Data handling cannot be performed on low-level detector signals, so electronic processing of the signal is necessary to convert it into a more manageable form.

Figure 18 shows the various processes the detector signal undergoes within a typical gas chromatograph.

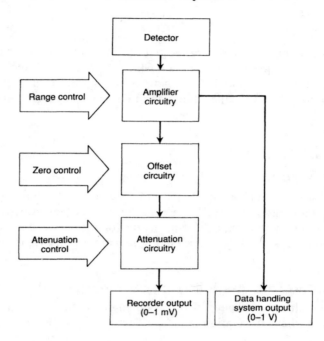

Figure 18. Steps involved in the production of analogue output signals.

The first step is to amplify the raw detector signal by many orders of magnitude to provide a more manageable signal. Some detectors (for instance an FID) will have a greater dynamic range than the amplifier, in which case, the effective gain of the amplifier may be adjusted by a 'range' switch to suit the size of the chromatographic peaks being produced.

A 'zero' or 'offset' control adjusts the output from the amplifier to 'position' a chromatogram correctly on a recorder or printer/plotter. On modern chromatographs, the signal offset is automatically reset prior to each run ('autozero').

Gas chromatographs usually provide two signal outputs for each detector: a low voltage (typically 0–1 mV) signal for analogue recorders and a higher voltage (typically 0–1 V) signal for connection to a digital data system. To plot a chromatogram on an analogue recorder, which has a fixed sensitivity, some control is required to ensure that the output signal is correctly scaled to suit this sensitivity. This is provided by an 'attenuation' control. Note that the attenuation is applied after signal amplification. No attenuation is normally provided on the integrator output as this facility is usually available within the data processing system itself.

4.4 Approaches to data handling

There are essentially two practical ways of generating results from the analogue signals delivered by the chromatograph:

(a) The recorder output can be connected to a suitable chart recorder and a chromatogram is traced on chart paper. Manual measurements can be made from the trace using a ruler to give retention times and peak heights. Peak areas can be derived by using gridded chart paper and estimating the number of squares under a peak, by cutting out each peak and weighing it, by measuring the peak width and height and calculating half their product, or by using a device called a planimeter. All these techniques are labour-intensive, are time consuming, require much manual transposition of data, and are prone to human error—they are hardly used in a modern laboratory except for diagnostic or confirmatory purposes.

(b) A much superior approach to data handling is to use a purpose-designed microprocessor-based system. Most laboratories now use such systems because of their low cost, speed, and accuracy. These devices come in many forms: personal integrators, workstations, or laboratory automation systems.

4.5 Microprocessor-based data handling systems

Although the software on a modern data handling system is highly sophisticated, it does need important input from the user in order to process data correctly. Data handling systems, therefore, should be regarded as analytical tools and it is the responsibility of the user to understand and optimize their operation. The general approach to data handling is similar on many systems but the mechanics of operation will vary widely. The details given here relate to a 'typical' data handling system.

4.5.1 Design fundamentals

Data handling systems are essentially computers containing some specialized electronics. Although they are produced in several physical forms, the general build is normally similar to that shown in *Figure 19*.

Central to the operation is the central processing unit (CPU) which is responsible for most operations within the system. It is driven by a software programme that contains all the CPU instructions necessary to perform the chromatographic data handling functions. The CPU has access to a block of random access memory (RAM) which can be used for temporary storage of chromatographic data and programme variables. The CPU communicates with the chromatograph through three interfaces: an analogue signal input (for the chromatographic signal), a binary coded decimal input (for vial numbers etc.), and a synchronization signal interface (for start/stop/ready

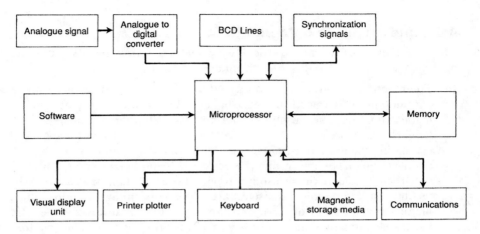

Figure 19. Schematic of a typical digital data handling system.

status signals). The more powerful systems will interface with more than one chromatograph and allow simultaneous asynchronous operation. The analogue signal cannot be accessed directly by the CPU so a specialized circuit is required to effect an analogue to digital conversion (ADC). This circuit may be housed within the data station, in an external interface box or even within the chromatograph itself. The CPU also is linked to a number of devices responsible for communicating with the outside world. These include a keyboard (for user instructions), a printer/plotter (for reports and chromatograms), a visual display unit (VDU, for review of status), magnetic media (floppy disks, hard disks, tapes, etc. for data storage), and communication interfaces for transmitting data to and from external computers (networking).

4.5.2 Data acquisition

As the analogue signal enters the data system, it is converted into its digital equivalent by the ADC circuit. The analogue signal is a continuous signal. To convert this into a digital format, discrete readings must be taken at regular intervals. An 'integrating ADC' is used for chromatography as, instead of taking readings at instants in time, it measures the mean signal over time intervals. The output from an integrating ADC will be a sequence of values representing what can be considered as 'time slices' taken from the original analogue signal. Time slice values are normally calculated with respect to a negative reference voltage to allow for a negative drift in the analogue signal. The ADC should possess a dynamic range in excess of that of the analogue signal coming from the chromatograph and should have a very high degree of linearity. The optimum sampling rate is determined by the software integration algorithm employed and will be related to the widths of the peaks to be processed. Typically, rates of up to 100 time slices per second are used for GC.

4.5.3 Generating analytical results from time slice data

The time slice values are processed by the CPU according to instructions provided by the software programme. *Figure 20* summarizes these processes for a typical system.

The first step is to store the time slice data in RAM. If there is sufficient RAM to store the whole chromatogram, then the data can be interactively reprocessed to enable optimization of the controlling parameters. The stored data can be used to produce chromatograms on a printer/plotter.

The stored time slice values may then be 'bunched' together to adjust the effective number of slices per chromatographic peak. Many peak processing algorithms depend on a set number of slices per peak for optimum performance.

The bunched slices are then normally subjected to digital smoothing to reduce the effects of noise and to remove any 'spikes'.

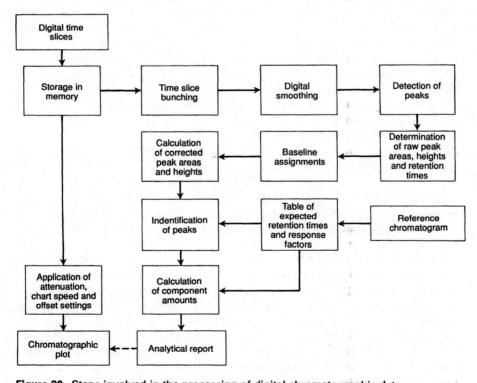

Figure 20. Steps involved in the processing of digital chromatographic data.

Andrew Tipler

4.5.4 Peak detection

The processed slices are then examined for the presence of peaks. Peak detection involves the discrimination of peaks from the background signal noise. The location of the peak start, crest, and end is normally necessary to establish the presence of a peak. There are various methods of establishing these points—most involve the comparison of some signal attribute such as rate of change (first differential), the rate of change (second differential), or area accumulation (integral) against a user-adjustable threshold.

Once a peak has been detected, the peak retention time can be computed from the slice number at the peak crest, the raw peak area can be calculated by summation of the slices between the peak start and peak end, and the raw peak height is equivalent to the value of the slice at the peak crest.

4.5.5 Baseline assignment

The next step is to predict the position of the background signal underneath the peak. This can be achieved, using the same algorithms applied in peak detection, by establishing baseline points on the signal where peaks are absent and interpolating a predicted background signal between these points. This baseline can then be subtracted from the raw peak areas and heights to give corrected values. In situations where the true baseline is drifting or is disturbed by solvent peaks, valve switching, etc., most systems provide tools to control baseline assignment locally using a table of *timed events*. Where VDU interaction is involved, it is usually possible to 'draw' the baseline manually on the screen. Most systems are able to plot the chromatogram with the applied baseline shown to confirm correct assignment.

4.5.6 Generating a peak table

At this stage, the software is able to construct a peak data table containing a list of retention times of detected peaks and their corrected areas (or heights). This data, in itself, is usually insufficient for analytical purposes, as no component identification or quantification has yet been performed, but it can be useful for development and diagnostic purposes and most data stations allow access to the peak data table through an 'AREA%' and/or 'HEIGHT%' report.

4.5.7 Peak identification

To perform peak identifications, a reference mixture containing known components is chromatographed and the retention times of the peaks of interest are recorded in a component list. The retention times of peaks in sample chromatograms (run under identical conditions) are then compared with this component list and examined for the presence of peaks at the expected retention times. This matching process is not exact, as retention times will vary from run to run, so an adjustable matching tolerance is usually provided. Better performance is achieved if peaks are matched using relative retention

66

times where all retention times are divided by the retention time of an easily identifiable reference peak which is present in all chromatograms. For qualitative analyses, peak identification is often all that is needed.

4.5.8 Quantification

For quantitative results, calibration by chromatography is required. A standard mixture containing known amounts of components is chromatographed, the peak area (or height) for each component is determined, and a calculated response factor is entered into the component list. Response factors are calculated as follows:

$$\text{Response factor} = \frac{\text{Peak area (or height)}}{\text{Component amount}}$$

The reference chromatogram (see Section 4.5.7) and calibration chromatogram are normally combined in a single chromatogram. The appropriate response factor is applied to each identified peak area (or height) in a sample chromatogram to give results representing the quantitative composition of the sample. Response factors can be applied to peak areas (or heights) from sample chromatograms in several ways:

$$\text{External standard, component amount in sample} = \frac{\text{Peak area}}{\text{Response factor}} \quad \text{(a)}$$

This calculation is usually used in instances where a small number of components are being determined in relatively complex chromatograms (e.g. residue analysis). Although it is probably the easiest to use, the results are directly affected by any variations in the injection process.

$$\text{Normalization (\% component in sample)} = \frac{100 \cdot \text{Peak area}}{\text{Response factor} \cdot \dfrac{\text{Peak area}_i}{\text{Response factor}_i}} \quad \text{(b)}$$

This calculation is used for simple mixtures where all components are apparent in the chromatogram (e.g. gas analysis). It compensates for variations in the injection process as all chromatographic peaks are included in the calculation.

$$\text{Internal standard, component amount in sample} = \frac{\text{Peak area} \cdot \text{Int Std Response factor}}{\text{Response factor} \cdot \text{Int Std Peak area}_i} \quad \text{(c)}$$

With this approach, all samples and calibration mixtures are fortified with the same amount of a compound which, when chromatographed, elutes close to the peaks of interest but does not coelute with any other peak. It is, therefore, suitable only for relatively simple chromatograms (e.g. quality control analyses). This calculation effectively compensates for any variations in the injection process.

A single response factor is only suitable for analyses where the peak size is directly proportional to the amount of analyte present. To provide better calibrations, most data stations support multi-level determinations (e.g. linear through origin, linear with offset, quadratic, cubic and logarithmic) involving chromatography of standard mixtures covering a range of concentrations followed by a statistical curve fit to the results.

4.5.9 Getting results

Once the quantitative data is available, the data system is able to print analytical reports summarizing this information (*Figure 21* shows a typical example). Many data handling systems have reporting options which allow report formats to be customized to suit the analysis. The information generated during the data processing, for instance peak retention times, peak names, and baseline assignments, can be used to annotate chromatographic plots to aid interpretation (*Figure 22* shows a typical example). Many data handling systems allow further processing of the data, either by in-built functions or by the use of programming languages such as BASIC, to suit the requirements of specific analyses. Despite the use of sophisticated software in modern systems, review and interpretation of the results by the human chromatographer remains a very important part of the analytical process. These devices are only tools and, as such, errors can occur which will remain undetected unless efforts are made to locate them.

4.5.10 Data traceability

As most modern data handling systems are able to store the raw data (digital time slices) and reports in digital files, some consideration must be given to how to preserve such data so that it can be reviewed and reprocessed at some later date. Many laboratories must work under regimes such as 'good laboratory practice' (LC Resources, Lafayette, CA) which specifies key requirements for data storage. One such requirement is the need to *trace* the final results back to the raw data and the data handling parameter settings so that if results need to be checked, the raw data can be reprocessed, replotted, and re-reported to produce identical results. It is therefore important that any file or printout includes references to the sample details, raw data file, the date of analysis, the data handling method, the chromatographic method, and the name of the analyst.

4.5.11 Data archive and communications

Because the chromatographic data can be represented in a digital form, this information can be stored on magnetic disk media for later review or reprocessing. If the data handling system possesses a disk drive, then these operations can be performed locally. Most systems have a digital communications interface which allows transmission of the data to a remote computer where they can be archived and reprocessed. The remote computer could be a

Peak #	Time [min]	Area [uV*sec]	Norm. Area [%]	Component Name
1	0.304	29.00	0.00	
2	8.148	11323.70	1.57	
3	8.798	12908.19	1.79	ethane
4	9.549	10655.74	1.48	ethylene
5	11.923	19310.01	2.68	propane
6	18.467	19512.10	2.71	propylene
7	21.224	23185.87	3.22	iso-butane
8	22.364	25777.57	3.58	n-butane
9	23.674	9918.08	1.38	acetylene
10	25.173	0.19	0.00	
11	26.350	178.82	0.03	
12	27.146	2532.31	0.35	
13	27.293	16745.47	2.32	trans-2-butene
14	27.790	19320.23	2.68	iso-butene
15	28.572	1072.09	0.15	1-butene
16	29.223	13858.52	1.92	cis-2-butene
17	30.377	617.71	0.09	
18	30.620	15450.06	2.14	cyclopentane
19	30.699	47133.33	6.54	iso-pentane
20	31.254	1221.17	0.17	
21	31.521	39512.70	5.48	n-pentane
22	32.951	3059.67	0.42	dichloroethylene
23	34.097	18025.59	2.50	2-methyl-2-butene
24	34.295	3968.78	0.55	cyclopentene
25	34.446	30237.39	4.20	trans-2-pentene
26	35.122	32250.94	4.47	3-methyl-1-butene
27	35.450	28647.43	3.97	1-pentene
28	35.656	1859.80	0.26	
29	36.024	24656.26	3.42	cis-2-pentene
30	37.378	60003.38	8.32	2,2-dimethylbutane
31	38.051	37702.66	5.23	3-methylpentane
32	38.157	49711.50	6.90	2-methylpentane
33	38.319	44960.26	6.24	2,3-dimethylbutane
34	39.539	38085.43	5.28	isoprene
35	39.730	182.66	0.03	
36	40.526	702.44	0.10	
37	41.086	430.80	0.06	
38	41.160	48393.24	6.71	4-methyl-1-pentene
39	41.631	90.67	0.01	
40	41.773	7495.82	1.04	2-methyl-1-pentene
41	44.106	85.66	0.01	
		720813.19	100.00	

Figure 21. Typical example of an analytical data report.

simple personal computer or it could be another data handling system forming part of a local area network (LAN) or it could be the central computer in a laboratory information management system (LIMS). Many remote systems also provide instrument control facilities to enable chromatographic methods to be stored and set up remotely.

These options are available for most data handling systems and their selection usually depends on the scale of the work being performed, although,

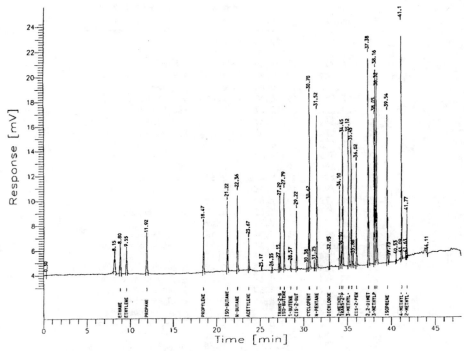

Figure 22. Example of an annotated chromatographic plot.

with the seemingly ever-decreasing cost of computer hardware, even small laboratories may find these systems affordable.

References

1. Grob, K. (1986). *Classical split and splitless injection in capillary gas chromatography*. Huethig Verlag, Heidelberg.
2. Grob, K., Fröhlich, D., Schilling, B., Neukom, H. P., and Nägeli, P. (1984). *J. Chromatogr.*, **295**, 55.
3. Hinshaw, J. (1992). *LC-GC Int.*, **5**, 14.
4. Grob, K. (1991). *On-column injection in capillary gas chromatography*. Huethig Verlag, Heidelberg.
5. Ioffe, B. V. and Vitenberg, A. G. (1984). *Head space analysis and related methods in gas chromatography*. Wiley, New York.
6. Bautz, D. E., Dolan, J. W., and Snyder, L. R. (1991). *J. Chromatogr.*, **541**, 1.

3

Development, technology, and utilization of capillary columns for gas chromatography

PETER A. DAWES

1. Introduction

The concept of using hollow tubes as chromatography separation columns stemmed from a theoretical mathematical investigation and later experimental work by Marcel Golay in 1957. The more correct terminology for this invention is open tubular capillary column but the shortened and more general terminology of 'capillary column' has found its way into common use.

The most obvious benefit of capillary columns over packed columns was the increased resolving capacity. An examination of the van Deemter equation will yield several reasons why a capillary column can have higher resolving power than a packed column. A large contributing factor is, however, simply the increased length.

The gas flow in packed columns is restricted by the small particles of packing, which means that only short low resolution columns may be used because the pressure drop across the column becomes too great. An open tubular capillary column has no obstruction in the centre of the tube which means that the carrier gas (mobile phase) is not greatly restricted and columns of much greater lengths and resolving capacity may be used.

Advantages of capillary columns over conventional packed columns are:

- greater resolving power
- increased speed of analysis
- greater sensitivity
- capability of eluting a greater range of components
- less interference from contaminants
- reproducibility between columns

Capillary gas chromatography is the preferred chromatographic technique, with greater resolution and speed of analysis, providing that the molecules of solute to be analysed are thermally stable, inert in terms of reacting with the capillary column, and have sufficient volatility.

2. Types of capillary column

Several types of capillary column (*Figure 1*) have been developed as follows.

(a) Wall-coated open tubular (WCOT) is the most common type of capillary column now used. The thin layer of polymer stationary phase is coated on to the wall of the capillary tube.

(b) Porous layer open tubular (PLOT) columns have a relatively thick layer of particles of a solid absorbent material such as aluminium oxide or a porous polymer coated on to the capillary tube wall.

(c) Support-coated open tubular (SCOT) columns have liquid stationary phase coated on to particles of a solid support material. These particles are then coated on to the capillary tube wall. SCOT columns were widely used but are rarely used today and will not be further discussed.

(d) Micropacked capillary columns are not open tubular columns but simply packed columns of capillary dimensions and will not be further considered.

3. Evolution of the modern capillary column

Capillary GC was available for many years before it became widely used. The critical factors that led to the wider acceptance of capillary gas chromatography were the introduction of fused silica capillary fabrication by Dandeneau and Zerenner of Hewlett Packard, new silylation methods for deactivation,

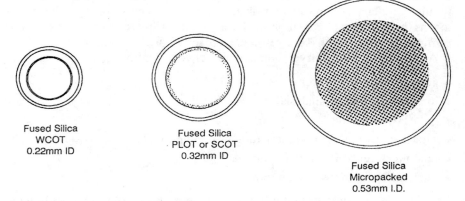

Fused Silica
WCOT
0.22mm ID

Fused Silica
PLOT or SCOT
0.32mm ID

Fused Silica
Micropacked
0.53mm I.D.

Figure 1. Types of capillary column.

stationary phase immobilization through cross-linking and more refined quantitative sample introduction methods.

3.1 Column support material

Fused silica is a high purity, synthetically produced glass with very low levels of metal contaminants. It is inherently straight and flexible due to the thin glass wall which is typically 0.05 mm thick.

Fused silica has greater strength than other glasses but in small capillary tubing it is only strong while it is free from imperfections. Defects from dust or moisture on the surface form nucleation sites for crack growth. Immediately the fused silica material is drawn in a high temperature furnace from a large preform tube down to a small capillary, the fused silica must be protected.

The most common and, to date, most satisfactory material for protecting the fused silica is polyimide. A thin film of polyimide is coated and then cured on to the surface of the fused silica to form a strong hermetic seal.

For applications where the maximum temperature of the polyimide (370°C) would be exceeded, aluminium-coated fused silica has been developed which can be used up to temperatures of 500°C, if permitted by the stationary phase coating on the inside of the capillary column.

One of the great benefits of a fused silica capillary column is the ease with which its thin wall permits low volume connections and insertions into injectors and detectors to be made. This is critical when dealing with the small carrier gas flows required in capillary gas chromatography. This advantage alone has made it possible to design more practical equipment for connections between columns, better detection transfer systems, multidimensional columns, outlet splitters, inlet splitter systems, and other refinements.

3.2 Stationary phase criteria and development

Stationary phases must meet certain criteria to be considered satisfactory for use in wall-coated open tubular capillary columns. These requirements are considered below.

3.2.1 Thermal stability

At the high temperatures often required in gas chromatographic analysis, the phase must not undergo thermal decomposition.

3.2.2 Physical stability

It must be possible to coat the stationary phase film uniformly over the inner wall of the capillary. The film must be physically stable as the column is heated.

3.2.3 Cross-linking

Although it is not essential for a stationary phase to be cross-linked and bonded to the fused silica capillary wall, nearly all capillary columns are now cross-linked. Cross-linked stationary phases have greater thermal stability and are non-extractable in most solvents.

3.2.4 Partition with solute molecules

The solutes must be capable of being in equilibrium between the mobile (gas) phase and the stationary phase at a useful temperature.

Most stationary phases have a temperature below which they no longer are able to partition the solute molecules with the carrier gas. Gas–liquid chromatography is no longer occurring and the peaks become *very* broad and poorly resolved. As the temperature is programmed above the minimum temperature the peak shapes return to normal. If the minimum temperature is too high, application of the column is limited. In addition the stationary phase must be capable of operating at temperatures high enough to be able to elute the compounds of interest without undergoing a breakdown or change in its characteristics. *Table 1* shows some typical minimum and maximum temperatures.

3.2.5 Chemical inertness to solute molecules

The solute molecules must not undergo a chemical reaction with the stationary phase or any impurities in it, and thus be lost during the analysis.

3.2.6 Phase selectivity

There are several modes of interaction and effects between the stationary phase and solute molecules, as listed below. The stronger the interaction between the solute molecule and the stationary phase the greater the retention of the solute molecule.

i. Dispersive interactions

Dispersive interactions result from van der Waals forces and tend to be weak and non-specific. These interactions tend to give separation simply by boiling point differences.

Table 1. Minimum and maximum operating temperatures of common stationary phases

Crosslinked stationary phase	Minimum temperature (°C)	Maximum temperature (°C)
Dimethylsiloxane (BP1)	−60	320
5% Diphenyl dimethyl siloxane (BP5)	−60	320
Dimethyl silarylene (BPX5)	−80	370
14% Cyanopropylphenyl dimethylsiloxane (BP10)	−20	270
50% Cyanopropylphenyl dimethylsiloxane (BP225)	40	250
Polyethylene glycol (BP20)	20	250
Cyanopropyl silarylene (BPX70)	25	270
Dimethylsiloxane–carborane copolymer (HT5)	10	460

ii. Dipole–dipole interactions
These interactions occur when the solute molecule and the stationary phase have a permanent dipole.

iii. Dipole-induced dipole interactions
These interactions arise where the dipole in a solute molecule induces a transient dipole in the stationary phase which usually contain aromatic moieties.

iv. Acid–base interactions
These interactions arise where electron pairs on the oxygen in polyethylene glycol phase or the lone electron pair on the nitrogen of cyanosiloxane stationary phases function as donors.

v. Molecular shape separations
These separations can be carried out by chirality using the range of cyclodextrin stationary and liquid crystal stationary phases that are now becoming available.

More information on the separation mechanisms of stationary phases is available (1, 2).

3.2.7 Polymer reproducibility
Stationary phases must be synthesized very reproducibly to ensure that a separation can be reproduced on different batches of the material.

3.3 Types of stationary phases for partitioning
There are several types of stationary phases for partitioning as described in the following sections.

3.3.1 Silicones
Silicones are the most commonly used stationary phases for partition GC (see *Figures 2a, b*, and *c*). The polymer backbone has good flexibility which allows high diffusion rates of the solute molecules into the polymer. Generally, siloxanes have ideal properties with good thermal stability, only a slight decrease in viscosity with increased temperature and good film-forming properties. They resist oxidation and very importantly they can be synthesized with a variety of pendant organic groups to provide a range of selectivities.

3.3.2 Polyethylene glycol
Polyethylene glycol (PEG) stationary phase (*Figure 3*) is the most commonly used stationary phase material after the siloxanes. PEG is a moderately polar material and for many years was the most polar stationary phase available on fused silica capillary columns due to difficulties coating and cross-linking polar siloxane stationary phases.

LIVERPOOL
JOHN MOORES UNIVERSITY
AVRIL ROBARTS LRC
TEL. 0151 231 4022

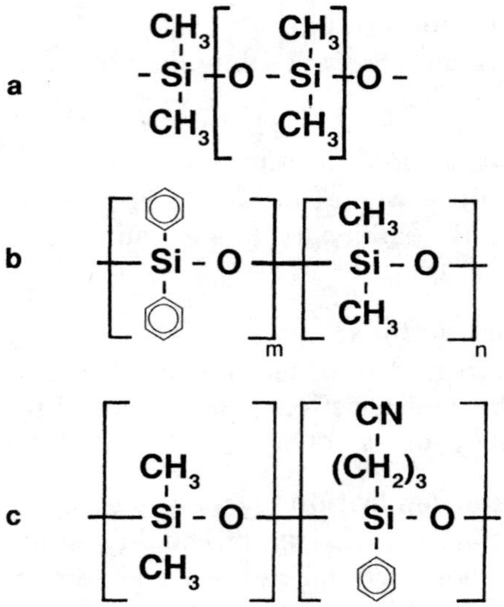

a

$$CH_3 \begin{bmatrix} CH_3 \\ | \\ -Si-O-Si-O- \\ | \quad\quad | \\ CH_3 \quad CH_3 \end{bmatrix}$$

b

$$\begin{bmatrix} \bigcirc \\ | \\ -Si-O- \\ | \\ \bigcirc \end{bmatrix}_m \begin{bmatrix} CH_3 \\ | \\ -Si-O- \\ | \\ CH_3 \end{bmatrix}_n$$

c

$$\begin{bmatrix} CH_3 \\ | \\ -Si-O- \\ | \\ CH_3 \end{bmatrix} \begin{bmatrix} CN \\ | \\ (CH_2)_3 \\ | \\ -Si-O- \\ | \\ \bigcirc \end{bmatrix}$$

Figure 2. Structures of the most commonly used siloxane stationary phases. (a) Dimethylsiloxane; (b) diphenyldimethylsiloxane; (c) cyanopropylphenylsiloxane.

The PEG materials have a limited maximum temperature and have a reputation of being very susceptible to damage by moisture and oxygen. These deficiencies have largely been overcome by the cross-linked PEG phases now available.

For the analysis of acidic solutes, particularly the free fatty acids, an acid-treated version of PEG phase is used.

$$-\!\!\left[CH_2 - CH_2 - O \right]_m$$

Figure 3. Structure of polyethylene glycol.

3.3.3 Modified siloxane polymers

The siloxane polymers may be modified to improve performance characteristics further. By substituting carborane and benzene rings into the backbone of the siloxane (see *Figure 4*), enhanced thermal stability and selectivity can be achieved.

Fused silica capillary columns coated with carborane-modified siloxane are capable of operation up to 480°C. This enables the chromatography of compounds of lower volatility than has previously been possible.

Silarylene stationary phases have higher thermal stability than ordinary siloxane stationary phases but are designed to have other characteristics very

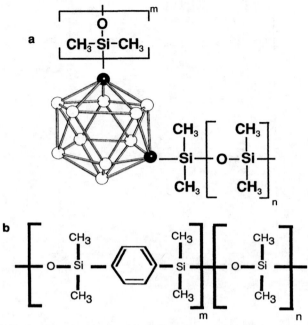

Figure 4. Structures of high temperature stationary phases. (a) Siloxane-carborane; (b) silarylene.

similar so that they may be used for the same applications with the advantage of reduced bleed (discussed in Section 4.7).

3.3.4 Solid adsorption chromatography

Capillary columns are now able to compete with packed columns in the analysis of volatile compounds and free gases due to the availability of adsorption PLOT columns. The main stationary phase materials used are: molecular sieve, aluminium oxide, and a number of porous polymer materials which are copolymers of styrene and divinylbenzene.

The porous polymer PLOT columns have been shown to be very versatile for the separation of a large number of volatile gases, very light hydrocarbons, and sulphur compounds.

Molecular sieve type columns make the separation of some free gases possible and alumina columns are useful for the analysis of volatile hydrocarbon isomers up to C_6. A fuller explanation of which stationary phase is most suitable for a particular analysis is available (3).

3.3.5 Chiral phases

Developments in cyclodextrin stationary phases have made the separation of many types of enantiomers possible (*Figure 5*). The cyclodextrin is a cyclic

Peter A. Dawes

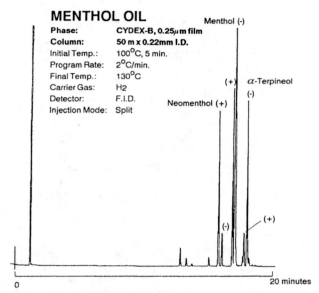

MENTHOL OIL

Phase:	**CYDEX-B, 0.25μm film**
Column:	**50 m x 0.22mm I.D.**
Initial Temp.:	100°C, 5 min.
Program Rate:	2°C/min.
Final Temp.:	130°C
Carrier Gas:	H2
Detector:	F.I.D.
Injection Mode:	Split

Figure 5. Enantiomer selective separations.

oligomer (*Figure 6*) which is substituted into a conventional siloxane stationary phase. Organic molecules of correct size and shape can interact more strongly with the cavity in the cyclodextrin and be more strongly retained on the capillary column. The cyclodextrin oligmer can be modified in many ways to make the phase selective to particular stereoisomers.

3.4 Manufacturing methods

A full range of capillary column types and configurations are readily available from commercial manufacturers. Although there is a great deal of information on capillary column manufacturing methods in the literature, the best

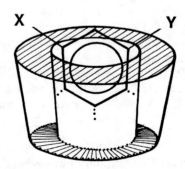

Figure 6. Structural unit of a cyclodextrin stationary phase.

78

Table 2. Commercially available stationary phases and equivalents

Cross-linked stationary phase	Commercial equivalent
Dimethylsiloxane	BP1, DB1, HP1, SE30, OV1, CPSil5
5% Diphenyldimethylsiloxane	BP5, DB5, HP2, SE54
14% Cyanopropylphenyldimethylsiloxane	BP10, DB1701, OV1701
50% Trifluoropropylmethylsiloxane	OV210, DB-210, QF-1
50% Cyanopropylphenyldimethylsiloxane	BP225, OV225
Polyethylene glycol	BP20, DB-WAX, CW20M
Cyanopropylsilarylene	BPX70
Dimethylsiloxane–carborane copolymer	HT5
Dimethylsilarylene	BPX5

technology is in the hands of the manufacturers who can produce capillary columns of excellent quality (see *Table 2*).

Some chromatographers are able to make their own capillary columns, which enables them to optimize the capillary columns for their separation requirements. The literature contains many procedures for the various steps in producing a capillary column. A comprehensive general guide to producing capillary columns is available (4).

The following sections are an overview of the steps involved in making a capillary column.

3.4.1 Fused silica surface modification

The surface chemistry of the fused silica capillary column can vary dramatically from one supplier to another and even significantly from batch to batch. The chemistry of the fused silica surface must first be adjusted to make it compatible for the following treatment steps.

3.4.2 Deactivation

Despite the high purity of fused silica, it can still exhibit a substantial degree of catalytic and adsorptive behaviour towards some solute molecules. The deactivation procedure must be compatible with the stationary phase to be employed.

3.4.3 Coating

The static coating method, in its various forms, is the preferred technique for applying the polymer stationary precisely and evenly on the wall of the capillary. The main steps are as follows.

(a) An accurately known concentration of the stationary phase is dissolved in a volatile solvent.

(b) The capillary is filled with the dilute solution and one end of the capillary is sealed.

(c) A vacuum is applied to the other end of the capillary causing the solvent to gradually evaporate, leaving the stationary phase coated evenly and in a known thickness on the wall.

3.4.4 Cross-linking

When the polymer is coated on to the column the polymer is a relatively low molecular weight oil or gum. Cross-linking of this layer forms an insoluble and physically stable polymer. Cross-linking is usually achieved through a free radical initiated reaction. The free radicals for cross-linking reactions may be generated through the use of peroxides, azo compounds, and gamma-radiation. Silanol-terminated prepolymers are now also typically used. The use of silanols is attractive from the point of view that potentially contaminating materials, or materials with undesirable reaction products, do not have to be introduced into the capillary column. Through thermal treatment, autocross-linking of the polymer is achieved. After cross-linking it is usual to wash the stationary phase with a volatile solvent and condition the column with carrier gas at an elevated temperature.

4. Column performance evaluation

As the technology for manufacturing capillary columns improves, the criteria for judging column performance becomes more stringent. Generally, standard tests are devised to ensure capillary columns are capable of performing as wide a range of applications as possible. These tests are useful for comparative purposes but in practice the only important criterion is how well the column performs the application for which it will be used. This section discusses some of the parameters that can be used to quantify the performance of a capillary column.

To determine the parameters for calculating column performance characteristics, a means of accurately measuring retention times and peak widths is necessary. Values for these parameters can be determined by some computer data acquisition software but without this facility a chart recorder is used. The chart speed per unit time needs to be known for determining the retention times of peaks and the width of the peaks in the same time units. A magnifying eyepiece with a graduated scale and a ruler measuring in the same units is all that is required. Rather than converting everything to minutes or seconds the calculations may be carried out in length units.

4.1 Resolving power

The resolving power of a capillary column can be thought of in terms of the number of resolved peaks that will fit into the analysis time. The narrower the peaks, the greater the peak capacity of the analysis and the higher the resolution of the column. This approach does not take into account factors such as the selectivity of the stationary phase.

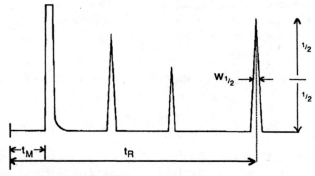

Figure 7. Parameters for the determination of efficiency.

In a chromatographic column, each solute band will broaden as it passes through the column. The measurement of resolution is simply a measure of how much the bands of solute have spread out as they proceed through the column.

4.2 Determination of efficiency and performance parameters

Examples of the determination of the parameters are given in the following section.

4.2.1 Retention parameters

- t_R = 242 mm (for chart speed 10 mm/min, equivalent to 24.2 min)
- t_M = 37.8 mm (equivalent to 3.78 min), elution time for unretained component.
- $w_{1/2}$ = 2.73 mm (equivalent to 0.273 min)

The data are used to compute values for the following efficiency and performance parameters (see *Figure 7*).

4.2.2 Theoretical plates (N_{Th})

$$N_{Th} = 5.54 \ (t_R/w_{1/2})^2$$
$$N_{Th} = 5.54 \ (242/2.73)^2$$
$$= 43\,500 \text{ theoretical plates}$$

4.2.3 Effective plates (N_{EFF})

$$N_{EFF} = 5.54 \ (t_R - t_M)^2/w_{1/2}^2$$
$$N_{EFF} = 5.54 \ [(242 - 37.8)/2.73]^2$$
$$= 31\,000 \text{ effective plates}$$

Theoretical and effective efficiencies may only be determined where the chromatogram is performed isothermally. Measurements of theoretical and effective plates are meaningless without also considering the capacity ratio of the peak on which the calculations have been performed. Values of N_{Th} and N_{EFF} are highly dependent on the capacity ratio for a component peak that has a short retention time.

4.2.4 Capacity ratio (also known as partition ratio or factor)

$$k' = (t_R - t_M)/t_M$$
$$k' = (242 - 37.8)/37.8$$
$$= 5.40$$

Determination of the capacity ratio of a peak at a particular temperature is essential for monitoring the condition of a column. A decrease in capacity ratio indicates a loss of stationary phase.

The effective plates and the theoretical plates are related by the equation:

$$N_{Th} = ((k' + 1)^2/k'^2)N_{EFF}$$
$$N_{Th} = [(5.40 + 1)^2/(5.40^2)].31\,000$$
$$= 43\,545 \text{ theoretical plates.}$$

4.2.5 Coating efficiency

Coating efficiency is useful for determining how well the resolving power of a column approaches its theoretical limit. The minimum height of a theoretical plate predicted by the theory is expressed as a percentage of the actual length of a theoretical plate.

The actual height equivalent of a theoretical plate (HETP) is obtained by working out the distance of a theoretical plate in the capillary column. The shorter the length occupied by a theoretical plate, the greater the efficiency of the column.

$$HETP_{actual} = L/N_{Th}$$

where L is the length of the capillary column in millimetres.

Using the results from above for a 12 metre capillary column:

$$HETP_{actual} = 12\,000/43\,500$$
$$= 0.276 \text{ mm.}$$

The minimum HETP, giving the greatest theoretically possible efficiency, is calculated for the peak on which the theoretical plates have been measured.

$$HETP_{min} = r_o \sqrt{(1 + 6k' + 11k'^2)/3} (1 + k')^2$$

where r_o is the internal radius of the capillary column in millimetres.

For a 0.22 mm internal diameter (i.d.) capillary column, $r_o = 0.11$ mm

$$HETP_{min} = 0.11 \sqrt{[1 + 6(5.40) + 11(5.40)^2]/3[(1 + 5.40)^2]}$$
$$= 0.11 \sqrt{(354/123)}$$
$$= 0.187 \text{ mm.}$$

The coating efficiency (CE) of the column is then given as a percentage of its theoretical maximum.

$$CE = HETP_{min}/HETP_{actual} \times 100$$
$$CE = [0.187/0.276] \times 100$$
$$= 67.8\%.$$

Coating efficiencies of commercial non-polar stationary phases such as the dimethylsiloxanes are typically at least 80%, with values approaching 100% being very common. The more polar stationary phases, particularly the cyanopropylsiloxane stationary phases, tend to have lower coating efficiencies, in the order of 70%, due to greater difficulties stabilizing these materials uniformly on to the fused silica wall.

4.2.6 Trennzahl separation number

The separation number or Trennzahl number (TZ) is a measure of how many peaks will fit with baseline–baseline separation between two peaks of a homologous series. It is also useful in that it can be used on a temperature-programmed chromatogram. The more peaks that can be placed between the peaks the greater the resolution of the column.

$$TZ = (t_{R(x + 1)} - t_{R(x)})/(w_{1/2(x + 1)} + w_{1/2(x)}) - 1.$$

Using the values shown in *Figure 8*:

$$TZ = (242 - 202)/(2.73 + 2.16) - 1$$
$$= 7.18.$$

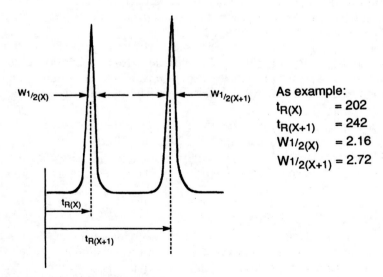

As example:
$t_{R(X)}$ = 202
$t_{R(X+1)}$ = 242
$W_{1/2(X)}$ = 2.16
$W_{1/2(X+1)}$ = 2.72

Figure 8. Parameters for calculating Trennzahl number.

Peter A. Dawes

4.2.7 Resolution
The resolution, R, between two peaks is the degree of separation between the two peaks.

$$R = 2(t_{R(x + 1)} - t_{R(x)})/1.699\ (w_{1/2(x + 1)} + w_{1/2(x)})$$
$$R = 2(242 - 202)/1.699(2.72 + 2.16)$$
$$= 9.65.$$

Resolution is also related to Trennzahl by:

$$R = 1.777(TZ + 1).$$

4.3 Kóvats' retention index
In evaluating the characteristics of a capillary column for a particular application, ensuring that it has the correct partitioning properties, is even more important than the overall resolving power of the column. Column manufacturers aim to make columns as reproducible as possible so that peaks are not shifted relative to each other in a chromatogram performed under identical conditions. Such shifts would make separation and/or identification of components difficult.

To make valid comparisons of Kóvats' retention indices between columns, the temperature, the stationary phase, and the phase ratio must be the same.

In calculating retention indices, the corrected retention time should be used. Corrected retention time $(t'_{R(x)})$ is the actual elution time of a component, $(t_{R(x)})$, less the time for elution of an unretained component, t_m.

The Kóvats' retention index relates the retention time of a particular compound to the retention time of a (hypothetical) n-paraffin. *Figure 9* shows the parameters that must be measured to determine retention indices and set of values to be used as an example.

An n-paraffin is given a retention index which is a product of its carbon number and 100. By logarithmic interpolation from the elution time of the compound of interest between the n-paraffins, a retention index relating to the carbon numbers of the n-paraffins can be calculated (see Section 7, Chapter 1).

An example of the determination of retention index is as follows (putting values for $t'_{R(N)}$, $t'_{R(A)}$, and $t'_{R(NH)}$ of 60, 202, and 242 respectively):

$$I_A = 100 \times 11 + 100 \times 4[(\log 202 - \log 60)/(\log 242 - \log 60)]$$
$$= 1100 + 400[0.527/0.606]$$
$$= 1448.$$

4.4 Speed of analysis
Excess resolution is of little or no benefit in itself but may be used to increase the speed of an analysis. Where excess resolution is available through the use of optimum selectivity or simply high resolving power, in terms of efficiency, there is potential to reduce analysis time. This can be achieved through faster

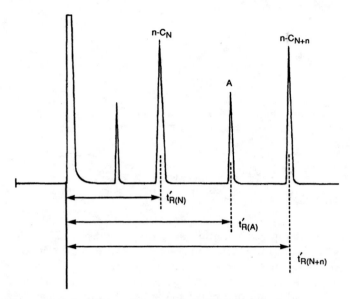

Figure 9. Measurement of parameters for calculating retention index.

carrier gas velocity, higher temperatures, or fast programme rates. The speed of analysis is simply measured as the time in which the necessary analysis can be achieved.

4.5 Sample capacity

If the amount of solute being chromatographed is too great, the stationary phase can approach saturation, affecting the equilibrium between stationary phase and gas phase. A portion of the solute molecules cannot dissolve into the stationary phase and remains in the mobile phase. It is carried further into the column until partitioning may again occur. Gradually, a small proportion of the solute gets ahead of the main solute band causing peaks to have a leading edge. The back of the peak is usually not distorted at all. The distortion (broadening) of the peak causes a loss of resolution (see *Figure 10*).

With sharp capillary column peaks it is not always easy to observe whether a peak has a leading edge, even though a loss in resolution in the area has occurred. There are two ways of determining precisely whether a peak is distorted.

(a) by observation of the peak *as* it is being traced. If the ascent of the peak is slower than the descent back to the baseline, the peak has a leading edge.

(b) using peak skew to quantify how much a peak has been distorted. Calculation of peak skew is explained in Section 4.6. A peak skew value of less than unity indicates a leading edge peak.

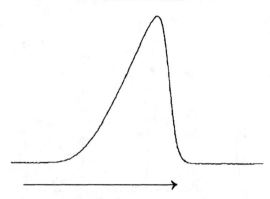

Figure 10. Overloaded, leading edge peak shape.

The amount of sample a column can chromatograph without overloading depends not only on the total sample size, but also on the amount of compounds in a particular region of a chromatogram.

There are two important characteristics when considering the possibility of peaks in a chromatogram being overloaded.

(a) The more stationary phase per unit length of a capillary column, the more solute molecule can be partitioned between the stationary phase and the mobile phase. The sample capacity can be taken as being proportional to the cross-sectional area of the stationary phase on the wall of the column. A greater thickness of stationary phase or larger column inside diameter gives greater sample capacity.

$$\text{Phase Cross-Section} = \pi.D.d_f$$

where D is the capillary internal diameter (μm) and d_f is the stationary phase film thickness (μm).

(b) The solubility of particular solutes in the stationary phase are important as a non-polar stationary phase will have a large capacity for a non-polar solute molecule. Conversely, a non-polar stationary phase will be overloaded relatively easily by a polar solute molecule.

Table 3 gives an indication of the approximate levels of sample capacity for different column inside diameters and stationary film thicknesses. This data should be used only as a guide because of other factors such as solubility variation of different solutes in particular stationary phases.

4.6 Inertness

It is of no value using a capillary column with high resolution if it is not possible to elute the solute molecules through the chromatographic system because they react in some way with active components in the system.

Table 3. Guide to sample capacity (ng) for different film thicknesses and inside diameters

Film thickness (μm)	Internal diameter (mm)			
	0.10	0.22	0.32	0.53
0.10	8	20	28	–
0.25	–	50	70	–
0.50	–	100	140	230
1.0	–	190	280	460
5.0	–	–	1200	2300

There are three categories of activity as follows.

(a) Reversible adsorption leads to peaks that may have the correct overall peak area but the peaks are observed as having tails, i.e. the downside section of a peak returns more slowly to the baseline.

(b) Non-reversible adsorption can give perfect peak shape but a portion of the solute remains adsorbed on to active sites within the system. The sample loss will give problems in quantification.

(c) Breakdown occurs where solute molecules, through an interaction with active components or thermal instability, breakdown or undergo a rearrangement.

Through advances in capillary column technology and manufacturing methods, the activity of columns has been reduced so that it is now possible to chromatograph solutes that were not considered feasible in the past.

There are many tests (of various sensitivities) for evaluating the activity of a chromatographic system. To the practising chromatographer it is only important to verify that the system will reliably chromatograph the solutes of interest. Tests to evaluate the system suitability for a particular analysis may be developed or standard tests can be used to give an indication of column and system performance.

The best known and most comprehensive test is the so-called Grob test (see *Figure 11*), which not only evaluates the activity of a capillary column but also parameters such as the resolving capacity of the column, its film thickness, and polarity. It is a complex test and needs to be carried out under the conditions specified. The methodology for this test has been reported (5) with additional notes elsewhere (6).

As column manufacturing techniques improve, more stringent tests for activity have been developed with more sensitive components. The molecules act as test probes for various types of activity, such as acids, bases, and alcohols. The test shown in *Figure 12a/b* is a general one for column and system inertness.

87

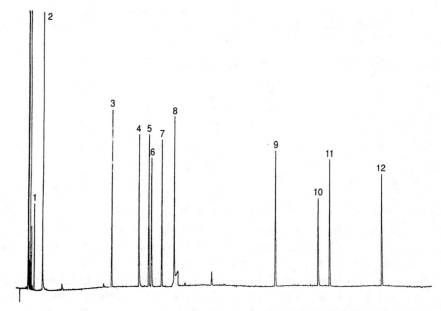

Figure 11. Standardized Grob test. Test mix: column, 25 QC2/BP1—0.25; length, 25 m; i.d., 0.22 mm; stationary phase, cross-linked methyl silicone BP1; film thickness, 0.25 µm; initial temperature, 40 °C; programme rate, 1 °C/min; final temperature, 145 °C; detector, FID; sensitivity, 1×10^{-11} AFS; carrier gas, helium; velocity 25 cm/sec; injection mode, split. **1**, 2,3-Butanediol; **2**, *n*-decane; **3**, 1-octanol; **4**, nonanol; **5**, 2,6-dimethylphenol; **6**, *n*-undecane; **7**, 2,3-dimethylaniline; **8**, 2-ethylhexanoic acid; **9**, a_{10} methyl ester; **10**, dicyclohexyl amine; **11**, C_{11} methyl ester; **12**, C_{12} methyl ester.

Once a level of performance has been established on a test, the system including the capillary column can be reassessed as needed. The activity of a column may change as it is used. Factors affecting column activity over time are:

(a) contamination of the column from the sample

(b) contamination of the column from carrier gas

(c) damage to the stationary phase polymer from high temperature operation

(d) damage to the front section of the column due to particular sample introduction techniques or sample types

For remedies to increased activity on a capillary column see Section 6.2.

Chromatographic conditions can have a dramatic effect on the extent to which solute molecules interact with active sites in the capillary column. The interaction of reactive components with the column may be reduced by shortening the time the component is in the column. Providing a de-

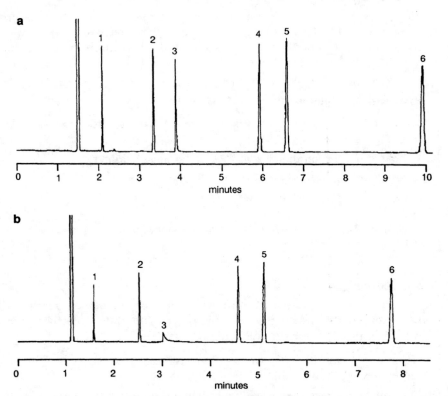

Figure 12. Chromatograms of (a) column A, showing good peak shape and inertness and (b) column B, showing some absorption and activity. Phase, BP5, 0.25 μm film; column, 25 m × 0.22 mm i.d.; temperature, isothermal at 140°C; detector, FID; sensitivity, 32 × 10^{-12} AFS; injection mode, split 60:1. Components were at 2–5 ng. **1**, *n*-Decane; **2**, 4-chlorophenol; **3**, *n*-decylamine; **4**, undecanol; **5**, biphenyl; **6**, *n*-pentadecane.

sired separation can be obtained, low interaction times can be achieved as follows:

- higher carrier gas velocity
- shorter column
- higher temperature
- higher temperature programme rate

Where there is activity towards a solute it is more difficult to work with small amounts of that solute because a larger proportion of the particular solute is lost. The effects of activity can often be masked by chromatographing larger amounts of the material.

Where there is reversible adsorption, a measure of column activity can be

Peter A. Dawes

made for the purposes of monitoring a system's performance by measuring the degree to which a particular peak is distorted. To calculate peak skew on a chart recorder, a high chart speed is required when the particular peak is eluting from the column (*Protocol 1*).

Protocol 1. Measurement of peak skew

1. From the apex of the peak drop a vertical line to the base line as shown in *Figure 13*.
2. Take the point on this line at 10% of the peak height.
3. From the centre line measure w_f and w_b
4. Determine skew using Skew[a] $= w_b/w_f$.

[a] Skew values greater than unity indicate that the peak is tailing which may be due to reversible adsorption.

4.7 Bleed

Stationary phase bleed arises from the degradation of the stationary phase allowing the volatile fragments from the polymer to elute from the column.
  Bleed is a problem for the following reasons.

(a) It limits the sensitivity of the analysis. Without some system of baseline compensation the baseline will move off scale.
(b) An unstable baseline or rising baseline can introduce errors in data collection and integration.
(c) Bleed will contaminate some types of sensitive detectors.
(d) Background bleed can limit the sensitivity of mass spectrometer detectors.

  Bleed from capillary columns is not simply material contaminating the phase eluting from the column. The mass spectra of column bleed from a

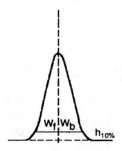

Figure 13. Measurement of peak skew.

90

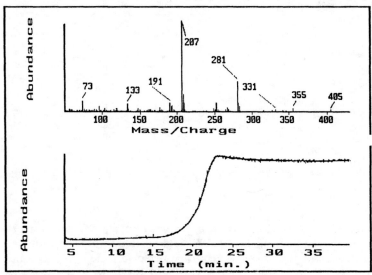

Figure 14. Profile and mass spectrum of bleed.

polysiloxane stationary phase (*Figure 14*) indicates two major ions, at m/z 207 and 281. These ions are the three and four member cyclic siloxane fragments shown in *Figure 15*.

The conditions that can lead to the depolymerization of the stationary phase and thus bleed are as follows.

(a) Catalytic breakdown from sites such as metal in the capillary column. Fused silica capillary columns have very low metal impurities.

(b) Strong acid or base catalysed breakdown of the polymer. This is more to do with what is injected into the column and can have a devastating effect on the stationary phase polymer.

(c) Stationary phases with impurities and precursors remaining from the synthesis or cross-linking. Such species tend to continuously react to cause breakdown of the polymer.

(d) Oxygen present in even small quantities in the carrier gas from impure gas from its source or gas that has been contaminated by leaking joints in the gas system or even inappropriate tubing and fittings (see Section 6.2).

Every polymer has an ultimate thermal limit at which the covalent bonds will be broken. There are two temperature limits to stationary phases that can be considered. The *maximum continuous temperature* (as shown in *Table 1*) has been defined as the temperature at which no change in the column efficiency, retention index, capacity ratio (k'), or activity occur after the column has been held at that temperature for an extended period (for

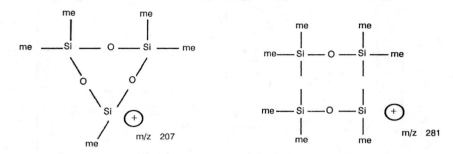

Figure 15. Major mass spectral fragments resulting from the bleed of a dimethylsiloxane stationary phase.

example, 72 hours). The *maximum programme temperature* for cross-linked stationary phases are usually 30°C higher than the maximum continuous temperature, but this is more arbitrary because it depends as much on how long the chromatographer expects the column to last. Non-cross-linked stationary phases should not be heated above the specified maximum temperature as permanent damage will rapidly occur. Cross-linked stationary phases do not tend to fail suddenly and can be heated to very high temperatures for short periods without damage.

4.7.1 What does not constitute stationary phase bleed?

When a column is temperature programmed and the baseline is observed to rise followed by a fall after the maximum temperature has been reached as shown in *Figure 16*, breakdown of the stationary phase is not likely to be the problem. Also if the mass spectrum of the material eluting from the column can be recorded, it is not likely to exhibit the *m/z* 207 and 281 ions which are characteristic of polymer breakdown. This effect is usually due to the column being contaminated by material from the sample or from some other source (see Section 6.2 for remedies to this problem).

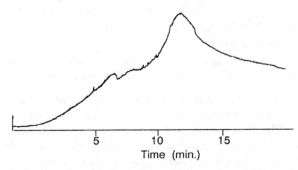

Figure 16. Baseline shift on temperature programming of a contaminated column.

4.7.2 Measuring column bleed

It is useful to quantify the bleed from the column. Traditionally, the baseline shift was simply related back to the units of measure of the detector signal. For a flame ionization detector (FID) most instruments have a maximum sensitivity setting where 1×10^{-12} amps is equivalent to full-scale deflection on the chart recorder output. This setting is of some practical use but there needs to be an awareness of variations in detector sensitivity when making comparisons. It can be useful to relate the bleed measured on a detector to the signal on the detector from a given compound. For example, the bleed is equivalent to the peak height of a quantity (ng) of $n - C_{20}$. Clearly, there are problems since this does not allow comparisons between columns of different configurations to be made. Here the peak widths will not be equivalent and consequently different peak heights will be observed.

Bleed from a capillary column is approximately proportional to the amount of stationary phase in the capillary column. Longer and thicker film columns will show proportionally more bleed.

Table 4 can be used as an indication of what would be appropriate bleed from a polysiloxane stationary phase column at its maximum continuous temperature for given inside diameters and film thicknesses. These values are for 25 metre columns. The bleed is also approximately proportional to the length of the column. Longer columns will have proportionally more bleed.

Table 4. Expected bleed levels at 320 °C from 25 metre dimethyl-siloxane columns with different internal diameters

i.d. (mm)	0.10	0.15	0.22	0.32	0.53
d_f (μm)	0.1	0.34	0.25	0.5	1.0
Bleed (pA)	2	9	10	29	96

5. Column selection

Such is the capability of capillary gas chromatography that many of the analyses carried out successfully are not optimized. There are benefits in optimizing the selection of a capillary column for an analysis:

- improved sensitivity
- greater speed of analysis
- ability to utilize a lower cost column
- reliability of the analysis by not 'stressing' the column

The effect of each of the four column parameters to be considered needs to be understood. The order of importance in making the selection can be given as:

- stationary phase
- internal diameter
- film thickness (or phase ratio)
- length

5.1 Stationary phase

The selectivity of stationary phases is by far the most powerful resolving mechanism available to the chromatographer. The resolving power or number of plates of a column are insignificant compared to a well-used stationary phase selectivity.

When choosing a stationary phase for an analysis the best place to start is usually the published literature and applications information from commercial suppliers.

With the high performance attainable in capillary gas chromatography, the majority of analyses can be performed on non-polar stationary phases, such as dimethylsiloxane or 5% phenylmethylsiloxane. It is also considered preferable to use a non-polar stationary phase where possible because these have higher maximum temperatures and are more durable. Non-polar phases rely on dispersive interactions and so the separation tends to be according to the volatility of the sample components.

Where other functional groups are substituted for methyl groups on the siloxane polymer backbone, other separation mechanisms occur (see Section 3.2). The 5% diphenyldimethylsiloxane stationary phase has an added degree of selectivity and is very useful for a wide range of applications, especially when aromatic groups are involved in the solute molecules.

Even though a separation as shown in *Figure 17* may be achieved, there

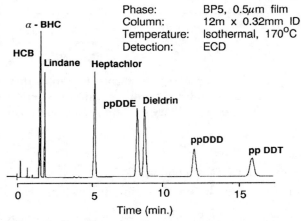

Figure 17. Separation of pp-DDE and dieldrin on a non-polar stationary phase. Phase, BP5, 0.5 μm film; column, 12 m × 0.32 mm i.d.; temperature, isothermal 170°C; detection, ECD.

may be advantages in trying the same analysis on a more polar column. The conditions for the chromatogram have been optimized on the BP5 (5% diphenyldimethylsiloxane) to achieve the separation in the shortest time. The limiting factor in this case was the separation of the DDE and dieldrin peaks. When the same analysis was carried out on a BP10 column (7% cyanopropyl, 7% phenylmethylsiloxane) the separation of all seven peaks was achieved and the resolution of the DDE and dieldrin was greater. Since excess resolution is of no value, the analysis time may be reduced, as indicated in *Figure 18*, to provide the minimum desired resolution between any of the peaks. The limiting peaks are still the DDE and the dieldrin, but with the BP10 moderately polar cyanopropylphenylsiloxane stationary phase the separation is achieved in a greatly reduced time.

In conclusion, optimization of the stationary phase selection can give dramatic decreases in analysis time.

5.2 Internal diameter

The efficiency or number of separating plates is inversely proportional to the internal diameter of the column. Small internal diameter columns also have

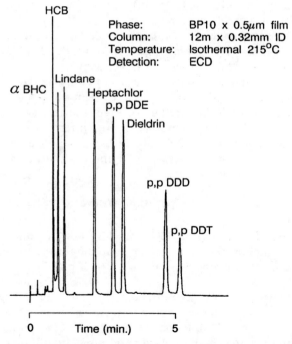

Figure 18. Separation of pp-DDE and dieldrin on a moderately polar stationary phase. Phase, BP10, 0.5 μm film; column, 12 m × 0.32 mm i.d.; temperature, isothermal 215°C; detection, ECD.

greater sensitivity, as the peaks should be sharper providing a greater detector response. There are several other aspects to internal diameter selection that should also be considered as follows:

- resolution required
- required speed of analysis
- sensitivity required
- ease of operation
- chromatograph system capabilities

Each of the internal diameter columns commercially available are discussed below.

5.2.1 Wide-bore (0.53 mm) columns

Wide-bore 0.53 mm i.d. columns have been important in the general acceptance of capillary gas chromatography because they allow easy transition from packed columns to capillary columns. They do not demand a high degree of expertise to operate and required modifications to the gas chromatograph injection and detection system are minimal. Usually the only requirements are a simple insert adapter that fits into a packed column inlet system and a similar adapter to the detector.

The large carrier gas flow rates, of the order of 2 ml/min to 30 ml/min, makes connections easier, with less possibilities for dead volume, and enables 0.53 mm i.d. columns to be operated in the same way as a packed column. The large gas flow rates also makes it easier to interface the columns to some sample introduction and detector systems. Columns of 0.53 mm i.d. have the largest sample capacity of the capillary columns (particularly when thick films are used), which makes the columns forgiving in terms of overloading.

If a chromatographer is already using, and is experienced with, smaller i.d. capillary columns there are few cases where there is justification in going back to the 0.53 mm i.d. columns.

5.2.2 Medium-and narrow-bore (0.32 mm and 0.22 mm i.d.) columns

Medium-and narrow-bore (i.d. 0.22 and 0.32 mm) columns are the most commonly used capillary columns and are suitable for the majority of applications. They offer better sensitivity than the 0.53 mm i.d. columns and considerably higher resolution.

In practical terms, the 0.32 mm i.d. columns would be first choice for many analyses. There is very little resolution difference between 0.32 and 0.22 mm i.d., and the 0.32 mm i.d. columns are more flexible in terms of the sample injection techniques that may be used. The most quantitative of the sample injection techniques, on-column injection, works well over a wide range of conditions and so is relatively uncomplicated with 0.32 mm i.d. columns. The

0.22 mm i.d. columns have traditionally been used when the maximum resolution is required.

5.2.3 Microbore (0.15 mm) columns

If a 50 metre column is to be used, 0.15 mm is the minimum practical internal diameter before the carrier gas pressure across the column becomes too high to be applicable. This type of column, therefore, has the maximum resolving power conventionally possible on a single column. The 0.15 mm i.d. capillary columns are also ideally suited for use with modern bench top quadrupole mass spectrometers which have limited carrier gas pumping capacity. Higher carrier gas velocities may be used in the capillary column without affecting the pumping capacity of the mass spectrometer and lowering the sensitivity.

5.2.4 Microbore (0.10 mm) columns

The efficiency of a capillary column with a small internal diameter is greater but it is generally not worthwhile to achieve higher resolving power. The real benefit of microbore columns is in fast analysis but not necessarily in high resolution analysis.

For small internal diameter columns considerable care is required in setting up the system to achieve good sample introduction without broadening the peaks in the process. The sample capacity of 0.10 mm i.d. columns is also very limited and only small sample sizes may be introduced if overloading is to be avoided. *Table 5* summarizes the characteristics of different i.d. capillary columns.

5.3 Film thickness

A more meaningful and useful parameter to consider when considering the thickness of the stationary phase is the phase ratio (β). This is the ratio of the volume of, mobile phase (gas) to the volume of stationary phase (liquid) in the capillary column given by

$$\beta = D/4.d_f$$

where D = column i.d. (μm) and d_f = stationary phase thickness (μm).

Table 5. Summary of characteristics of columns of various internal diameters

Function	Internal diameter (mm)				
	0.10	0.15	0.22	0.32	0.53
Resolution	****[a]	****	***	***	*
Sensitivity	****	****	***	**	*
Sample loading	*	**	***	***	****
Analysis speed	****	****	***	**	*
Operation ease	*	**	**	***	****

[a] Denotes degrees of performance/capability.

The capacity ratio and the retention of the solute molecules relate strongly to the phase ratio. The greater the film thickness, the lower the phase ratio and the greater the retention. Even though two columns may possess different internal diameters and film thicknesses, if they have the same stationary phase and phase ratio, then under the same temperature conditions they should behave identically.

Thick films of stationary phase (low phase ratio) should be used where volatile components are to be analysed. The greater retention of volatile components gives better opportunity to achieve separation. In addition, thick films of stationary phase permit the analysis of volatile components at higher and more easily manageable temperatures, and can allow cryogenic cooling to be avoided.

The sample capacity of a capillary column, which is the amount of sample before it becomes overloaded, giving a loss of resolution, is also directly related to the amount of stationary phase on the capillary column wall. Thick films can tolerate larger amounts of sample.

Thin films of stationary phase (high phase ratio) retain solute molecules less and are, therefore, useful for allowing the elution of non-volatile components at lower temperatures, which is of advantage in terms of column bleed and column longevity.

5.4 Length

Because resolution is only proportional to the square root of the efficiency and therefore the length of the column, doubling the length of the column gives only about 40% improvement in resolution. In achieving this increase in resolving power there are several disadvantages, e.g. doubling the length leads to:

- doubled analysis time
- decreased sensitivity
- doubled column bleed
- reduced inertness due to a greater time allowed for the sample to interact with active sites in the column

It is far more rewarding to consider carefully the stationary phase type and phase ratio when setting up an analysis than the length of the column. Longer columns should only be used where high resolving power is required, such as with a complex wide boiling range sample where it is difficult to optimize across the whole range of the solutes to be analysed.

6. Care of a capillary column

Advances in capillary column technology have improved the durability, but there are still some guidelines that need to be followed to obtain long and reliable service.

6.1 Protection of fused silica fabric and installation hints

The polyimide layer protecting the fused silica is a tough material but is only in the order of 15 to 20 μm thick. Damage to the polyimide layer by abrasive materials should be avoided.

Fused silica tubing should only be cut by first scratching it with a hard material like a diamond tip pencil, sapphire scribe, a sharp piece of zirconia, or carborundum. There are a number of products available for this purpose which will cut through the polyimide coating, putting a small mark in the fused silica surface. If this is done properly the fused silica will break cleanly and straight.

If fused silica is broken incorrectly, shattering of the fused silica may occur, depositing particles of material inside the column giving problems with activity and peak distortion.

The performance of a capillary column will be destroyed by particles such as ferrule material entering the column. When installing a ferrule (particularly graphite) onto a capillary column, the entrance of the column should be inverted and a few centimetres of the column removed after the ferrule has been positioned. Graphite broken from the ferrule can travel several metres into a large inside diameter capillary column and will be very difficult to dislodge.

6.2 Column contamination

Cross-linked stationary phases are very robust but will be damaged severely by strong acids and bases and severely contaminated by non-volatile or reactive materials. Symptoms from a deteriorating column are:

- activity
- increased bleed
- unstable baseline
- broadened peaks

Some faults are fatal to the performance of the column, but in general the performance of the capillary column can be recovered. For these four symptoms the treatment is much the same. However, the treatment will only be beneficial if the column is at fault and not another part of the system, such as the injection or detection system.

Activity towards sensitive sample components is usually the first sign of column deterioration. The activity of a column may change as the column is being used. Factors that will affect the column activity over time are:

- contamination of the column from sample
- contamination of the column by carrier gas

- damage to the polymer by high temperature operation
- damage to the front section of the column due to the sample and sample introduction technique

The column can be restored as indicated in *Protocol 2* or washed as described below.

Protocol 2. Restoration of column performance

The steps included will not suit all situations, but are useful as a general procedure to follow.

1. Condition the column at its maximum continuous temperature and observe the detector signal. The baseline should eventually settle and not drift up or down.

2. If the baseline is still unstable or the problem persists remove approximately 0.5 m from the injection end of the column. This removes non-volatile material or a damaged section at the front of the column.

3. If the performance has still not improved wash column to remove contaminating material (cross-linked and bonded phase only).

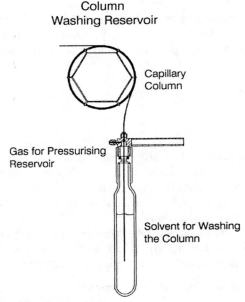

Figure 19. Column washing reservoir.

One end of the capillary column is placed into the reservoir containing the selected solvent as shown in *Figure 19*. The vessel is pressurized, pushing the solvent through the column. Usually a pressure in the order of 15 lb/in^2 (100 kPa) is sufficient.

When selecting a solvent to wash a column there should be some consideration of the samples that have been analysed on the column and which solvent is likely to be the most appropriate for removing the material. In practice, washing with pentane where a non-polar solvent would be better, or methylene chloride where a more polar solvent is needed, covers most eventualities.

After washing, all solvent should be removed from the column before connecting it it to the detector by heating the column during carrier gas flow. After washing, it is useful to condition the column briefly at its maximum temperature for a short period or until a stable baseline is achieved.

The capacity ratio as discussed in Section 4.2 can be used to monitor changes in the film thickness and therefore condition of the column. It is essential that the test is carried out at precisely the same temperature each time, and there should be an awareness that there may be significant temperature calibration differences between gas chromatographs. A reduction in the capacity ratio indicates that stationary phase has been lost during use.

A further symptom of the deterioration of a stationary phase coating is the appearance of a series of peaks as shown in *Figure 20* on temperature programming. This could be an indication that the stationary phase is undergoing thermal breakdown.

6.3 Column operation

When using a capillary column there are a few hints that can help the column give longer more reliable service.

A capillary column should never be heated for any significant period

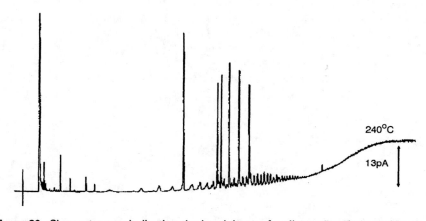

240°C

13pA

Figure 20. Chromatogram indicating the breakdown of a siloxane stationary phase.

without carrier gas flowing through the column. The greatest danger is air diffusing into the column, particularly from the detector end, causing oxidation of the stationary phase.

If a capillary column is going to be unused for an extended period, it is best to remove it from the chromatograph and seal the ends. When the column is reinstalled in the instrument a very short conditioning period should be enough to restore its condition.

It is not recommended that the column remains in the chromatograph with carrier gas flowing through the column unless the column is also heated. At room temperature, capillary columns will concentrate any impurities on to the stationary phase, which at best is an inconvenience when having to condition the column to remove the material and at worst it may permanently damage the phase coating.

To extract the maximum performance and lifetime from a capillary column, it is essential that the carrier gas is of high purity, with oxygen scrubbers and possibly organic scrubbers. It has been shown that much of the oxygen that is introduced into a capillary column comes from leaking fittings and inappropriate polymer components in the gas delivery system.

A more comprehensive treatment of capillary column troubleshooting is given elsewhere (7).

References

1. Burns, W. and Hawkes, S. J. (1977). *J. Chromatogr. Sci.*, **15**, 185.
2. For a range of references see Golovnya, R. V. and Polanuer, B. M. (1990). *J. Chromatogr.* **517**, 51.
3. De Zeeuw, J., De Nijs, R. C. M., Zwiep, D., and Peene, J. A. (1991). *Am. Lab.*, **23**(9), 44.
4. Grob, K. (1986). *Making and manipulating capillary columns for gas chromatography*. Huethig, Heidelberg.
5. Grob, K. Jr., Grob, K., and Grob, G. (1978). *J. Chromatogr.*, **156**, 1.
6. Grob, K., Grob, G., and Grob, K. Jr. (1981). *J. Chromatogr.*, **219**, 13.
7. SGE (1985). *Capillary GC operating hints booklet*. SGE International, Australia.

4

Applications of packed and capillary GC

PETER J. BAUGH

1. Introduction

Although it can be said that capillary columns (high performance, HP) have superceded packed columns[a] (low performance, LP) in GC applications and industrial uses of GC, the latter for many years has remained the preferred technique for quality control and chemical process monitoring. The reason has not necessarily been a reluctance to change but purely because LP columns could, with high capacity, achieve the performance adequately enough to satisfy the requirements. In such applications it was only necessary to determine the major components and impurities ($< 1\%$) could be ignored.

A large section of industrial uses of GC is devoted to solvents and volatile/gaseous product monitoring which requires profiling and analysis of highly volatile organic compounds (HVOs) and purgable volatile organic compounds (VOCs). For HVOs, it remains that gas solid chromatography (GSC) using the LP mode is preferred, although surface-coated capillary columns containing Al_2O_3 or Porapak Q have now been perfected to compete in the analysis of both permanent gases and HVOs. On the other hand, for VOCs the film thickness of stationary phase available for capillary columns is now adequate to provide an increase in retention necessary for separation and analysis to give the HP mode ascendency over the LP mode.

Other specialized uses in which capillary columns now excel lie in high-temperature GC (HTGC) applications which have always presented problems for the LP mode because of the excessive bleed at the maximum operating temperature (MAOT) or above. For analysis of compounds of low volatility, it was necessary to employ short stainless steel columns (1.5 ft × 1/16 in) with low stationary phase (SP) composition (1–3%).

Since the advent of the polysiloxane-carborane (HT5 from SGE) and more recently BTX5 (also SGE) phases which can be used up to 450°C and 375°C,

---

[a] It is worth noting for other reasons that with supercritical fluid chromatography (SFC), the packed column mode is more versatile because modifiers can be added at a later stage in the pressure programming to induce chromatography of polar analytes (1).

Peter J. Baugh

respectively, it is now possible to utilize conventional non-polar capillary columns with intermediate film thickness (d_f) to elute compounds of low volatility, such as hydrocarbons up to C_{100} and polywaxes.

A literature search has revealed an intercomparison of LP and high-pressure (HPGC) (2). The areas covered include theory (2), comparisons of retention data (Kóvats' indices) (3–6), detector optimization and sensitivity (7, 8), quantification (9), and specific applications involving toxicology, volatile organic analysis (VOA) (10), volatile oils (3), drugs (5), fatty acids (11), pesticides such as polychlorobiphenyls (PCBs), organochlorines (OCs) (4, 12), and residues (13).

This chapter highlights the applications and use of techniques with LP and HPGC columns in a comparative way in selected instances, and also, in isolation where appropriate. Specific applications of LP, narrow/medium, and wide bore columns (see also refs 14–17) with special phases are also included. Other chapters (notably Chapter 3) include a detailed discussion of column technology and performance. Although there is some unavoidable overlap, the emphasis in this chapter is on illustrating the chromatography of a variety of analytes using the most suitable column or column combination (Section 3 on Multidimensional GC, MDGC).

Microbore columns which have been used to a limited extent for fast analysis in HPGC[a] have not been applied in main stream GC and their applications are not considered here.

2. Experimental considerations

As indicated in the introduction, only limited details of the installation of columns and other related experimental considerations will be included to avoid unnecessary overlap with other chapters.

2.1 Column requirements

Column requirements are discussed in detail in Chapters 2 and 3. This chapter briefly covers the inlet and outlet end coupling of all columns utilized.

2.1.1 Packed columns

Regarding packed columns, the coupling is straightforward and the fittings used are dependent on the internal diameter (i.d.) of the column (i.d. 1/4 in or 1/8 in, the latter for narrow-bore packed columns). Usually graphite ferrules (OGF4-4 or OGF8-8, SGE) are suitable unless the detector is mass spectrometric, for which graphitized vespel is employed to reduce the problems of air leaks at high temperature in the GC column–MS interface. With

[a] A high-pressure accessory coupled to the GC inlet is required for high pressure GC in which pressures up to 300 lb/in^2 are employed. In addition, a higher viscosity carrier gas, such as N_2 is used to minimize leaks which would arise with He or H_2.

the high flow rates in LPGC there are few problems with dead volumes at the injector or detector, which would result in loss of performance.

2.1.2 Narrow and medium bore capillary columns

With the advent of fused silica open tubular (FSOT) columns (*ca.* 1980) and when the problem of fragility had been overcome by applying polyimide coating, the coupling of capillary columns was facilitated.

Considerable patience and skill was formerly required for installing boro-silicate columns, which had to be straightened at the entrance and exit prior to coupling to the injector and detector. Being thinner-walled and more delicate than conventional packed columns, the borosilicate columns could be easily fractured, thus affecting the ability to maintain the GC up-time (more critical in GC/MS applications, see Chapter 11 for other details).

The FSOT columns, however, because of their flexibility and straightness, can be more precisely aligned in the injector and detector (viz. directly coupled through to the ion source of a mass spectrometer).

The graphite ferrules (OGF16-004/005, SGE for 0.22 mm and 0.32 mm, i.d. columns, respectively) are satisfactory for GC with conventional detectors, but care is required to avoid flaking of the graphite, which can cause a restriction in the flow through the column and in the extreme case, blockage. It is important that the column ends are cut as near to 90° as possible to the axis along the column, using a diamond tool. Poorly machined ends can lead to loss in sensitivity in the GC or MS detector. Less problems are encountered in this respect if make-up gas is employed in large-volume packed column detectors in LPGC systems adapted for capillary use, which also assists in optimizing the sensitivity. The ferrules used for GC–MS are of the graphitized vespel type (GVF16-004 and 005 for 0.22 and 0.32 mm i.d., respectively).

2.1.3 Wide-bore or megabore capillary columns

LPGC systems are easily adapted for applications of wide-bore (0.53 mm i.d.) capillary columns in place of packed columns. Fittings are available from SGE (glass-lined tubing, glt) or Alltech Associates (glass) which allow alignment of the capillary column in the injector and detector (with or without make-up tee, SGE only). Standard 10 μl microlitre syringes with 47 gauge stainless steel needles can be employed to allow direct injection, as with packed columns.

Make-up gas is not absolutely necessary, although advisable (see Section 4.3), since flow rates approach those used for packed columns. Until recently, using an MS as detector a jet separator was required for both LP and wide-bore open tubular (WBOT) columns to divert the carrier gas because of the inability of the vacuum system to cope with high carrier gas flow rates. Although still required on older GC–MS systems, this problem has been overcome for WBOTCs by reconfiguring the ion source/pump geometry to enable high vacuum and thus the sensitivity to be maintained.

The use of wide-bore capillary columns has been criticized recently by Grob and Frech (18). They state that, although a case can be made out for the replacement of packed by wide bore columns, in their opinion it is evident that in all applications the use of medium or narrow bore columns would be preferred.

Larger bore columns were used in the early 1970s to a limited extent. With the advent of FSOT columns about 12 years ago, 0.53 mm i.d. columns also found favour, although they are more fragile than standard capillaries. The advantages over packed columns are considered in terms of increased speed of analysis, improved inertness, increased column capacity, low elution temperatures, high carrier gas flow rates, and modified packed column instruments.

Grob and Frech (18) have used SE54 stationary phase with d_f adjusted to give equal β values for MBOT and WBOT columns (0.15 μm for 0.32 mm i.d.; 0.25 μm for 0.53 mm i.d. columns). Separation efficiencies were measured by Trennzahl number (TZ) determined at fixed temperature (55 °C) using C_{11}–C_{15} *n*-alkanes in *n*-hexane. Analyses were duplicated using H_2 and He as carrier gases. From their data it is evident that WBOTCs do not compare favourably with the narrower bore OTCs in performance (see *Table 1*).

However, many workers have changed from LP columns to WBOTCs to minimize the reconfiguration and cost of conversion of old GC injectors to the split/splitless mode required for HPGC using narrow or medium bore OTCs.

3. Multidimensional gas chromatography

Multidimensional gas chromatography (MDGC) has specific applications which make use of particular configurations which involve combinations of packed with capillary or capillary with capillary columns. Recent texts by Schomburg (19) and Clement (20) describe the technique and its applications in some detail. Several configurations are considered to highlight particular applications.

Table 1. Comparison of narrow, medium, and wide bore open tubular columns. TZ values at different carrier gas flow rates[a]

Column dimensions		Flow rate (ml/min)			
i.d. (mm)	**length** (m)	2	6	12	20
0.53	10	18	15	9.5	7.5
	25	29.5	24	18	14
0.32	5	17.5	12.5	9	6.5
	10	28	17	13	10
	25	41	29	22	18
0.27	10	30	19	14	10

[a] Reproduced with kind permission from ref. 18, 1988 © Int. Sci. Commun.

The objectives of MDGC are as follows:

(a) short analysis times by partial analysis of complex mixtures, i.e. when components with long retention are unimportant with respect to the aim of the analysis can be back-flushed from the first column by reversing the direction of the carrier gas flow

(b) higher resolution for the separation of selected groups of compounds transferred from the first column

(c) increase of the mass flows or concentration of trace components at the peak maximum in order to achieve higher S/N ratios

(d) multiple sets of retention and peak area data from both or several separations executed within the same coupled system (NB separation of target compounds in a complex mixture can be improved)

(e) good separation and reliable determination of small peaks eluted on the tail of a large peak (solvent) can be achieved by the heart cut technique[a]

In most instances the flow direction of the carrier gas in the first column is reversed for backflushing and venting of unwanted species between the injector and the first column or with elution in the normal flow direction between the columns. A monitor detector is required between columns.

Column or flow switching can be effected either with a multiport valve, or valveless switching which is the preferred technique. Column switching in MDGC should be executed automatically according to a previously set programme. Instrumentation should also be equipped with two separate ovens for each column to be operated at the optimal temperature on programme.

A number of modes exist for different applications and are illustrated in *Figure 1* and exemplified in the sections following.

3.1 Capillary–capillary column coupling with differing film thickness of SP

The application used to illustrate this mode is the analysis of a crude oil fraction, C_4–C_{30}. The first column allows reasonable retention for the low volatile compounds, but the volatile components are not separated and pass into the second column with a thicker phase while the low volatile compounds are detected in a monitor detector between the two columns and are not passed to the second column. *Figure 2* illustrates the chromatograms obtained for the analysis of C_4 to C_{12} and C_{13} to C_{30}.

3.2 Capillary–capillary column coupling with consecutive non-polar and polar SPs

An example of an application of this mode is the analysis of coal-derived gasoline fraction. Here the first capillary column (SP-PEG) is used to retard

---

[a] The transfer of eluate cuts may cause instrumental problems.

Peter J. Baugh

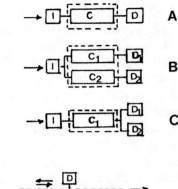

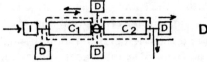

Figure 1. MDGC configurations. (A) Single column/detector; (B) parallel columns/ detectors; (C) single column/dual detector; (D) serial columns/multidetector. (Adapted from ref. 19 with kind permission 1990 © VCH Publishers.)

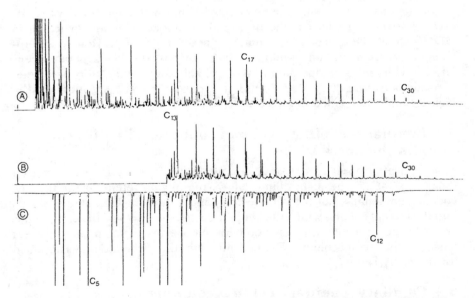

Figure 2. Analysis of a crude oil fraction using two capillary columns with different film thickness of stationary phase. (A) Chromatogram of whole sample; (B) cut after elution of more volatile compo. ents ($<C_{12}$); (C) separation of cut compounds (C_4 to C_{12}). Column 1: 30 m SE52, d_f 0.1 μm. Column 2: 50 m OV1, d_f 1.0 μm. Conditions A and B: 5 min at 50°C, 50 to 300°C at 8°C/min; Condition C: 5 min at 0°C, 0 to 50°C at 10°C/min, 50 to 250°C at 4°C/min. Carrier gas, hydrogen 0.11 and 0.08 MPa. Detectors, FID. (Reproduced from ref. 19 with kind permission 1990 © VCH Publishers.)

polar components, such as aromatic hydrocarbons, ketones, alcohols, and nitriles, but not separate non-polar compounds which pass unresolved to the second column coated with a non-polar methylpolysiloxane SP.

3.3 Capillary column and FID/ECD as dual detectors

In this mode the application involves a complex mixture containing electron affinic and non-affinic compounds, such as chlorohydrocarbons and paraffins or PAHs, respectively, with the single capillary column fitted with a split tee to enable simultaneous detection and analysis of the two types of components by ECD and FID. A more complicated configuration has been described by Schomburg (19) and involves the use of double oven GC with polar (Silar 10 CP) and non-polar (methylpolysiloxane) columns. One ECD is employed as a monitor detector between the two ovens and another ECD is placed in parallel with a second type of detector—FID (to monitor *n*-alkanes added as Retention Index standards). The configuration was designed specifically for the optimized separation of a complex mixture of chlorohydrocarbons, e.g. PCBs, PCDDs, and PCDFs.

3.4 Packed-capillary coupling

The mode can be used to improve the separation of the solvent and other major components using a packed column which may be overloaded on capillary column (efficiency reduces rapidly with overloading). The shape of the major peak is not so broadened by overloading with packed columns.

Focusing within the inlet may be required because of excessive width of the band which is cut. This can be achieved as follows:

- cold trapping
- dissolution of the solutes in the SP

An example is the analysis of an aqueous solution of phenols. A short pre-column containing Tenax (lipophilic polymer) provides the conditions for very short retention of highly polar water and retardation of more lipophilic compounds such as phenols. An injector between the two columns can be used to transfer and separate the solvent and phenols on the packed column. After the water has been eluted and vented, the trapped phenols are back-flushed to the capillary column in the split mode (required because the carrier gas flow through the packed column is much higher than that through the capillary column). Cold trapping then occurs to focus the phenols which are then rapidly vaporized by heating the trapping section of the column to begin the chromatographic separation.

4. Applications of packed and capillary columns

In this section the applications of the column types are described and compared in selected instances to illustrate the differences in performance and capability of the columns for the analysis of particular analytes.

4.1 Specific applications for packed columns

Traditionally, packed column GC is simpler to arrange and set up than capillary GC, when considering a stationary phase and its preparation and conditioning for particular applications. This arises because of the ability to apply the SP easily to the solid support and modify it appropriately to the particular chromatography required. The contact with the glass surface is minimized because the SP is held on the solid support and, furthermore, deactivation of the glass can be effected using a suitable silylating reagent to limit the effect of adsorption on the inner wall of the column.

Consider, as an example, the problems associated with the analysis of amines on capillary columns. The general difficulty in the chromatography lies in the absorptivity leading to tailing due to the low concentration of surface OH (silanol) groups.

As an example, packed columns can be selectively modified for analysis of amines. Simple treatment with KOH reduces the adsorption to a minimum, allowing good peak shape and optimum low performance chromatography. *Protocol 1* describes the procedure for analysis of volatile amines on packed columns.

Protocol 1. Analysis of volatile amines on packed columns

1. Use the materials listed below for preparing the column:
   - borosilicate glass column, 2 m × 2 mm, with standard ground-glass ends
   - SP Apiezon L or PEG, 4–10%, w/w, solution in DCM
   - Solid support—chromasorb P (80/100) adequate weight to provide a SP/chromasorb[a] composition of 4%, w/w
   - HMDS or TMCS as deactivation reagent
2. Prepare the column as follows.
   (a) Deactivate the borosilicate glass column by treatment with HMDS (50 μl) at 200°C or TMCS in toluene (10%, v/v) for 8 h.
   (b) Treat support with an appropriate volume of 1–5%, w/w, KOH and dry under a nitrogen stream to remove water.
   (c) Prepare the SP/solid support by adding the solution of the SP in DCM to the solid support and evaporating the solvent under a gentle vacuum or a countercurrent of dry nitrogen.
   (d) Add the free flowing SP/solid support mix to the column (plugged at exit by deactivated glass wool) ensuring uniform distribution by tapping during addition (a moderate vacuum applied to the column exit can assist in this process).

---

3. Use typical GC conditions/temperatures and times listed:

- carrier gas/flow rate N_2/20–40 ml/min
- injector temperature   200°C
- initial temperature   70°C
- ramp rate   4°C/min or isothermal (dependent on mid-range boiling point of analytes)
- final temperature   190°C
- initial time   5 min/0°C min
- final time   5 min conditional on boiling range of analytes (not less than 200°C)

[a] 4% Carbowax 20M R + 0.8% KOH on Graphpac R-GC, 60/80 commercially available from Alltech.

The amine mixture can be injected from a headspace system (for example originating from a biological source, such as detritus material) using a gas-tight syringe, or in suitable solvent if this does not present any interference with the analysis. *Figure 3* illustrates the analysis of a standard amine mixture on a KOH-treated packed column. Note that volatile acids can be analysed on 10% Alltech AT[TM]-1200 + 1% H_3PO_4 on Chromasorb W-AW, 80/100.

4.2 Comparison of applications of packed and capillary columns

A number of examples are used below to illustrate the comparative applications of packed and capillary columns and reference can be made to a recent text (20).

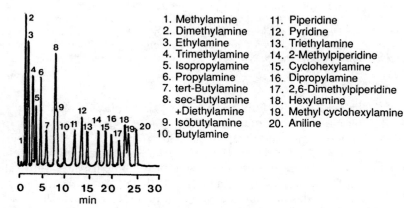

1. Methylamine
2. Dimethylamine
3. Ethylamine
4. Trimethylamine
5. Isopropylamine
6. Propylamine
7. tert-Butylamine
8. sec-Butylamine +Diethylamine
9. Isobutylamine
10. Butylamine
11. Piperidine
12. Pyridine
13. Triethylamine
14. 2-Methylpiperidine
15. Cyclohexylamine
16. Dipropylamine
17. 2,6-Dimethylpiperidine
18. Hexylamine
19. Methyl cyclohexylamine
20. Aniline

Figure 3. Analysis of volatile amines on a KOH treated packed column. (Reproduced with kind permission from Alltech Associates Inc.)

4.2.1 Fatty acid methyl ester (FAME)

Traditionally, fatty acid (FFA or FAME) analyses were performed on packed columns with a polar phase (FFAP or carbowax 20M). This is still the requirement for food or edible additive/product analysis specification, which requires precise elution order of $C_{18_{0,1,2, \text{ and } 3}}$ acids/esters with base line separation. However, with the availability of capillary columns and the progress in (automated) injection, column deactivation, and coating techniques, the practical usefulness of OTCs has been realized. It is worth pointing out that, in terms of theoretical plates per metre, there is little difference between the two modes. It is the increase in length of OTCs that determines the much greater efficiencies and thus performance achievable.

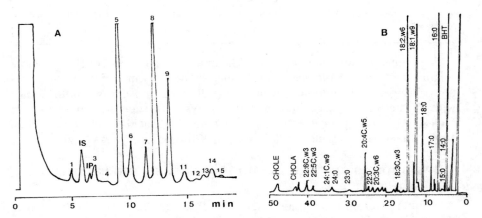

Figure 4. Comparison of packed and capillary GC for the analysis of FAME. (A) Packed column; (B) capillary column (conditions outlined in *Protocol 2*). 1, 12:0, IS 13:0; 3, 14:0; 4, 14:1, w5; 5, 16:0; 6, 16:1, w7; 7, 18:0; 8, 18:1, w9; 9, 18:2, w6; 11, 18:3, w3; 12, 22:0; 13, 22:1, w9; 14, 20:4; 15, 24:0. (Reproduced, with kind permission, from (A) Hockel, M., Dunges, W., Holzer, A., Brockerhoff, P. and Rathgen, G. H. (1980). *J. Chromatogr.*, **221**, 205; and (B) Muskiet, F. A. J., van Doormaal, J. J., Martini, A., Wolthers, B. G. and van der Slik, W. (1983). *J. Chromatogr.*, **278**, 231. © Elsevier Science Publishers BV.)

Practical considerations limit the length of LP to about 6 m (head pressure and d_f), whereas the small diameter of HP and the permeability of capillary columns allow column lengths up to *ca.* 500 m.

For fatty acid methyl ester (FAME) analysis using capillary columns, lengths of 25–60 m are usually used. A typical comparison is illustrated in *Figure 4* for FAME analysis obtained from FFA originating from pooled serum and human plasma analysed by LP and HPGC, respectively. The comparative conditions and details of the method are outlined in the procedure in *Protocol 2*.

Protocol 2. Comparison of LP and HPGC for the analysis of FFA
as FAME derivatives

1. Use LP column and conditions as follows:
   - column: 10%, w/w, SILAR 10CP (CP88)
   - support: Chromasorb W HP (100–120 mesh)
   - carrier gas: N_2
   - flow rate: 20–40 ml/min
   - detector: FID

2. Set up GC temperature programme:
   - initial temperature   250°C
   - initial time   2 min
   - ramp rate   10–15°C/min
   - final temperature   280°C
   - final time   5 min

3. Use capillary column and GC conditions as follows:
   - column: CP-SIL-88 (CP88), dimensions 25 m × 0.25 mm; d_f 0.25 μm
   - carrier gas: He
   - flow rate: 1 ml/min; 25–30 cm/sec
   - detector: FID
   - injector: split, ratio 100:1

4. Set up the GC temperature programme:
   - injector   25–300°C
   - ramp rate   10–15°C/min
   - final   300°C
   - initial time   2 min
   - final time   5 min

5. Use the derivatization method described in Chapter 5, p. 146, Section
   4.3.1 for FFA with
   - FFA mixture, each component   50 μg
   - acetone/ether   1 ml

It is worth noting that, because of the thermal instability of both analytes
and SP, high-boiling-point lipids (TMS DG, DS, and CE) are normally determined by using short WCOT narrow bore HP of 5–10 m length. Hydrogen

can be used as the carrier gas to assist in the reduction of analysis time. The use of short columns is also exemplified by HTGC applications in Section 5.

4.2.2 Pesticides

HPGC has had a major impact in residue analysis (21) in that improved resolution of a pesticide from coextractives and, more importantly, the resolution of active and inactive isomers of certain pesticides is attainable. An example is the pyrethroid insecticide, cypermethrin, which comprises four pairs of enantiomers designated *cis*-1, *cis*-2, *trans*-1, and *trans*-2. With HPGC, all four isomer pairs can be resolved, whereas with LPGC only partial resolution is achieved. Another advantage is the relative inertness of capillary columns but a drawback is the limited capacity, which can be a serious problem for a sample containing large amounts of co-extractives. It is possible to overcome this with column switching techniques to remove interfering materials (see Section 3 on MDGC).

Figure 5A and *B* show the chromatographic profiles obtained with LP and

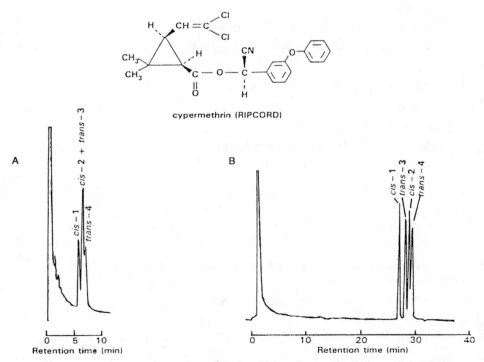

Figure 5. Analysis and separation of cypermethrin isomers. (A) Packed column; (B) capillary column (conditions summarized in *Protocol 4*). 1, *cis*-1; 2, *trans*-3; 3, *cis*-2; 4, *trans*-4. (Reproduced from ref. 21 with kind permission 1985 © Elsevier Science Publishers BV.)

114

HPGC, respectively, and the GC conditions for LP and HPGC used in the analysis of commercial cypermethrin pesticide are outlined below.

i. *Analysis of commercial cypermethrin by LP and HPGC*
(a) Use LPGC conditions as follows:
  - column: 1.5 m × 0.4 cm glass
  - SP: GEXE 60 (1.2% m/m)
  - support: Gas Chrom Q (100/120)
  - temperature: 225°C, isothermal
  - carrier gas: N_2
  - detector: FID
  - sample: 8 μl × 0.02 μg/l
  - flow rate: 20–40 ml/min
(b) Use HPGC conditions as follows:
  - column: HP SE54, 25 m × 0.3 cm ID, d_f 0.25 μm
  - temperature: 230°C
  - carrier: He
  - detector: ECD
  - injector: split, ratio 11:1
  - sample: 2 μl × 0.2 μg/l
  - flow rate: 1 ml/min

4.3 Application of wide-bore capillary columns
4.3.1 Optimization of GC–ECD for the analysis of a narrow range of organochlorines (OCs) and pyrethroids under isothermal temperature conditions

Several illustrations of the capability of wide-bore capillary columns with stationary phase of 15% cyanopropylphenyl dimethylsiloxane, d_f 1.0 μm (SGE or Alltech) are given whereby improved separation and sensitivity can be obtained under isothermal conditions, making appropriate use of carrier and make-up gas together with optimized flow rates. A less expensive GC (e.g. Varian 3710 or Gowmak) can be employed and, furthermore, with ECD as the detector, quite sensitive analysis can be achieved for a narrow range of OCs, viz. 'drins together with pyrethroids.

The combination of helium and nitrogen as carrier and make-up gases is far superior to nitrogen alone as the carrier with no make-up or nitrogen employed in both capacities. The use of helium as the carrier gas enhances the performance and therefore the sensitivity. The use of nitrogen as make-up is twofold leading to:

(a) maintenance of the sensitivity in ECD detection through an increase in flow rate through the detector (the volume is designed for packed column operation and flow rates)

(b) an increase in pressure at the column exit which reduces the pressure drop across the column (greater for WBOTC than medium or narrow bore OTC) and consequently the flow rate (FR) through the column (lower FR increases efficiency). The longer retention time resulting allows a greater separation of components eluting under isothermal conditions.

In operation of WBOTC, lower flow rates are experienced through the column. Thus the use of make-up gas also allows a high flow through the detector zone, which is required when thermal cleaning[a] of the detector (ca. 420°C overnight) is necessary to reduce background. There is a build-up of involatile material due to the contamination of the detector after prolonged use.

Protocol 3 summarizes the procedure for optimising the GC operation, detector exit, and column flow rates for analysis of a selected range of CHCs and pyrethroids.

[a] Graphite not vespel ferrules should be used because of the necessity of raising the detector zone temperature to 400–420°C during thermal cleaning.

Protocol 3. Procedure for the optimization of GC–ECD operation for analysis of OCs and pyrethroids using a WBOT column

1. Use injector and detector adapters as indicated in SGE hints for chromatographers for modifying packed column GC for WBOTC operation.

2. Set up GC with helium as carrier gas and nitrogen as make-up gas[a] with head pressures of 7 and >40 lb/in², respectively.

3. Check the flow rate through the column with make-up valve shut using a bubble flowmeter attached to the exit port of the ECD detector and adjust the flow to 6–10 ml/min.

4. Open the make-up valve to allow nitrogen to enter the detector zone and monitor the increase in FR as in step 3. Adjust the flow through the detector to obtain a combined FR above 20 ml/min.

5. Set the injector, column, and detector temperatures to 250, 240, and 320°C, respectively.[b]

6. Inject 1 μl of a mixture of 'drins and pyrethroids (100 parts per billion, p.p.b. = 100 pg/μl injected) and adjust the make-up gas pressure until the optimum response is obtained. Determine the effect of altering the carrier gas flow which, if too low, will lead to longer than desired retention times particularly for pyrethroids (the FR through WBOTC will be more restricted than for packed columns and therefore cannot be varied over as wide a range).

[a] The detector adapter with a make-up tee is required if the GC has no make-up gas inlet (denoted by auxiliary inlets 1 or 2 at rear of instrument).
[b] *Warning*, Do not change the He cylinder gas without cooling the column. The detector with make-up gas flow on can remain at high temperature. Also, do not change the nitrogen cylinder gas while detector is hot unless the carrier gas is flowing through the column.

Figure 6 illustrates the chromatography obtained under a variety of conditions, e.g. nitrogen as carrier with no make-up, helium as carrier and nitrogen as make-up, optimized isothermal temperature and flow rates (column and make-up). As can be seen with nitrogen as carrier and no make-up, the separation of mirex (M), permethrin (P), and cyfluthrin (C) is limited with poor resolution and the elution is rapid. With helium and nitrogen in combination but FR through column low (< 4 ml/min), M is eluted but C and P have long t_r values. Increasing the FR by a factor of 1.5 leads to the elution of all three analytes with baseline separation and the profile of cis and trans P (23% cis) with Resolution (R_s = 1.5). The isomer profile for C (four components, see Section 4.2.2), although showing incomplete resolution, is more easily identifiable than for nitrogen as carrier gas (important in locating C in a higher background situation, dirty matrix). Further improvement in the chromatographic efficiency could be made by using a 0.5 m length of WBOTC as retention gap coupled to a 0.3 mm i.d. OTC with a lower d_f (hot on-column injection, see Section 4.3.2). This technique is applicable to other detection modes, i.e. FID with appropriate alteration in the measurement of column and make-up flow rates.

4.3.2 Application of WBOTC–MBOTC with hot on-column injection for the analysis of pesticides

Although not widely used, hot on-column (HOC) injection provides an alternative to cold on-column (COC) injection for the analysis of less volatile compounds. Because of the use of WBOTC (deactivated) as a pre-column it has the advantage of utilizing a standard 47-gauge needled syringe for injection, which requires less care in cleaning and presents no assembly difficulties (capillary needles are fragile, problems can arise in blockage within the syringe barrel, and air leaks can occur during loading). Thus maximal transfer of the solute to the analytical column is maintained. HOC can be compared to COC injection except that the sample is flash

117

Peter J. Baugh

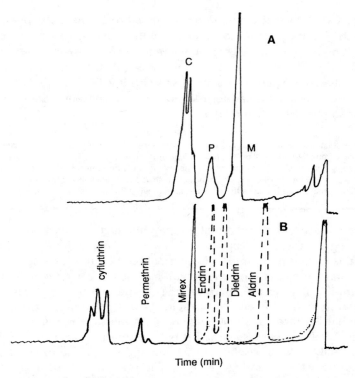

Time (min)

Figure 6. Optimization and analysis of pesticides on a wide-bore capillary column. (A) Mirex, permethrin and cyfluthrin (25 p.p.b.) carrier gas nitrogen, flow rate 7 ml/min, no make-up; (B) The *'drins* (10 p.p.b.), mirex, permethrin, and cyfluthrin (25 p.p.b.), carrier gas He at 5–6 ml/min with N_2 as make-up gas, total FR 20 ml/min (optimization as outlined in *Protocol 3*).

vaporized (splitless-direct as for WBOTC or packed column injection). This condition discounts the use of secondary cooling but the oven temperature should be cold enough to condense the solvent so that cold trapping and solvent effects will be enabled. However, because the flash vaporization occurs in a very small volume there is the possibility of backflow due to high solvent pressures.

A simple hot on-column injector which makes use of the SGE kits for:

(a) conversion of the injector/detector (see Section 2.1.3)

(b) providing a coupling of WBOTC as retention gap with dead volume union to a medium or narrow bore column

The full kit (SGE OCA-4/135 capillary injection adapter, part number 1034611) includes the injector conversion glass-lined tubing (GLT) which fits into the packed column injector port and 5 m of deactivated 0.53 m i.d.

118

capillary tubing. A packed column injector port, if fitted to the GC, can be conveniently used.

i. Column coupling

The WBOTC acts as an integral part of the injection device and also as a retention gap (or short column with a suitable phase as for MDGC). All solutes migrate even at low temperature until they reach the SP in the analytical column (0.32 or 0.22 mm) where they are trapped and reconcentrate. The coupling requires the use of a dead volume union to prevent leakage of carrier gas and to minimize loss in transfer of analytes from the WBOTC to the analytical column. There are two basic designs available as follows:

(a) glass fit connectors comprising a glass tube with a restriction which can accommodate capillary combinations of 0.53;0.32 or 0.53;0.22. The square cut column fits into the connector and the polyimide coating provides an air-tight seal to lock the column in position

(b) stainless steel butt connectors which use ferrules and lock nuts to hold the column ends together. With this type of connector it is possible to fit the 0.32 mm analytical column inside the 0.53 mm retention gap. This feature aids mass transfer and limits further dead volumes.

As with COC it is necessary to keep all parts of the injector clean; the conversion tube and pre-column can be cleaned with solvent. Also it is good practice before coupling the pre-column to the analytical column, with the injector assembled, to renew the septum, and then flush the pre-column with several hundred microlitres of solvent.

ii. Analysis of pesticides

The application of this injector technique to the analysis of pesticides, viz. permethrin (cis and trans isomers), is described here and compared with COC injection (J & W Scientific) using an MBOTC (0.32 mm i.d.) as a pre-column attached to 0.25 mm i.d. column. With COC the use of a pre-column of 0.32 mm i.d. prior to the narrow bore analytical column is preferred to minimize peak splitting (Figure 7). Figure 8 shows the analysis of permethrin isomers with mirex as internal standard using COC injection and a 60 cm pre-column of 0.32 mm i.d. It can be seen that, with COC, excessive tailing with DCM as the solvent is observed, whereas with diethyl ether the chromatography is considerably improved. It is apparent that COC injection with a pre-column is not entirely compatible with the use of DCM as the solvent widely employed for environmental analysis. The precolumn can be active and should be silylated or alternatively deactivated if suspected. Similar results are found with HOC for DCM, with diethyl ether again exhibiting reduced tailing with the pre-column fitted (Figure 9). Since the injection takes place through a septum there is the chance of contamination through transfer of

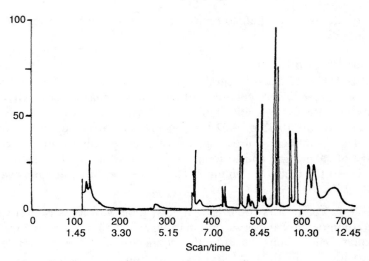

Figure 7. Cold on-column injection and analysis of mirex and permethrin (1 p.p.m.) on a narrow-bore capillary column with no pre-column showing peak splitting exaggerated by the presence of a mixture of *n*-alkanes (>C_{15} to C_{36}).

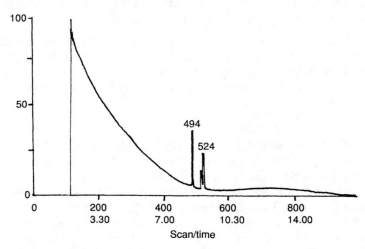

Figure 8. Cold on-column injection and analysis of permethrin and mirex in DCM (1 p.p.m.) on a narrow-bore capillary column with a medium bore pre-column (0.6 m) showing excessive tailing.

septum fragments to the WBOTC. The use of a pre-pierced septum could be an advantage. However, injection through this type of septum must be undertaken with a dome-tipped needled syringe.

In summary, HOC injection with WBOTC pre-column prior to MBOT analytical column gives comparable results to COC injection using MBOT

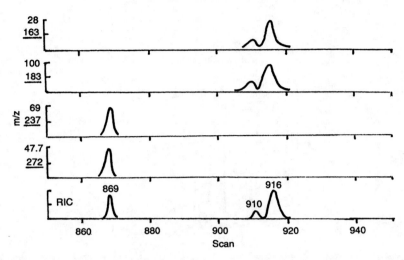

Figure 9. Hot on-column injection of permethrin and mirex in diethylether on to a MBOTC column using a WBOTC pre-column (0.5 m).

and a NBOT column in combination and avoids the problems associated with FSOT capillary needled syringes (P. J. Baugh and K. J. Larmer, unpublished results) (22).

5. High-temperature GC (HTGC) and analysis of high molecular weight low volatile analytes

HTGC has been developed for several reasons but there are a number of problems which require consideration as follows:

(a) injection discrimination

(b) thermal decomposition in the injector or column

(c) reaction with functional groups of the SP

(d) catalytic hydrogenation when H_2 is the carrier gas

With the development of cold on-column injection and high-temperature-resistant, catalyst-free, cross-linked SP, direct injection of compounds normally not easily chromatographable because of the high boiling points can be analysed. Direct injection of fatty acids can be conducted using HTGC. In addition, related materials may still require derivatization (TMS), e.g. phospholipid with substitution of the phosphobase.

Figure 10 illustrates the chromatographic profile of high boiling lipids on a 5 m column coated with SP2100 (CP5 equivalent) with a maximum operating temperature (MAOT) of 340 °C.

Peter J. Baugh

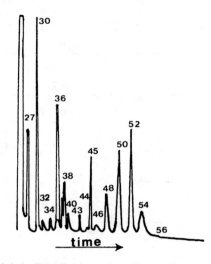

Figure 10. Profile of high boiling lipids analysed on a short non-polar capillary column (5 m). Very low density lipoprotein isolated from human plasma. Conversion to TMS ethers after dephosphorylation. Peaks: 27, chol-TMS; 30, tridecanoyl glycerol (IS); 34, TMS palmitoylsphingosine; 36, 38, 40, TMS-diacylglycerols, 34, 36 and 38 acyl carbons; 43, 45, 47, cholesterol esters with 16, 18 and 20 acyl carbons; 44, 46, 48, 50, 52, 54, 56, triacylglycerols with a total number of 44, 46, 48, 50, 52, 54, and 56 acyl carbons. (Reprinted with kind permission from Kuksis, A., Myher, J. J., Geher, K., Breckenridge, W. C., Jones, G. J. L. and Little, J. A. (1981). *J. Chromatogr.*, **224**, 1. 1981 © Elsevier Science Publishers BV.)

Compounds such as PAH, TG, and *n*-alkanes $> C_{30}$ can be separated at temperatures at which the vapour pressure is high enough to give rise to reasonably short t_r. The temperatures in the GC system and also those of the injector and detector must not be so low that condensation and/or adsorption of the low volatility polar solutes occur.

There are a number of further considerations. Capillary columns are preferred to packed columns because of the large ß values, low d_f, and in particular the application of short capillary columns with d_f 0.1 μm and SP of high stability (non-polar only, e.g. HT5 SGE, see also Chapter 3). Deactivation of glass or FSOT must also be effective above 300 °C. Connections of the column must be stable (graphite not vespel ferrules). The columns must exhibit low bleed to avoid too much background and the carrier gas must be completely oxygen-free.

The high MAOTs possible with the HT5 phase—480 °C programmed and 460 °C continuous—requires aluminium-clad FSOT columns (SGE; AL SIL™). HT5 exhibits virtually zero bleed at normal GC temperatures and is, therefore, eminently suitable for use with MS. In addition, using short columns (6–12 m) to limit the retention time, fast analysis of low volatile high RMM compounds can be achieved.

SGE have illustrated the use of HT5 in one of their application articles (23) and several examples are illustrated in the following sections.

5.1 Analysis of polyaromatic hydrocarbons

With its extended operating range of 10–480°C, analysis of large RMM range of polyaromatic hydrocarbons (PAHs) is possible. *Figure 11* illustrates the chromatogram for the analysis of PAHs from naphthalene (two-ring) to benzoperylene (six-ring). The extended MAOT of HT5[a] should enable separation of PAHs with 10–12-ring structures. The analysis details are summarized in *Protocol 4*.

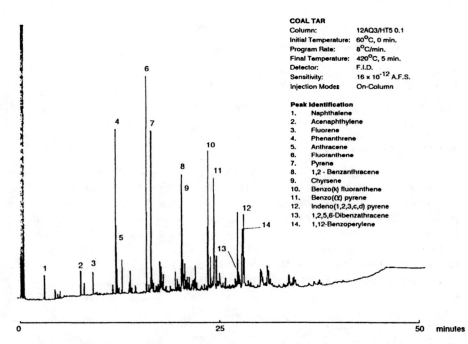

Figure 11. Analysis of PAHs over an extended temperature range on a narrow bore capillary column coated with HT5 stationary phase (SGE Ltd). 1, Naphthalene; 2, acenaphthylene; 3, fluorene; 4, phenanthrene; 5, anthracene; 6, fluoranthene; 7, pyrene; 8, 1,2-benzanthracene; 9, chrysene; 10, benzo(*k*)fluoranthene; 11, benzo(a)pyrene, 12, indeno (1,2,3,c,d) pyrene; 13, 1,2,5,6-debenzanthracene; 14, 1,2-benzoperylene. (Reproduced with kind permission from Dawes, P. and Cumbers, M. (1989). *Am. Lab.*, **21**, 18. © 1989 Inst. Sci. Comm.)

[a] As well as being capable of high-temperature operation, HT5 has selective properties suitable for the complete separation of alkyl substituted aromatics.

Protocol 4. HTGC conditions for separation of PAHs on HT5 SP coated FSOT columns

1. Extract the mixture of PAH from coal tar using
   - hexane
   - coal tar sample
2. Dilute the extract to *ca.* 10 ng/μl
3. Analyse on GC–FID using
   - column   12 m AQ3/HT5; d_f 0.1 μm
   - injection volume   1 μl
   - injection mode   cold on-column
   - initial temp.   60°C; 0°C/min
   - programme rate   8°C/min
   - final temp.   420°C
   - carrier gas H_2 or He
   - flow rate   *ca.* 1 ml/min

5.2 Analysis of triglycerides

A routine method for the analysis of triglycerides (TGs) by capillary GC was not available until the advent of HT5 SP. However, the greater thermal stability of HT5 allows separation of TGs according to their respective carbon number which is important in the food and confectionery industry. *Figure 12* illustrates the chromatography for analysis of TGs up to 54 carbon atoms and *Protocol 5* summarizes the procedure and GC conditions applicable. The analysis of TGs is limited to temperatures < 375°C due to the thermal instability of TGs at higher temperatures.

Protocol 5. Analysis of triglycerides in butter milk fat using short HT5-coated FSOT columns

1. Prepare a sample from butter milk/fat for extraction using:
   - milk/fat   10 ml
   - solvent: DCM/acetone   10 ml
2. Extract the sample (by liquid-liquid extraction or in ultrasonic bath) to obtain a final concentration in the range, 1–5 μg/ml.
3. Analyse on GC–FID using the following column and conditions
   - column   6 m AQ3/HT5; d_f 0.1 μm
   - injection volume   1 μl

- injection mode   cold on-column
- initial temperature   200°C, 0°C/min
- programme rate   10°C/min
- final temperature   370°C, 5 min
- carrier gas H_2 or He
- flow rate   1 ml/min (25–30 cm/sec)

5.3 Analysis of porphyrins

Traditionally, analysis of alkyl porphyrins has required preparation of low-boiling-point derivatives as no existing SP could operate at temperatures of 400°C and above. Because alkyl porphyrins are very thermally stable and temperatures up to 480°C can be employed with HT5 SP coated columns, the analysis is facile for both metal (M = TiO, Cu, VO, Pd) complexes and free bases. Free-base haemoglobin porphyrins are analysable in this way as

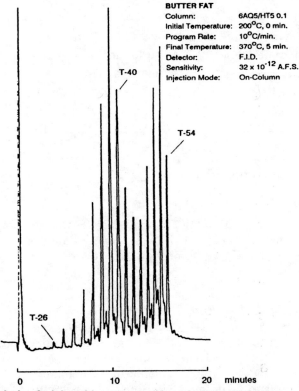

BUTTER FAT

Column:	6AQ5/HT5 0.1
Initial Temperature:	200°C, 0 min.
Program Rate:	10°C/min.
Final Temperature:	370°C, 5 min.
Detector:	F.I.D.
Sensitivity:	32×10^{-12} A.F.S.
Injection Mode:	On-Column

Figure 12. Analysis of triglycerides extracted from butter milk fat using a short NBOTC column coated with HT5. (Reproduced with kind permission as Figure 11.)

illustrated in *Figure 13*. *Protocol 6* summarizes the analysis procedure and conditions.

Protocol 6. Analysis of free-base haemoglobin porphyrins using HT5-coated FSOT columns

1. Prepare a stock solution (100 μg/ml) of the free base porphyrin as follows:

   - porphyrin   10 mg
   - DCM or hexane   100 ml

2. Dilute the stock solution to obtain a final concentration in the μg/ml range.

3. Analyse on GC-FID using the following column and conditions:

   - column   12 m AQ3/HT5; d_f 0.15 μm
   - injection volume   1 μl
   - injection mode   cold on-column
   - initial temperature   260°C, 0°C/min
   - programme rate   10°C/min
   - final temperature   400°C, 5 min
   - carrier gas   He
   - flow rate   1 ml/min

5.4 Other analytes

5.4.1 FAME

Because of the higher MAOT, the analysis of FAME can be extended to higher RMM methyl esters. At the same time because of the selectivity of HT5 it is possible to resolve isomers of target FAME, e.g. $C_{18:1}$ and $C_{18:2}$ *cis* and *trans* isomers.

5.4.2 Crude oil and wax

The wide temperature range of HT5 also enables the analysis of high RMM waxes and hydrocarbons, C_4 to C_{100}, to be made achievable in relatively short analysis times, viz. 45 min. *Figure 14* illustrates the separation of components of a crude oil and wax mixture eluting in the temperature range, 10–480°C. The results compare favourably with those obtainable by SFC (1) but at a considerably reduced cost and less complexity in instrumentation.

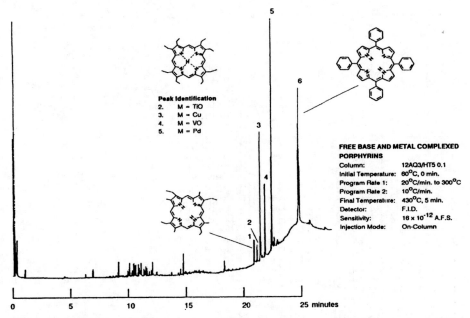

Peak Identification
2. M = TiO
3. M = Cu
4. M = VO
5. M = Pd

FREE BASE AND METAL COMPLEXED PORPHYRINS

Column:	12AQ3/HT5 0.1
Initial Temperature:	60°C, 0 min.
Program Rate 1:	20°C/min. to 300°C
Program Rate 2:	10°C/min.
Final Temperature:	430°C, 5 min.
Detector:	F.I.D.
Sensitivity:	16×10^{-12} A.F.S.
Injection Mode:	On-Column

Figure 13. Analysis of alkyl porphyrins at high temperature on a NBOTC coated with HT5. (Reproduced with kind permission as Figure 11).

6. Gas–solid chromatography (GSC) and gas–liquid chromatography (GLC) for the analysis of highly volatile organics (HVOs)

Gas–solid chromatography (GSC) can be conducted using either packed or wide-bore capillary columns. Applications of both types of column together with thick film capillary GLC to the analysis of gaseous and highly volatile compounds are described in the following sections.

6.1 GSC using packed columns

Packed columns using supports of the Porapak type—Porapak polydivinyl polymer operated with a ramp rate of 6.5°C/min in the temperature range 50–200°C using H_2 as the carrier gas show the resolution of C_3 and C_4 isomers to be insufficient and separation requires long times analysis time (t_R = 23 min).

The GC operation is described below. The column has a suitably high content of SP, i.e., a low ß value and low separation efficiency. Porapak Q and Chromasorb N can also be applied for such separations and subambient temperature assists in further improvement of resolution.

127

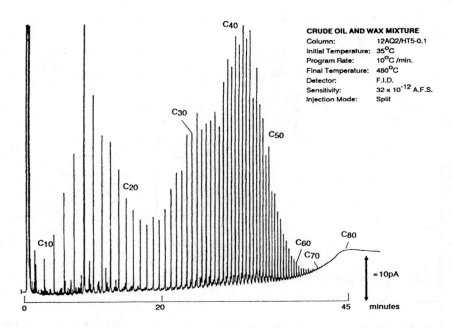

Figure 14. Analysis of a crude oil and wax mixture on HT5 using an extended temperature range, 10 to 480°C. (Reproduced with kind permission as Figure 11.)

6.1.1 Analysis of C1 to C5 hydrocarbons on Porapak N

Analyse the sample CH_4, CO_2, C_2H_4, C_2H_6, C_2H_2 (propane/propene), isobutane using the following column and conditions

- column support   Porapak N
- column dimensions   2 m × 3 mm
- temperatures
    initial   50°C
    final   200°C
    ramp rate   6.5°C/min
- carrier gas H_2 at 0.11 mPa
- detector   FID
- analysis time   23 min

6.2 GSC capillary columns

The support-coated open tubular (SCOT) columns are used for this type of analysis. The inner surface of the capillary column (i.d. 0.4–0.53 mm) is

128

coated with a fine particle Al_2O_3. Generally, the highest selectivity for separation of saturated and unsaturated HCs of low carbon number can be achieved by adsorption chromatography on silica gel, alumina, or molecular sieves. The intermolecular interaction is much stronger than for GLC for which the process is partition.

Compounds of high molecular weight can only be eluted at high-temperature at which labile compounds decompose. There is no SP which will bleed at high temperature and therefore GSC columns are of long life. Ramping is required to elute higher RMM in reasonable analysis times. The details of the analysis is contained below and *Figure 15* illustrates the chromatographic profile.

6.2.1 Analysis of C_1 to C_5 HCs on GSC capillary coated with Al_2O_3

Analyse the sample as in Section 6.3.1 using the following column and conditions:

- column
  - fabric   glass or FSOT
  - support Al_2O_3 (KCl deactivated)
  - dimensions   60 m × 0.4 mm
  - d_f 0.8 μm
- temperature
  - initial   100°C for 15 min
  - ramp rate   6°C/min
  - final   180°C
- carrier gas N_2 at 0.15 Mpa
- detector   FID
- analysis time   16 min

6.3 GLC using thick film capillary columns

GC capillary columns can be successfully applied to the separation of C_1 to C_5 HCs only if the SP d_f is large or by operating at subambient temperatures and temperature programming. Even here the resolution of 1-butane and isobutene is poor. This separation can be achieved with highly polar SP but this SP cannot be fixed to glass or FSOT columns. Typical details for analysis on thick film capillary columns are provided below.

6.3.1 Analysis of C_1 to C_5 HCs on a thick film capillary column

Analyse the sample of gaseous and volatile C_1 to C_5 hydrocarbons using the following column and conditions:

Peter J. Baugh

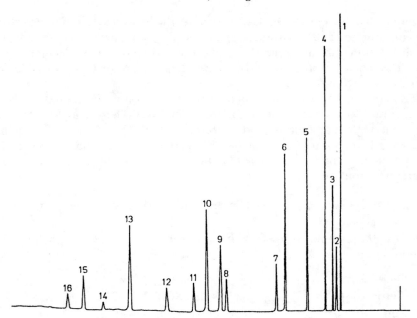

Figure 15. Analysis of HVOs (C_1–C_5) by GSC on a capillary column coated with alumina. 1, Methane; 2, ethane; 3, ethene; 4, propane; 5, propene; 6, isobutene; 7, *n*-butane; 8, *trans*-2-butene; 9, 1-butene, 10, isobutene; 11, *cis*-2-butene; 12, isopentane; 13, 1,3 butadiene; 14, *trans*-2-pentene; 15, 1-pentene; 15, *cis*-pentene. (Reproduced from ref. 19 with kind permission 1990 © VCH Publishers.)

- column
    - fabric   alkali glass or FSOT 30 m × 0.32 mm
    - SP   OV-1 methylpolysiloxane
    - d_f 0.8–1.0 μm
- temperature   −10 to 20°C ballistically
- detector   FID
- carrier gas N_2 at 0.025 MPa
- analysis time   19 min

7. Conclusions

A number of illustrations of the applications of packed columns, wide-bore, medium and narrow capillary columns have been presented. The coverage is not and cannot be exhaustive. However, in all areas there are extensive and excellent examples of the chromatography achievable with LP and HPGC provided in the manufacturers/suppliers catalogues, notably SGE, Alltech,

Chrompack, J & W Scientific, and Jones Chromatography, to which the new or non-specialist user should look as the first step towards their particular requirements. Furthermore, the applications laboratories (where available) and those of the GC instrument makers are always willing to give sound advice on GC columns, injection techniques, and instrumental requirements and will carry out preliminary examination of analytes requiring separation and profiling by GC to determine the most appropriate column and conditions for the chromatography.

References

1. Smith, R. D. (Ed.) (1988). *Supercritical fluid chromatography*. RSC Monograph, RSC, London.
2. Ettre, L. S. (1984). *Chromatographia*, **18**, 477.
3. Betts, T. J. (1984). *J. Chromatogr.*, **294**, 370.
4. Fehringer, N. V. and Walters, S. M. (1984). *J. Assoc. Off. Anal. Chem.*, **67**, 91.
5. Japp, M., Gill, R., and Osselton, M. D. (1987). *J. Forensic Sci.*, **32**, 1574.
6. Korol, A., Solodchenko, N. N., and Ermakov, A. A. (1983). *J. High Resolut. Chromatogr. Chromatogr. Commun.*, **6**, 279.
7. Novak, J. and Roth, M. (1984). *J. Chromatogr.*, **292**, 149.
8. Wells, G. (1983). *J. Chromatogr.*, **270**, 135.
9. Kominar, R., Onuska, F. I., and Terry, K. A. (1985). *J. High Resolut. Chromatogr. Chromatogr. Commun.*, **8**, 585.
10. Clark, R. R. and Zalikowski, J. A. (1987). *Proc. Water Qual. Technol. Conf.*, p. 15.
11. Rovei, V., Sanjuan, M., and Hrdina, P. D. (1980). *J. Chromatogr.*, **182**, 349.
12. Gutierrez, A., Garci, McIntyre, A. E., Lester, J. N., and Perry, R. (1983). *Environ. Technol. Lett.*, **4**, 521.
13. Zenon-Roland, L., Agnessens, R., and Nangniot, P. (1984). *J. High Resolut. Chromatogr Chromatogr. Commun.*, **7**, 480.
14. Grant, D. W. (1986). *Lab. Pract.*, **35**, 59.
15. Berg, J. R. (1987). *Liquid Chromatogr.—Gas Chromatogr.*, **5**, 206, 233.
16. Herraiz, T., Reglero, G., Herraiz, M., Alonso, K., and Cabezudo, M. D. (1987). *J. Chromatogr.*, **388**, 325.
17. Bogusz, M., Bialka, J., Gierz, J., and Klys, M. (1986). *J. Anal. Toxicol.*, **10**, 125.
18. Grob, K. and Frech, P. (1988) *Int. Lab.*, p. 18.
19. Schomburg, G. (1990). *Gas chromatography. A practical course.* VCH, Weinheim.
20. Clement, R. E., ed. (1990). *Gas chromatography. Biochemical, biomedical, and clinical applications*, J. Wiley & Sons, New York.
21. Roberts, T. R. (1985). *Trends in Anal. Chem.*, **4**, 5.
22. Baugh, P. J. and McDonald, W. (1989). *Anal. Proc.*, 219; McDonald, W. (1988). M.Sc. dissertation, University of Salford.
23. SGE (1988). *HT5, a new high temperature stationary phase for capillary GC*, GC3/88, Publication Part No. 5000227. SGE International, Ringwood.

5

Chemical derivatization in gas chromatography

DAVID G. WATSON

1. Introduction

Gas chromatography (GC) of volatile or non-polar compounds may be carried out without derivatizing the sample, indeed derivatives of compounds such as hydrocarbons or halogenated hydrocarbons cannot easily be prepared. It is possible to analyse polar compounds such as carboxylic acids and amines, without prior derivatization, on polar GC phases such as those based on polyethylene glycol. However, derivatization is useful in many instances where it may:

(a) increase the volatility and decrease the polarity of polar compounds;

(b) stabilize compounds which are unstable at the temperatures required for GC;

(c) improve the separation of groups of compounds on GC;

(d) yield information with regard to the number and type of functionalities present in mixtures of unknown compounds;

(e) improve the behaviour of compounds towards selective detectors such as electron capture or nitrogen selective detectors and mass spectrometry.

There are a number of drawbacks in derivatization:

(a) the derivatizing agent may be difficult to remove and interfere in the analysis and this is particularly disadvantageous when the purity of a compound is being assessed by GC;

(b) the derivatization conditions may cause unintended chemical changes in a compound for example dehydration;

(c) the derivatization step increases the time required for analysis.

For these reasons, GC with derivatization is less frequently employed in quality control applications, where the purity of a single substance or the components in a formulation are being determined, than for instance high pressure liquid chromatography where derivatization is usually not necessary.

For quantitative accuracy in a derivatization procedure, it is best that reaction of the analyte with the derivatizing reagent is complete and that an internal standard is used. An internal standard may largely compensate for losses of analyte incurred during derivatization, particularly if it is a close analogue of the substance being derivatized. This is particularly important for compounds with a degree of instability, for example corticosteroids, catecholamines, and other compounds prone to thermal elimination of water or oxidation.

Derivatization reactions are usually simple chemical reactions which are likely to occur in nearly quantitative yield such as acylation, alkylation, or silylation. Two major books on chemical derivatization reactions have been published (1, 2). Another useful textbook provides information on derivatization of many compounds as well as examples of applications in capillary (GC) (3).

2. Apparatus

Derivatization reactions require relatively simple apparatus. The most generally useful apparatus is described below.

2.1 Sample containers and reaction vessels

Several types of sample containers and reaction vessels are described below.

2.1.1 Thick walled Reacti-Vials (Pierce Chemical Co.) or V-Vials (Aldrich Chemical Co.)

The most useful vials have volumes of 0.3 and 1 ml. The vials are fitted with Teflon-faced caps. These tapered vials enable the concentration of the sample to a few microlitres and the use of few microlitres of reagent in derivatization procedures.

2.1.2 Glass sample tubes *ca.* 3.5 ml capacity with aluminium-lined screw caps

These tubes are used for handling sample solutions. It is best to avoid contact between the sample and plastic caps since plasticizer, leached from the caps by the solvent in which the samples are dissolved, may produce additional chromatographic peaks.

2.1.3 Glass sample tubes *ca.* 4.5 ml capacity with Teflon-lined screw caps

These tubes are used for storing reagents removed in small amounts from stock bottles. Sample tubes with open-top caps and Teflon-faced silicone rubber septa may also be obtained but the advantages of avoiding contact of the reagent with the atmosphere may be outweighed by the fact that it is difficult to get such septa to seal perfectly once they have been punctured with a syringe.

2.2 Heating and evaporation

2.2.1 Aluminium heating blocks

Aluminium blocks with holes drilled to fit Reacti-vials, sample tubes, and a thermometer are suitable for heating and evaporation. These may be bought either undrilled or predrilled (Pierce Chemical Co.), or may be constructed by a workshop from billeted aluminium. A typical dimension for a block is 15 × 10 × 6 cm with holes cut to a depth of 3 cm.

2.2.2 A simple thermostated hot plate

This hot plate may be used for heating. In general reaction temperatures do not have to be exact. More expensive hot plates or dry block heaters will give greater accuracy in temperature control but it is rare that this is necessary.

2.2.3 Apparatus for blowing down the sample

This apparatus usually comprises a nitrogen gas cylinder with a pressure-regulating head to which rubber tubing may be attached. A Pasteur pipette may then be inserted into the end of the rubber tubing to provide a stream of gas which may then be used to evaporate samples. The gas flow may be crudely regulated using a screw clip to pinch the rubber tubing. If evaporation of several samples at once is required, a simple multiport blowing-down apparatus, consisting of a glass tube sealed at one end with several openings along its length to which rubber tubing may be attached, may be constructed by a glassblower. Pasteur pipettes may be inserted into the rubber tubing and these may be individually regulated using screw clips to pinch the rubber tubing. Alternatively Teflon glass taps, such as those used in constructing chromatography columns, may be used instead of screw clips to provide a finer control of gas flow. This type of apparatus may also be bought from a number of suppliers. The advantage of using Pasteur pipettes as opposed to the steel needles used in most commercially available apparatus is that they are disposable and may be changed between samples thus avoiding cross contamination.

2.3 Sample and reagent handling

Many of the reagents used in derivatization are highly reactive and contact with air and moisture must be kept to a minimum. The reagents are also often irritant and corrosive and must be handled in a fume cupboard.

The preferred method of transferring measured volumes of sample solutions or reagents is by use of a microlitre syringe. These syringes are available from a number of different manufacturers and perhaps the most useful volumes are 10 μl, 50 μl, 100 μl and 500 μl. Small volume glass pipettes are a cheaper alternative, but require more skill in their use and allow more contact of reagents with the air. Contact of reagents and solvents with plastic automatic pipette tips is best avoided, especially when low amounts of analyte are being determined. Syringe barrels may become sticky with repeated use and may be washed by filling the barrel with a solution of strong detergent such as Decon[R] followed by thorough rinsing in water and methanol.

Pasteur pipettes are useful for transfer of sample solutions and may be plugged with cotton wool and filled with a drying agent such as anhydrous sodium sulphate when drying of sample solutions is required. They may also be used to contain short columns of chromatographic reagents which may be used to remove derivatization reagents.

2.4 Removal of derivatizing reagents.

The simplest procedure for removing excess of derivatization reagent is by its evaporation under a stream of nitrogen. For involatile reagents a chromatographic filtration step may be used. Strongly adsorbent chromatographic materials such as silica gel or alumina are usually not suitable for removal of reagents since they may degrade the product or irreversibly adsorb it. Sephadex LH20 or Sephadex LH60 may be used for removal of reagents, these chromatographic materials are lipophilic but polar and will readily remove polar reagents.

3. Standard procedures in derivatization

Many of the procedures described in the protocols have repetitive elements in them and certain standard procedures may be defined at the outset and are as follows.

(a) When volumes of reagent are small, reactions are carried out in 0.3 ml or 1 ml capacity Reacti-Vials or V-vials. When volumes of solvents or reagents are greater, such as in aqueous phase reactions, then 3.5 ml screw-top sample tubes with aluminium-lined caps are used.

(b) The reagents or solvent are evaporated under a stream of nitrogen gas with the sample maintained at 60–80°C in a heating block. Obviously, less volatile reagents require heating at higher temperatures for their efficient removal. If the sample is volatile, evaporation at a low temperature for a longer time may be required or it may be better to inject it without removing the reagents.

(c) Drying is carried out by passing the sample through *ca.* 3 cm of anhydrous sodium sulphate contained in a Pasteur pipette plugged with cotton wool. Anhydrous magnesium sulphate may be used as an alternative.

(d) Dissolution of the derivatized sample prior to analysis is carried out by treating the sample with 2 ml of solvent for capillary column GC using the splitless injection mode (the volume may be adjusted if a split injection is used) or 100 μl for packed column GC. Inject 1 μl of the product solution; 200 μg of material is chosen as a suitable starting point for the development of a method, since in most circumstances the derivatized compound should be clearly seen in relation to any interfering peaks from reagent residues.

(e) Removal of excess reagents is carried out by passage through a short column of Sephadex LH20. The sample is passed through a short column prepared by introducing Sephadex LH20 suspended in EtOAc/hexane (1/1, v/v) into a Pasteur pipette plugged with cotton wool and allowing the solvent to drain out to leave a plug of *ca.* 3 cm of the adsorbent.

The Sephadex LH20 is prepared by suspending 50 g in 500 ml MeOH for 1 h and allowing it to swell. The suspension is filtered under vacuum through a sinter funnel and the collected LH20 is then washed with 500 ml EtOAc. The LH20 is removed from the sinter funnel and suspended in 250 ml of EtOAc/hexane (1:1, v/v), and the suspension allowed to settle. The solvent is decanted and replaced with a further 250 ml EtOAc/hexane (1:1, v/v). The suspension is then ready for use and used LH20 may be regenerated by following the above procedure.

4. Derivatization reactions involving one reagent

4.1 Silylation reactions

The theory behind silylation reactions has been comprehensively reviewed (4).

4.1.1 Trimethylsilyl (TMS) derivatives

These derivatives may be prepared from a wide range of functional groups including: hydroxyl, carboxylic acid, amine, amide, thiol, phosphate, hydroxime, and sulphonic acid. The reaction is of the general type:

ROH, RNH_2, $RCOOH$, $RNH.COR'$, RSH, $ROPO_3H$, RNH_2OH, RSO_3H $\xrightarrow{\text{Silylating reagent}}$ $ROTMS$, $RNHTMS$, $RCOOTMS$, $RNTMS.COR'$, $RSTMS$, $ROPO_3TMS$, $RN(OTMS)$, $RSO_3\,TMS$.

In some instances, e.g. TMS derivatives of sulphonic acids and phosphates, the stability of the derivative is very low. There are a number of TMS donor reagents available but some have wider ranges of application than others. It is difficult in most instances to see advantages of many of the available commercial cocktails over simply using the neat reagent.

Protocol 1. TMS derivatives of unhindered groups

1. Heat at 60°C for 15 min a mixture of the following:
   - analyte   200 μg
   - N,O-bis(trimethylsilyl)acetamide (BSA)   50 μl
2. Evaporate the BSA and dissolve the residue in EtOAc.
3. For analysis of carboxylic acids to improve stability of the derivative (5), carry out the reaction as in steps 1 and 2 but do not remove the BSA, and inject the excess of reagent with the sample, diluted with EtOAc to an appropriate volume, into the GC.
4. For compounds of poor solubility in BSA:
   (a) Heat at a higher temperature, e.g. 90°C for 20 min. If this fails to dissolve the analyte.
   (b) Dissolve initially in dry pyridine at 90°C, add BSA, and heat at 90°C for 20 min.

N,O-Bis(TMS)trifluoroacetamide (BSTFA) and N-TMS-N-methyltrifluoro-acetamide (MSTFA) may be substituted for BSA. These reagents are stronger silyl donors, which is important for less reactive groups such as thiols and amides, but are not as good solvents. They are also more volatile, which is important where they are being used to derivatize volatile compounds, e.g. short-chain fatty acids.

Trimethylsilylimidazole (TMSIM) is the most powerful silylating reagent available and it may be used to prepare TMS derivatives from sterically hindered alcohols such as steroids with a side chain at C-17 and 17-hydroxygroup. It may also be used in the silylation of carboxylic acids, but not in the silylation of amines.

Protocol 2. TMS derivatives of hindered alcohols

1. Heat at 80°C for 20 min the following mixture:
   - analyte   200 μg
   - trimethylsilylimidazole (TMSIM)   50 μl
2. Cool the mixture and dilute with 100 μl EtOAc, mix thoroughly, add 1 ml hexane.
3. Pass the solution through a short column of Sephadex LH20, thus removing the involatile TMSIM which would leave residues on the column and in the detector.
4. Evaporate the eluent and dissolve the residue in EtOAc.

The procedures described in *Protocols 1* and *2* should enable silylation of most compounds to be carried out, although a large number of different silylation procedures have been described in the literature. Another procedure that has been widely used is the incorporation of a small amount of a catalyst to increase the rate of silylation of a hindered group or unreactive group. In most instances it is better to use TMSIM in silylation of hindered groups since the silylation catalysts trimethylchlorosilane (TMCS) and trimethylbromosilane (TMBS) may produce traces of mineral acid which may catalyse degradation of an analyte, e.g. the dehydration of a tertiary alcohol. A typical procedure employing a catalyst is as follows:

(a) Heat at 60 °C for 30 min the following mixture:
- analyte   200 μg
- BSA   50 μl
- TMCS or TMBS   1 μl

(b) Evaporate the reagents and dissolve the residue in EtOAc.

4.1.2 Tertiarybutyldimethylsilyl (TBDMS) derivatives

The general equation for the reaction is identical to that involved in TMS formation. There is not a wide range of commercially available silylating reagents for introduction of this group. The effectiveness of different reagents for introducing the TBDMS group has been studied (6). The advantage of TBDMS ethers is that they are much more stable than TMS ethers (see *Protocols 3* and *4*).

Protocol 3. TBDMS derivatives of unhindered groups

1. Heat at 60 °C for 30 min the following mixture:
   - analyte   200 μg
   - a DMF solution containing 1 M tertiarybutyldimethylsilyl chloride (TBDMSCl) and 1 M imidazole   100 μl

2. Cool the mixture, dilute with 100 μl EtOAc, mix thoroughly, and add 1 ml hexane.

3. Pass the mixture through a short column of Sephadex LH20 removing excess reagents (alternatively the products may be extracted into 200 μl hexane).

4. Evaporate the eluent and dissolve the residue in EtOAc.

Using this procedure TBDMS derivatives may be prepared from alcohols and phenols. The derivatives are stable to moisture or chromatography on silica gel. Derivatives of carboxylic acids may also be prepared by this procedure, although there are likely to be losses of analyte due to the imidazole

139

acting as an acid scavenger. The procedure is not suitable for the preparation of TBDMS derivatives of amines.

There are few good procedures for the direct silylation of Krebs cycle acids. The following procedure minimizes the risk of the formation of enol silyl ethers and yields stable products.

Protocol 4. TBDMS derivatives of Krebs cycle acids

1. Heat at 60°C for 1 h a mixture of the following:

   - analyte   200 μg
   - a pyridine solution containing 0.5 M TBDMSCl and 1.25 M imidazole   100 μl

2. Cool the mixture and dilute with 100 μl EtOAc, mix thoroughly, and add 1 ml hexane.

3. Pass the mixture through a short column of Sephadex LH20.

4. Evaporate the eluent and dissolve the residue in EtOAc.

Di-isopropylethylamine may be used to replace imidazole if the TBDMS group is being introduced at a hindered hydroxyl (7). 1,8-Diazabicyclo-(5,4,0)undec-7-ene (DBU) shows promise as an alternative base catalyst for conversion of amines, thiols, and carboxylic acids to TBDMS derivatives (8).

MTBSTFA is a commercially available analogue of MSTFA which is used to form TMS derivatives. The advantage of MTBSTFA is that it is volatile but it is of low reactivity and only reacts with alcohols when there is carboxyl group in the same molecule (D. G. Watson, unpublished data). However, it is useful for the formation of TBDMS derivatives of acids which are more stable than the corresponding TMS derivatives and has also been used to prepare TBDMS derivatives of dipeptides (9). The following procedure describes TBDMS formation using *N*-tertiarybutylsilyl-*N*-methyltrifluoracetamide (MTBSTFA).

(a) Heat at 90°C for 1 h a mixture of the following:
   - analyte   200 μg
   - MTBSTFA   100 μl
   - TMCS   1 μl

(b) Evaporate the reagents and dissolve the residue in EtOAc.

4.1.3 Other silylating reagents

Isopropyldimethylsilyl chloride and ethyldimethylsilyl chloride are both commercially available although derivatives such as the *N,O*-(bis)silylacetamides are not. The chlorides may be used to derivatize compounds by substituting them for TBDMSCl in *Protocol 3*. A number of silylating reagents have been prepared from these compounds and have been used in derivatization reactions (10).

4.2 Acylation reactions

The reaction is of the general type:

$$\text{ROH, RNH}_2, \text{RNHOH} \xrightarrow{\text{R'COX}} \text{ROCOR', RNHCOR', RNHOCOR'}$$

where R'COX represents an acid anhydride, chloride or imidazole.

4.2.1 Acetate formation

The following procedure may be applied to unhindered alcohols, phenols, amines, and monosaccharides.

Protocol 5. Formation of acetates from unhindered groups

1. Heat at 60°C for 15 min a mixture of the following
   - analyte   200 μg
   - pyridine   30 μl
   - acetic anhydride   30 μl
2. Evaporate the reagents and dissolve the residue in EtOAc.

The following procedure is applicable to hindered alcohols.

Protocol 6. Formation of acetates from hindered groups

1. Leave the following mixture at room temperature for 2 h (gentle heating might be required initially to dissolve the analyte).
   - analyte   200 μg
   - acetic anhydride containing 1% *p*-toluene sulphonic acid   50 μl
2. Evaporate the reagents and dissolve the residue in 1 ml methylene chloride.
3. Wash with 1 ml of 1 M Na_2CO_3 and remove and dry the organic layer.
4. Evaporate and dissolve the residue in EtOAc.

Monosaccharides may be analysed as their alditol acetates, the advantage of converting monosaccharides to their alditol acetates is that, whereas the unreduced sugars as their acetates (or TMS derivatives) yield four peaks upon derivatization due to the formation of derivatives of the α- and β-anomers of their pyranose and furanose forms, as their alditol acetates they yield a single peak.

Protocol 7. Formation of alditol peracetates from monosaccharides

1. Leave the following solution at room temperature for 1 h:
   - monosaccharide   1 mg
   - 1 M ammonia solution   0.5 ml
   - $NaBH_4$ 10 mg
2. Carefully add 100 µl glacial acetic acid.
3. Evaporate the solution and dissolve the residue in 1 ml MeOH containing 10% glacial acetic acid.
4. Evaporate the MeOH/acetic acid, repeat this three more times adding further 1 ml amounts of MeOH/acetic acid in order to remove boronic acid (which would interfere in the final acetylation step) as its methyl ester. Remove residual acetic acid by adding a further 2 ml MeOH alone and evaporating.
5. Suspend the final residue in 0.5 ml acetic anhydride and heat for 1 h at 90°C.
6. Evaporate the reagent and take up the residue in 1 ml water and extract with 2 ml EtOAc.
7. Remove and dry the organic layer.

The following procedure (*Protocol 8*) employs the fact that acid anhydrides (or chlorides) react more rapidly with phenols or amines than with water. The method is particularly useful for the analysis of biogenic amines such as dopamine which are amphoteric and cannot be extracted directly from biological fluids.

Protocol 8. Aqueous phase acetylation

1. Prepare the following solution:
   - analyte   200 µg
   - water   1 ml
2. Add the following:
   - saturated Na_2CO_3 100 µl
   - acetic anhydride   50 µl
3. Shake the mixture vigorously for 1 min
4. Extract the product into 2 ml EtOAc and dry the extract.
5. Evaporate to dryness and dissolve the residue in EtOAc.

In the above procedure propionic anhydride may be used instead of acetic anhydride and, since it reacts more slowly with water, only 10 μl is required. The procedure may also be applied to the derivatization of amines such as adrenaline or phenols such as 3-methoxy-4-hydroxyphenylethylene glycol. In these cases the final residue obtained above is treated with acetic anhydride/pyridine (*Protocol 5*).

4.2.2 Fluorinated acylating groups

Trifluoroacetic anhydride (TFAA), pentafluoropropionic anhydride (PFPA), and heptafluorobutyryl anhydride (HFBA) may all be used. These reagents are applicable to the derivatization of alcohols (apart from tertiary alcohols which they tend to dehydrate), phenols, amines, amides, and sugars. In the case of alcohols and phenols, the derivatives formed are susceptible to hydrolysis but are quite stable in the case of amines. The fluoroacylated derivative of a compound has considerably shorter retention time compared with its acetate or trimethylsilyl derivative. The derivatives are also highly electron-capturing which makes them useful for electron capture detection (ECD). All these reagents are poor solvents but their high reactivity means that often a compound will gradually dissolve in the anhydride as it is acylated. A typical procedure is described below.

(a) Heat at 70°C for 15 min in a vial with a close fitting cap the following mixture:

- analyte   200 μg
- acetonitrile   100 μl
- fluoracyl anhydride   100 μl

(b) Cool the solution before removing the vial cap and evaporating the reagents and dissolving the residue in EtOAc.

The ECD response it usually greatest for HFB derivatives, but a disadvantage of these derivatives is that the lower volatility of HFBA makes it more difficult to remove residual traces of reagent which may cause interference when analysis using ECD is carried out. If an amine is being analysed it may be better to dissolve the final product in MeOH, converting residual traces of acid or anhydride to the more volatile methyl ester which will readily evaporate. Addition of MeOH may also be used when an amine group is present in the molecule which also contains hydroxyl groups; in this case the hydroxyl groups are hydrolysed by the MeOH leaving the acylating group on the amine; the free hydroxyl groups may then be acetylated (*Protocol 5*) or trimethylsilylated (*Protocol 1*).

TFA-imidazole, PFP-imidazole, and HFB-imidazole may be used to acylate compounds that are sensitive to traces of acid. These reagents acylate the same groups as the anhydrides and are also recommended for the acylation of indole amines (see *Protocol 9* for an example of the use of TFA-imidazole).

Protocol 9. Fluoroacylation with fluoroacyl imidazoles

1. Heat at 90°C for 2 h the following mixture:
   - analyte   200 μg
   - fluoroacyl imidazole   100 μl
2. Cool the mixture and dilute with 100 μl EtOAc, mix thoroughly, and add 1 ml hexane. Pass the mixture through a short column of Sephadex LH20 (alternatively the products may be extracted directly with hexane).
3. Evaporate the eluent and dissolve in EtOAc.

Aqueous phase acylation with heptafluorobutyryl chloride (HFBCl) has been used in the extraction derivatization of biogenic amines such as dopamine (11) and is particularly useful in the derivatization of polyamines such as putrescine (D. G. Watson, unpublished data). *Protocols 10* and *11* contain typical procedures for the use of HFBCl.

Protocol 10. Derivatization of polyamines with HFBCl

1. Prepare the following solution:
   - analyte   200 μg
   - MeOH   50 μl
2. Add the following:
   - 0.5 M aqueous NaOH   1 ml
   - HFBCl   50 μl
3. Immediately shake vigorously for 2 min.
4. Extract with 2 ml EtOAc, remove, and dry the organic layer and adjust to the appropriate volume for analysis.

Protocol 11. Derivatization of phenolic amines with HFBCl

1. Prepare the following solution:
   - analyte   200 μg
   - 1 M phosphate buffer, pH 7.5   200 μl
2. Add the following:
   - HFBCl   5 μl
3. Immediately shake vigorously for 5 min.
4. Extract the solution with 1 ml hexane.

5. Shake the organic layer with 2 × 200 μl water.

6. Dry the organic layer, evaporate the solvent, and dissolve the residue in EtOAc.

Pentafluorobenzoyl chloride can be used to derivatize amines and phenols. The derivatization is carried out in the aqueous phase and the reagent remains there upon extraction of the product (see *Protocols 12* and *13*). The derivatives have a very high ECD response.

Protocol 12. Derivatization of an amine with PFBCl

1. Prepare the following solution:
- analyte   200 μg
- MeOH   50 μl

2. Add the following:
- 0.5 M aqueous NaOH   1 ml
- PFBCl   5 μl

3. Shake vigorously for 5 min and extract with 2 ml diethyl ether.

4. Remove and dry the organic layer, evaporate the solvent, and dissolve the residue in EtOAc.

Protocol 13. Derivatization of a phenol with PFBCl

1. Prepare the following solution.
- analyte   200 μg
- MeOH   50 μl

2. Add the following:
- 1 M potassium phosphate buffer pH 8.0   1 ml
- PFBCl   5 μl

3. Shake vigorously for 5 min and extract with 2 ml diethyl ether.

4. Remove and dry the organic layer, evaporate the solvent, and dissolve the residue in EtOAc.

Protocols 12 and *13* may also be carried out using 3,5-ditrifluoromethyl-benzoyl chloride instead of PFBCl.

4.3 Alkylation reactions

Alkylation reactions may be used to derivatize carboxylic acids, amines, sulphonic acids, phosphonic acids, phosphates, barbiturates, uracils, purines, penicillins, thiols, and inorganic anions.

4.3.1 Methylation

The following procedure may be used to esterify carboxylic acids. The general equation for the reaction is:

$$RCOOH + R'OH \longrightarrow RCOOR' + H_2O$$

(a) Heat at 60°C for 30 min a mixture of the following:

- analyte   200 µg
- dry MeOH to which has been freshly added 3%, v/v, acetyl chloride   100 µl

(b) Evaporate the reagents and dissolve the residue in EtOAc.

Other alcohols, e.g. ethanol or isopropanol, may be used instead of MeOH.

The following procedure (*Protocol 14*) may also be used to methylate carboxylic acids.

Protocol 14. Methylation with boron trifluoride–MeOH complex

1. Heat at 60°C for 15 min a mixture of the following:
   - analyte   200 µg
   - MeOH containing 10% w/v BF_3 100 µl
2. Extract with 2 × 0.5 ml hexane, wash the organic layer with 2 × 1 ml of water, and dry.
3. Evaporate the solvent and dissolve the residue in EtOAc.

The general equation is as given above.

The following procedure (*Protocol 15*) may be used to hydrolyse and methylate the carboxylic acids in glycerides in single step.

Protocol 15. Methylation of the fatty acids from glycerides

1. Prepare the following solution:
   - analyte   1 g
   - petroleum ether   10 ml
2. Add the following:
   - 2 M methanolic KOH   1 ml
3. Shake vigorously for *ca.* 0.5 min and allow the glycerol layer to settle.
4. Make an appropriate dilution of the upper layer and analyse by GC.

Diazomethane may be used to methylate carboxylic acids, phenols, barbiturates, penicillins, amines, phosphonic acids, and sulphonic acids. Diazomethane may be conveniently prepared from 1-methyl-3-nitro-1-nitrosoguanidine (MNNG) by its reaction with 5 M sodium hydroxide in a mini-diazomethane generator available from Aldrich Chemical Co. Instructions are provided with the apparatus; an additional point is that a stronger solution of diazomethane is obtained if the apparatus is left immersed in the ice bath for 45 minutes after addition of sodium hydroxide solution to the MNNG. The same procedure may be used to generate diazoethane.

$$\text{RCOOH, RSO}_3\text{H, RPO}_3\text{H, ArOH} \xrightarrow{\text{CH}_2\text{N}} \text{RCOOMe, RSO}_3\text{Me, RPO}_3\text{Me, ArOMe.}$$

All acidic hydrogens in the molecule are replaced and thus ureides such as barbiturates or uracils incorporate two methyl groups which replace both amide protons as shown in *Structure I* for a barbiturate. The following procedure can be used for methylation with ethereal diazomethane.

Structure I. Barbiturate derivative.

(a) Leave the following mixture at room temperature for 10 min
- analyte   200 µg
- ethereal diazomethane   0.5 ml

(b) Evaporate the solvent and dissolve the residue in EtOAc.

Trimethylanilinium hydroxide (TMPAH) may be used to carry out 'flash heater' methylation of barbiturates, xanthines, phenols, and nucleotides. Tetramethylammonium hydroxide may be used instead of TMPAH. The disadvantage of these reagents is that they require a column dedicated to their use, since they leave a residue on the column that will methylate compounds injected subsequently, whether this is required or not. The following procedure describes methylation using TMPAH.

(a) Inject the following mixture into the GC with the injection port temperature set at 250°C:
- analyte   200 µg
- 0.1 M TMPAH in MeOH   100 µl (packed) or 2 ml (capillary)

The following procedure (*Protocol 16*) may be used in the methylation analysis of polysaccharides or in the direct analysis of oligosaccharides by GC;

alternative procedures exist for most other compounds. Methylsulphinylcarbanion is prepared by dissolving sodium hydride in DMSO (12).

Protocol 16. Methylation using methylsulphinylcarbanion/methyl iodide

1. Prepare the following solution in a 5 ml Reacti-Vial fitted with a Reacti-Vial magnetic stirrer and a Mininert push-button valve (Pierce Chemical Co.):

   - analyte   500 μg
   - dry DMSO   1 ml

2. Remove the cap and flush the vial with dry nitrogen or preferably argon and inject 1 ml 2 M methylsulphinylcarbanion solution via the Mininert valve.

3. Stir the solution for 30 min and then cool it in ice. Inject 1 ml methyl iodide and stir the mixture for a further 30 min.

4. Dilute the reaction mixture with 5 ml water and extract with 2 × 8 ml chloroform.

5. Remove the chloroform by rotary evaporation and dissolve the residue in 10 ml hexane. Wash the hexane layer with 3 × 10 ml water in separating funnel to remove residual DMSO and dry.

6. Remove the hexane by rotary evaporation and dissolve the residue in EtOAc.

4.3.2 Alkylation with benzyl halides

The general equation is as follows:

RCOOH, RNH$_2$, ArOH etc. + R'Br → RCOOR', RNR'$_2$, ArOR' + HBr

Pentafluorobenzyl bromide (PFBBr) can be used to prepare strongly electron-capturing derivatives of carboxylic acids, amines, barbiturates, uracils, thiocarbamides, phenols, and thiols (see *Protocol 17*). Ureides give structures of the type shown in *Structure II* for uracil; also both protons on a primary amine may be replaced by the PFB group.

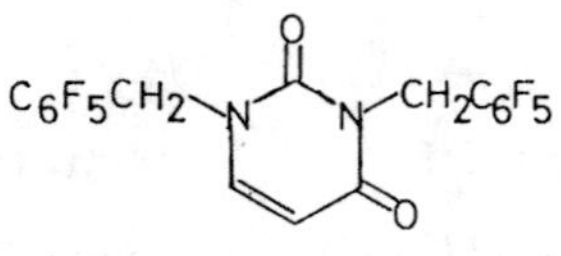

Structure II. Uracil derivative.

Protocol 17. Direct reaction with PFBBr

1. Prepare the following solution:
   - analyte   200 μg
   - acetonitrile   60 μl
2. Add the following:
   - PFBBr   10 μl
   - triethylamine   10 μl
3. Leave the solution at room temperature for 15 min then add 0.5 ml EtOAc, mix thoroughly, and add 0.5 ml hexane.
4. Leave the solution at room temperature *ca.* 15 min while a white precipitate forms and then wash the organic layer with 2 × 1 ml 0.5 M HCl.
5. Dry the organic layer, evaporate, and dissolve the residue in EtOAc.

This procedure may also be carried out using 3,5-ditrifluoromethylbenzyl bromide; the resultant derivatives have less tendency to form tailing chromatographic peaks than the equivalent PFB derivatives (13). PFBBr has also been used to derivatize inorganic anions (14).

The following procedure (*Protocol 18*) may be used to alkylate acids, phenols, and sulphonamides:

Protocol 18. Extractive alkylation with PFBBr

1. Prepare the following solution:
   - analyte   200 μg
   - methylene chloride   1 ml
2. Add the following:
   - aqueous 0.1 M tetrabutylammonium hydrogen sulphate in 0.2 M aqueous NaOH   1 ml
   - PFBBr   20 μl
3. Shake for 30 min at room temperature.
4. Remove the organic layer, dry, evaporate, and dissolve the residue in diethylether.

A recent procedure describes the use of octafluorotoluene for the alkylation of phenols (15).

David G. Watson

4.4 Condensation reactions

4.4.1 Oxime formation
The general equation for the reaction is:

$$RR'CO + R''ONH_2 \cdot HCl \longrightarrow RR'CNOR'' + H_2O.$$

Reaction may be carried out in either the organic or aqueous phase, and reaction of an aldehyde or ketone with hydroxylamine is usually followed by acetylation or TMS formation to derivatize the hydroxyl group on the oxime and other hydroxyl groups in the molecule (see *Protocols 19, 20,* and *21*).

Protocol 19. Reaction with hydroxylamine

1. Prepare the derivative by leaving for 12 h at room temperature a mixture of the following:

   - analyte   200 µg
   - a solution of hydroxylamine·HCl 20 mg/ml in pyridine   50 µl

2. After 12 h add 50 µl BSA and heat the solution for 15 min at 60 °C.

3. Evaporate the reagents and dissolve the residue in EtOAc.

Protocol 20. Formation of oximes from keto acids

1. Prepare the derivative by heating at 60 °C for 30 min a mixture of the following:

   - analyte   200 µg
   - 0.5 M aqueous NaOH   1 ml
   - a solution of hydroxylamine·HCl 25 mg/ml in water   1 ml

2. Cool, acidify with 6 M HCl and extract with 2 ml EtOAc.

3. Dry the organic layer, evaporate, and dissolve the residue in BSA. [a]

4. Heat for 30 min at 60 °C, do not remove the reagent, and dilute to an appropriate volume for analysis.

[a] MTBSTFA (See p. 140) may be used instead of BSA.

Protocol 21. Formation of alkyl oximes

1. Heat for 15 min at 60°C a mixture of the following:
   - analyte   200 μg
   - a solution of methoxylamine·HCl 80 mg/ml in pyridine   50 μl
2. Cool and add 100 μl EtOAc, mix, add 1 ml hexane, and wash solution with 2 × 1 ml 0.5 M HCl.
3. Dry the organic layer, evaporate, and dissolve the residue EtOAc.

Sterically-hindered ketones such as a corticosteroid with a keto group at C_{11} may require a longer reaction time at a higher temperature, e.g. 90°C for complete derivatization. Ethyloxime·HCl, tert-butyloxime·HCl or benzyloxime·HCl may be used instead of methoxylamine·HCl. However, the larger the alkyl group the less readily the oxime will react with a sterically-hindered ketone.

Reaction of a carbonyl compound with pentafluorobenzyl hydroxylamine·-HCl (PFBO·HCl) produces a highly electron-capturing derivative useful in detection by ECD (see *Protocol 22*). PFBO·HCl is relatively unreactive and some ketones react with it only slowly giving a low yield of product (16). It is not clear what factors govern whether or not derivatization will occur, e.g. a good yield of product is obtained with testosterone, whereas the yield of product from reaction of PFBO·HCl with the closely structurally-related androstene dione is very low.

Protocol 22. Formation of PFBO derivatives

1. Heat at 60°C for 15 min a mixture of the following:
   - analyte   200 μg
   - PFBO·HCl 100 mg/ml in pyridine   50 μl
2. Cool the solution and mix with 100 μl followed by 1 ml hexane.
3. Wash the solution with 2 × 1 ml 0.5 M HCl.
4. Dry, evaporate, and dissolve the residue in EtOAc.

For ketones that do not react readily with PFBO·HCl, reaction for several days at room temperature may improve the yield of derivative.

4.4.2 Condensation of amines with carbonyl compounds

The general equation for the reaction is:

$$RNH_2 + R'R''CO \rightarrow RNCR'R''.$$

David G. Watson

A typical procedure for reaction with low molecular weight ketones is as follows:

(a) Leave at room temperature for 2 hours a mixture of the following:
  - analyte   200 µg
  - acetone or other volatile ketone   100 µl or 2 ml

(b) Inject directly into the GC.

A typical procedure for reaction with pentaflurobenzaldehyde (PFBA) is as follows.

(a) Heat at 60°C for 1 h a mixture of the following:
  - analyte   200 µg
  - acetonitrile   50 µl
  - PFBA   25 µl

(b) Cool and dilute to an appropriate volume with EtOAc.

The derivative has a high ECD response but for sensitive work it is difficult to remove the PFBA which may interfere in the analysis.

4.5 Derivatives of miscellaneous types

4.5.1 Hexafluoroacetylacetone (HFAA) derivatives of guanidines

The derivative formed in this case is a bis(trifluoromethyl)pyrimidine shown in *Structure III*.

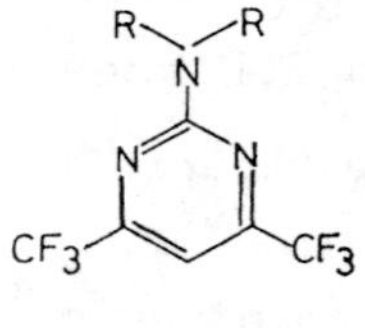

Structure III. Bis(trifluoromethyl)pyrimidine derivative.

Protocol 23 describes the general procedure for derivatization.

Protocol 23. Reaction with HFAA

1. Heat at 120°C for 1 h a mixture of the following:
  - analyte   200 µg
  - pyridine   50 µl
  - HFAA   50 µl

2. Cool the solution, add 1 ml ether followed by 3 ml HCl.

---

3. Shake the mixture, centrifuge, and remove and dry the organic layer.
4. Evaporate and dissolve the residue in EtOAc.

4.5.2 Derivatization of tertiary amines

The following procedure may be used to convert a tertiary amine to an electron-capturing carbamate (17). The general equation for the reaction is:

$$RN(CH_3)_2 + R'OCOCl \rightarrow RNCH_3OCOR'$$

and *Protocol 24* describes the derivatization procedure.

Protocol 24. Reaction of tertiary amines with pentafluorobenzyl-chloroformate (PFBCF)

1. Prepare the following solution:
   - tertiary amine free base   200 μg
   - heptane   200 μl
2. Add the following:
   - PFBCF   50 μl
   - anhydrous Na_2CO_3 10 μg
3. Heat at 100°C in a tightly capped vial for 1 h at 100°C.
4. Shake the heptane layer with 1 ml 1 M NaOH and analyse the organic layer by GC

In an amine, such as imipramine, one of the two methyl groups on the tertiary nitrogen is replaced by a pentafluorobenzylformate group.

4.5.3 Derivatization of quaternary amines

Since quaternary amines carry a charge they will not pass through a GC column. The following procedure (see *Protocol 25*) has been used in the analysis of acetylcholine (18) and it may have application to other quaternary amines. The equation for the reaction is:

$$(CH_3)_3N^+ (CH_2)_2OCOCH_3 \xrightarrow{\quad C_6H_5S^- \quad} (CH_3)_2N(CH_2)_2COOCH_3$$

Protocol 25. Demethylation of a quaternary amine

1. Heat in vial flushed with N_2 at 80°C for 30 min with shaking every 5 min a mixture of the following:
   - acetylcholine   200 μg
   - butanone containing 6 mg/ml sodium benzenethiolate   500 μl

---

Protocol 25. *Continued*

2. Cool and add 0.1 ml 0.5 M aqueous citric acid and 2 ml pentane.

3. Shake vigorously, centrifuge, and discard the upper layer.

4. Wash the aqueous layer with 2 × 1 ml pentane and then evaporate traces of pentane from the aqueous phase.

5. Add 50 μl $CHCl_3$ and 0.1 ml of a mixture of 2 M ammonium citrate/ 7.5 M ammonium hydroxide. Shake and centrifuge.

6. Remove and dilute the organic layer appropriately for GC analysis.

The procedure removes one of the methyl groups from the quaternary nitrogen converting the acetylcholine to a tertiary amine.

5. Mixed derivatives

5.1 Silyl-acyl and silyl-carbamate derivatives

Phenolic amines present problems in their extraction from an aqueous environment since they are amphoteric and ionized appreciably at all pH values. Difficulties in extraction are avoided by acylation or alkyl formylation of amine and phenolic groups in the aqueous phase. The product is then extracted into the organic phase and if the acyl or alkyl formyl groups attached to phenolic oxygens are particularly bulky they are selectively removed by shaking the organic phase with 10 M ammonia solution. The free phenolic and aliphatic groups in the molecule are then reacted with a silylating reagent, e.g. BSA (see *Protocol 26*). The removal of acyl group from phenolic oxygens is only necessary if the group is bulky and makes GC retention times very long or if, as in the case of alkyl formate groups, it is not stable to further derivatization after extraction.

Protocol 26. Aqueous phase derivatization of phenolic amines

1. Prepare the following solution:
   - phenolic amine   200 μg
   - 1 M potassium phosphate buffer pH 7.5   1 ml
2. Add either propionic anhydride   10 μl
   or acetic anhydride   40 μl
3. Shake vigorously for 2 min and extract the aqueous layer with 2 ml EtOAc, and remove and dry the organic layer.
4. Evaporate the solvent and react the residue with either a silylating reagent (*Protocol 1* or *Protocol 3*) or with acetic anhydride (*Protocol 5*).

Reaction with PFBCl may be used to prepare electron capturing derivatives of phenolic amines (see *Protocol 27*).

Protocol 27. PFB derivatives of phenolic amines

1. Prepare the following solution:
   - phenolic amine   200 μg
   - 1 M potassium phosphate buffer pH 7.5   1 ml
2. Add 5 μl PFBCl.   5 μl
3. Shake the mixture vigorously for 5 min and extract with EtOAc.
4. Extract with 2 ml EtOAc and remove the organic layer and shake with 0.5 ml 10 M ammonia solution for 5 min.
5. Remove, dry, and evaporate the organic layer.
6. React the residue with BSA (*Protocol 1*) but do not evaporate all the BSA, leave a few microlitres to facilitate chromatography of traces of acylating reagent and dissolve the residue in EtOAc.

3,5-Ditrifluoromethylbenzoyl chloride may be used instead of PFBCl and silyl groups such as TBDMS (*Protocol 3*) may be introduced instead of TMS (19). *Structure IV* shows the PFB/TMS derivative of adrenalin.

Structure IV. PBF/TMS derivative of adrenalin.

Alkyl chloroformates, e.g. methyl chloroformate, may be used in aqueous phase derivatizations as indicated in *Protocol 28*.

Protocol 28. Alkyl chloroformate TMS derivatives of phenolic amines

1. Prepare the following solution:
   - phenolic amine in 50 μl MeOH   200 μg
   - 1 M potassium phosphate buffer pH 7.5   1 ml
2. Add 30 μl methyl chloroformate and shake vigorously for 5 min.
3. Extract with 2 ml EtOAc, remove the organic layer and shake with 0.5 ml 10 M ammonia for 5 min.
4. Remove, dry, and evaporate the organic layer.
5. React the residue with BSA (*Protocol 1*).

Trichloroethylchloroformate may be used if an electron-capturing derivative is required and other silyl groups such as TBDMS (*Protocol 3*) may be
introduced instead of TMS.

Alcohol amines, including compounds such as ephedrine and the ß-
blocking drugs, can be extracted directly from alkaline solution and converted
to a single derivative, e.g. TMS or PFB. However, in the case where an
involatile derivatization reagent such as PFBCl is used, it is advantageous to
leave the excess derivatizing agent behind in the aqueous layer and derivatize
the aliphatic hydroxyl after extraction (see *Protocol 29*).

Protocol 29. PFB/TMS derivatives of alcohol/amines

1. Prepare the following solution:
   - alcohol amine in 50 μl MeOH   200 μg
   - 0.5 M aqueous KOH   0.5 ml
2. Add 5 μl PFBCl, shake vigorously for 5 min, and extract with ether.
3. Remove and dry the organic layer and evaporate the solvent.
4. React the residue with BSA (*Protocol 1*).

5.2 Acyl/acyl derivatives

The following procedure (*Protocol 30*) may be used to derivatize tryptamine,
serotonin, melatonin, and related structures. The derivative structure of the
derivative formed is shown for serotonin (see *Structure V*).

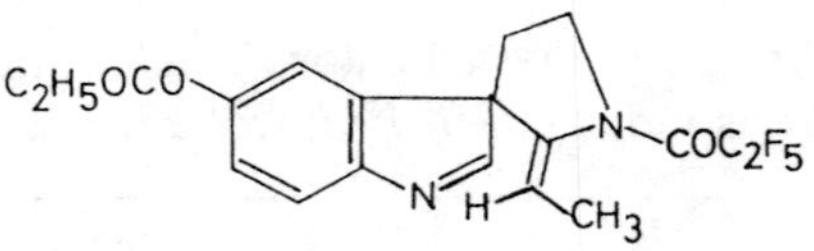

Structure V. Acyl/acyl derivative of serotonin.

Protocol 30. Spirocyclic derivatives of tryptamines

1. Prepare the following solution:
   - tryptamine   200 μg
   - 0.4 M perchloric acid   500 μl
2. Add:
   - saturated Na$_2$CO$_3$ 100 μl
   - pyridine   10 μl

---

3. Finally add 50 μl propionic anhydride and shake vigorously for 5 min.

4. Extract with 2 ml EtOAc, and remove and dry the organic layer.

5. Evaporate the solvent, dissolve the residue in 0.5 ml PFPA and allow the PFPA to evaporate while heating at 60°C in a vial without a cap.

6. Dissolve the residue in EtOAc.

The derivative chromatographs as two peaks due to the formation of two geometrical isomers in the course of derivatization (21).

5.3 Acyl/alkyl derivatives

The following type of derivative (see *Protocol 31*) has been widely applied to the analysis of amino acids where the nitrogen in the molecule is acylated and the carboxyl group alkylated, but it applies equally to hydroxy acids, non-steroidal anti-inflammatory drugs such as aspirin or fenamates, acidic meta-bolites of biogenic amines such as homovanillic acid and bile acids.

Protocol 31. Acyl and fluoroalkyl derivatives of amino acids and hydroxy acids

1. Heat at 60°C for 30 min a mixture of the following:

- analyte   200 μg
- methanol containing 3% acetyl chloride   100 μl

2. Evaporate the reagents, dissolve the residue in 100 μl trifluoroacetic anhydride, and heat at 60°C for 30 min.

3. Evaporate the reagents and dissolve the residue in EtOAc.

Propanol, isopropanol, and butanol may be used instead of methanol; and acetic anhydride, pentafluoropropionic anhydride, and heptafluorobutyric anhydride may be used instead of trifluoroacetic anhydride.

The following procedure provides derivatives of amino acids and hydroxy-acids which yield a high ECD response (21). It is also applicable to acylated amino acids such as hippuric acid, where the amide function becomes acylated during the derivatization giving improved chromatography. The acylation and alkylation are accomplished in a single step. Either trifluoroethanol (TFE) or hexafluoroisopropanol may be used.

(a) Heat at 100°C for 1 h the following mixture in a vial with a tightly fitting cap:

- analyte   200 μg
- TFE   5 μl
- PFPA   50 μl

(b) Evaporate the excess reagents and dissolve the residue in EtOAc.

David G. Watson

The following procedure (see *Protocol 32*) provides derivatives of phenolic acids such as the tyramine metabolite *p*-hydroxyphenylacetic acid (PHPA) or salicylic acid with a high ECD response (22). The structure of the derivative for PHPA is shown in *Structure VI*.

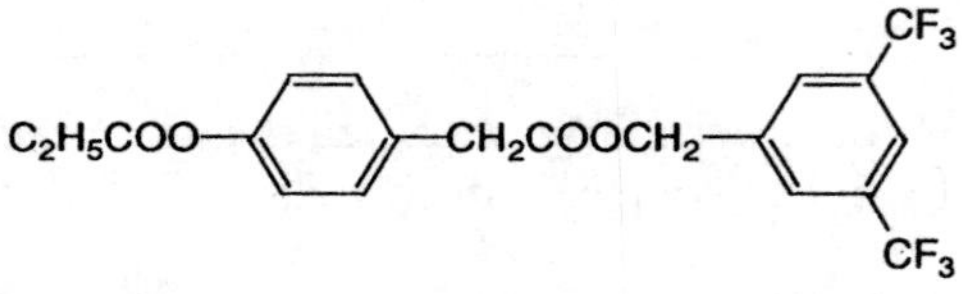

Structure VI. PHPA derivative.

Protocol 32. Ditrifluorobenzyl/propionyl/acetyl derivatives

1. Prepare the following solution:
   - analyte   200 μg
   - 1 M potassium phosphate buffer pH 7.5   1 ml

2. Add 10 μl propionic anhydride, shake the solution vigorously for 5 min, and extract with 2 ml EtOAc.

3. Remove, dry, and evaporate the organic layer making sure that traces of propionic acid are also evaporated and dissolve the residue in 50 μl acetonitrile.

4. Add: 10 μl 3,5-ditrifluorobenzyl bromide and 10 μl triethylamine and leave the solution at room temperature for 15 min.

5. Add, with thorough mixing, 500 μl EtOAc followed by 500 μl hexane, and leave for 15 min while a white precipitate forms.

6. Wash the organic layer with 0.5 ml 0.5 M HCl and then dry.

7. Evaporate the solvent and, to derivatize any remaining aliphatic hydroxyl groups, react the residue with acetic anhydride/pyridine (*Protocol 5*).

5.4 Acyl/amide derivatives

The following is a useful procedure (*Protocol 33*) for derivatizing the sulphonic amino acid taurine (derivative shown in *Structure VII*) and it could be adapted both for other amino acids and sulphonic acids. An amide of a sulphonic acid is much more stable than an ester.

158

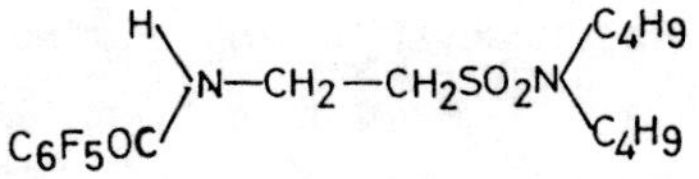

$$\underset{\substack{C_6F_5OC}}{\overset{\displaystyle H}{\diagdown}} N-CH_2-CH_2SO_2N \underset{\diagdown C_4H_9}{\overset{\diagup C_4H_9}{}}$$

Structure VII. Taurine derivative.

Protocol 33. PFB/sulphonamide derivatives

1. Prepare the following solution
   - taurine   200 μg
   - 0.25 M aqueous NaOH   1 ml
2. Add 20 μl PFBCl and shake the mixture for 2 min.
3. Adjust the pH of the reaction mixture pH 1–2 with 0.5 M HCl and wash with 3 × 3 ml ether.
4. Add 100 μl 10% tetrabutylammonium hydrogen sulphate in 0.2 M NaOH and extract with 2 ml methylene chloride by shaking for 3 min at room temperature.
5. Centrifuge for 1 min, transfer the organic layer to another tube, and evaporate.
6. Dissolve the residue in 50 μl thionyl chloride, heat at 80°C for 10 min, and evaporate the thionyl chloride.
7. Add 100 μl of di-*n*-butylamine in acetonitrile and leave the solution for 2 min at room temperature, acidify the reaction mixture with 1 ml 20% orthophosphoric acid, and extract with 2 × 3 ml hexane.
8. Remove the organic layer, evaporate and dissolve the residue in EtOAc.

5.5 Silyl/alkyl oxime and acyl/oxime derivatives

The following procedure (*Protocol 34*) is used for the analysis of corticosteroids, where oxime formation is used stabilize the steroid side chain prior to trimethylsilylation (23, 24).

Protocol 34. Methoxime trimethylsilyl derivatives

1. Heat at 60°C for 20 min the following mixture:
   - corticosteroid   200 μg
   - a pyridine solution containing 80 mg/ml
     methoxylamine·HCl   30 μl

Protocol 34. *Continued*

2. Then add 30 µl trimethylsilyl imidazole and heat for a further 20 min at 60 °C.

3. Cool the reaction mixture, add 100 µl EtOAc, mix thoroughly, and add 1 ml hexane.

4. Pass the solution through a short column of Sephadex LH20, evaporate the eluent, and dissolve in EtOAc.

The oximation of the two ketone groups in corticosteroids may potentially produce four isomers, i.e. possibly four GC peaks but often only one or two forms predominate. In the derivatization of steroids such as dexamethasone, which has a hindered 20-ketogroup, or cortisone, with a hindered 11-ketogroup, heating at 90 °C for 2 h is required in the oximation step. In the case of dexamethasone which contains a very hindered hydroxyl group at C_{17}, this is followed by heating for 3 h at 90 °C with TMS-IM to effect trimethylsilylation.

Protocol 34 may be used for other ketols and, if they do not contain hindered hydroxyl groups, BSA (*Protocol 1*) may be used as the silylation reagent in the second step. Monosaccharides may be analysed as trimethylsilyl/methoxime derivatives each sugar yielding two peaks due to separation of the syn and anti forms of the methoxime.

The following derivative is not an oxime but is formed by a similar procedure, the reaction converting the aldehyde group in the aldose to a nitrile. The derivative gives a single peak for aldoses via a simple procedure (see *Protocol 35*) and provides an alternative to the lengthy procedure used for formation of alditol acetates (*Protocol 7*).

Protocol 35. Wohl derivatives of aldoses

1. Heat together for 1 h at 60 °C the following mixture:

   - aldose   200 µg
   - a MeOH solution containing 1 mg hydroxylamine·HCl   200 µl
   - dry sodium acetate   2.5 mg

2. Evaporate the MeOH, add 200 µl acetic anhydride, and heat at 120 °C for 1 h.

3. Evaporate the reagents and shake the residue with 2 ml EtOAc to remove the product from the reagent residues.

The derivatives may be analysed by ECD as well as by flame ionization detection.

The following derivative (see *Protocol 36*) has been used in analysis of 18-hydroxysteroids. The structure for the derivative of 18-hydroxycorticosterone is shown in *Structure VIII*.

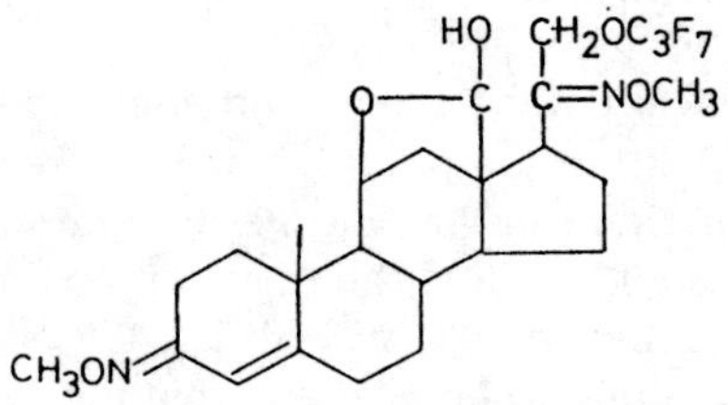

Structure VIII. 18-Hydroxycorticosterone derivative.

Protocol 36. Heptafluorobutyl/methoxime derivatives

1. Leave overnight at room temperature the following mixture:
   - analyte   200 μg
   - a pyridine solution containing 16 mg/ml
     methoxylamine·HCl   100 μl
2. Dilute the solution with a saturated aqueous NaCl and extract with 2 × 1 ml EtOAc.
3. Wash the extract with 1 ml 1 M HCl, dry, and evaporate the EtOAc.
4. Dissolve the residue in 50 μl 2:1 toluene/pyridine, add 100 μl hepta-fluorobutyric anhydride, and heat the solution at 60°C for 30 min.
5. Evaporate the reagents and dissolve the residue in EtOAc.

5.6 Derivatization procedures for prostaglandins

Prostaglandins may contain hydroxyl, carboxyl, and keto groups, and require three stages in derivatization as indicated in *Protocol 37*. The structure of a typical derivative is shown in *Structure IX*.

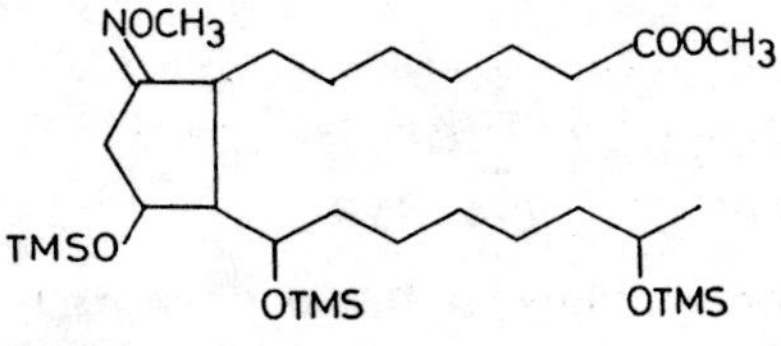

Structure IX. Prostaglandin derivative.

161

Protocol 37. Methyl/trimethylsilyl/methoxime derivatives

1. Leave at room temperature for 10 min the following mixture:
   - analyte in 20 μl MeOH   20 μg
   - ethereal diazomethane (see p. 147)   100 μl

2. Evaporate the solvent. If the compound does not contain a keto group proceed to step 3. If the compound contains a keto group dissolve the residue in 30 μl pyridine containing 30 mg/ml methoxylamine·HCl and heat for 10 min at 60°C. Dilute the reaction mixture with 500 μl EtOAc followed by 500 μl hexane. Wash with 2 × 0.5 ml 0.5 M HCl. Remove, dry, and evaporate the organic layer. Treat the residue as in step 3. If the compound does not contain a ketogroup proceed directly to step 3.

3. Add 30 μl BSA, heat for 10 min at 60°C, evaporate the reagents, and take up the residue in EtOAc (10 μl or 200 μl).

Pentafluorobenzyl hydroxylamine·HCl may be used instead of methyl-oxylamine·HCl to yield an electron-capturing derivative.

Another electron-capturing derivative of prostaglandins may be prepared as follows as described in *Protocol 38*.

Protocol 38. Pentafluorobenzyl ester/trimethylsilyl/oxime derivatives

1. Leave at room temperature for 15 min the following mixture:
   - analyte   20 μg
   - acetonitrile   20 μl
   - pentafluorobenzyl bromide (PFBBr)   2 μl
   - triethylamine   2 μl

2. Dilute with 500 μl EtOAc followed by 500 μl hexane, leave for 30 min, then wash with 2 × 0.5 ml 0.5 M HCl, and then dry the organic layer.

3. Evaporate the solvent and then treat the residue as in *Protocol 37* from step 2 onwards.

3,5-Ditrifluoromethylbenzyl bromide may be used instead of PFBBr.

6. Bifunctional and mixed bifunctional monofunctional derivatives

The advantage of bifunctional derivatizing agents is their selectivity; in a mixture of compounds, they will selectively derivatize those compounds with reactive functional groups either on neighbouring carbon atoms or separated by one carbon atom. An extensive review of this type of derivative has been made (25).

6.1 Bifunctional silylating agents

A study of this type of derivative used di-*tert*-butyldichlorosilane to prepare di-*tert*-butylsilylene derivatives of diols and hydroxyacids (26). The structure of a typical derivative is shown in *Structure X*. The derivatives are stable in

Structure X. Di-*tert*-butylsilylene derivative.

comparison with corresponding dimethylsilylene derivatives and provide a means of selectively analysing diols and hydroxyacids in complex mixtures. A silylating reagent reacting with both mono- and difunctionalities has been studied (27). The following outlines a typical procedure using this type of reagent.

(a) Heat at 80°C for 15 h the following mixture:

- analyte   200 μg
- acetonitrile   60 μl
- *N*-methylmorpholine   20 μl
- 1-hydroxybenzotriazole (dried *in vacuo* at 40°C prior to preparing stock solution 3 mg/ml in acetonitrile)   9 μg
- di-*tert*-butyldichlorosilane   3.5 μl

(b) Dilute the mixture to an appropriate volume for analysis with EtOAc.

6.2 Aldehydes and ketones as bifunctional derivatizing agents

Benzaldehyde, pentafluorobenzaldehyde, and hexafluoroacetone have all been used to prepare bifunctional derivatives. However, the most useful reagent of this type is dichlorotetrafluoroacetone (DCTFA). DCTFA selectively forms cyclic derivatives with amino acids (oxazolidinones) (28, 29) and α-hydroxyacids (30) (dioxolanones) under very mild conditions (see below for typical procedure) and the resultant derivatives are strongly electron-capturing and stable to chromatographic clean-up. Structures of typical

David G. Watson

Structure XI. Bifunctional derivative of an amino acid using dichlorotetrafluoroacetone.

Structure XII. Bifunctional derivative of an α-hydroxy acid using dichlorotetrafluoro-acetone.

derivatives are shown in *Structures XI* and *XII*. A typical procedure for a DCTFA derivative is as follows.

(a) Leave at room temperature for 15 min the following mixture:
- analyte   200 μg
- acetonitrile   30 μl
- DCTFA   10 μl
- pyridine   5 μl

(b) Evaporate the reagents and dissolve in EtOAc for analysis or, if derivatization of other functional groups is required, react it with BSA (*Protocol 1*) or acetic anhydride pyridine (*Protocol 5*).

DCTFA does not readily form cyclic derivatives with diols, although reaction with DCTFA followed by prolonged heating with acetic anhydride may result in formation of cyclic derivatives of these compounds. With ß-hydroxyamines, DCTFA either forms Schiff's bases or, with secondary amines, the initial product formed by reaction of DCTFA with the amino group does not cyclize.

6.3 Alkyl boronates as bifunctional derivatizing agents

Alkyl boronates are the most extensively used bifunctional derivatizing agents. They are formed under very mild reaction conditions and have good thermal and GC properties. They have certain disadvantages: they are hydrolytically unstable; they form partial derivatives with isolated functional groups, e.g. hemiesters with isolated hydroxyl groups, causing poor or non-existent chromatography; they may be unstable to further reactions required to derivatize other functional groups in the molecule, and excess alkyl boronic acid accumulates as a residue on the GC column and this may cause 'flash' derivatization of subsequent samples. Such reagent residues may be largely removed by injecting a few microlitres of 1,3-propanediol several times after work with boronates has finished. Despite these drawbacks, alkyl boronic acids have been found to be very useful as selective derivatizing agents in the analysis of mixtures, e.g. sesquiterpene vicinal diols (31). Alkyl-boronic acids will react with 1,2-diols, 1,3-diols, 1,2-hydroxyacids, 1,3-hydroxyacids, 1,2-hydroxyamines, 1,3-hydroxyamines, and aromatic compounds with orthosubstituted phenol/amine and phenol/carboxylic acid groups. Thus these derivatives may be applied in the analysis of steroids, lipids, nucleosides, carbohydrates, catecholamines, and prostaglandins. The alkylboronic acids: methane boronic, butane boronic acid, *tert*-butane-boronic acid, cyclohexane-boronic acid, benzene-boronic acid, 3,5-di-trifluoromethyl-benzene-boronic acid, and ferrocene-boronic acid have all been used in derivatization. The structure of a butane boronate derivative of a vicinal diol is shown in *Structure XIII*. A typical procedure is outlined below.

Structure XIII. Butane boronate derivative of a vicinal diol.

(a) Leave the following solution for 15 min at room temperature
- analyte   200 µg
- pyridine or EtOAc containing 100 µg of alkyl boronic acid   50 µl

(b) Evaporate the solvent and dissolve the residue in EtOAc.

This type of derivative may be used for simple diols including steroid diols and some sugars. Ferrocene boronates have been found to be particularly useful in the analysis of low molecular weight diols and hydroxyacids by GC-MS (32).

In most instances other functional groups in a molecule may be derivatized after the formation of a boronate without causing the derivative to break down, which is not the case for boronates formed with the dihydroxyacetone side chain of corticosteroids. *Protocol 39* is typical of the procedures employed.

Protocol 39. Acetyl and trimethylsilyl alkyl-boronates

1. For acetyl derivatives, dissolve the residue from (b) above in the following mixture:
   - dry pyridine   100 µl
   - acetic anhydride   20 µl

165

Protocol 39. *Continued*

2. Leave at room temperature for 2 h, then evaporate the reagents, and dissolve the residue in EtOAc.

3. For TMS derivatives, dissolve the residue from (b) above in the following mixture:

  • hexamethyldisilazane   100 μl
  • pyridine   100 μl

4. Leave at room temperature for 5 min, then evaporate the reagents, and dissolve the residue in EtOAc.

7. Derivatives used for separation of enantiomers

Enantiomers in a mixture can be separated by converting them to a pair of diastereoisomers via reaction with a single enantiomer of a chiral derivatization reagent.

7.1 Acylation with chiral reagents

The following procedure (*Protocol 40*) may be used in the chiral derivatization of amines and alcohols.

Protocol 40. Derivatization with R-(−)-2-phenylbutyryl chloride ((−)PBCl)

1. Leave for 1 h at room temperature the following mixture:

  • (−)-2-phenylbutyric acid   50 mg
  • freshly distilled thionyl chloride (gentle warming may
    be required to dissolve the acid)   1 ml

2. Evaporate the thionyl chloride and dissolve the residue in 1 ml dry toluene and add 50 μl of this solution to an equimolar amount of amine or alcohol (*ca.* 0.5 mg) and heat at 60 °C for 1 h.

3. Evaporate the solvent and dissolve the residue in EtOAc.

Chiral natural products such as (+)-chrysanthemic acid may also be used in this procedure as derivatizing agents (33). Amino acids may be derivatized by this procedure after prior conversion to their methyl esters (see p. 157). The reagent used in the following procedure, *N*-trifluoroacetyl-L-prolyl chloride (TPCl), is commercially available as its chloride and may be used to determine the enantiomeric composition of chiral amines.

(a) Leave for 15 min at room temperature the following mixture:
- analyte   200 µg
- chloroform   50 µl
- 0.1 M TPCl in chloroform   40 µl

(b) Evaporate the solvent and dissolve the residue in EtOAc.

The following method (*Protocol 41*) may be used to carry out analysis of enantiomers of phenolic amines such as adrenaline or phenylephrine (34).

Protocol 41. Derivatization with L-(−)-N-heptafluorobutyrylphenylalanyl chloride (HFBPALCl)

1. To prepare L-(−)-HFBPALCl stir the following mixture at room temperature until all the phenylalanine has dissolved (*ca.* 4 h):
   - L-(−)-phenylalanine   1 g
   - acetonitrile   50 ml
   - heptafluorobutyryl chloride (HFBCl)   2 g

2. Remove the solvent and excess HFBCl by rotary evaporation.

3. Dissolve 100 mg of the product-heptafluorobutyrylphenylalanine in 1 ml thionyl chloride by warming gently and leave the solution for 1 h at room temperature.

4. Evaporate the excess thionyl chloride and dissolve the residue in 2 ml EtOAc.

5. To carry out derivatization prepare the following solution:
   - phenolic amine   200 µg
   - 1 M potassium phosphate buffer pH 7.5   1 ml

6. Shake the solution vigorously with 50 µl of the solution prepared in step 4 for 10 min and then add 2 ml EtOAc.

7. Remove the organic layer and shake with 0.5 ml 10 M ammonia solution for 5 min, and then remove and dry the organic layer.

8. Evaporate the organic layer and then react the residue with BSA (*Protocol 1*) or TBDMSCl (*Protocol 3*).

HFBPALCl is best when freshly prepared from a stock of HFBPAL. Reaction with TBDMSCl yields products which are stable to chromatography on silica gel which may then be used to remove some of the reagent residues. The structure of a derivative prepared from *p*-synephrine is shown in *Structure XIV*.

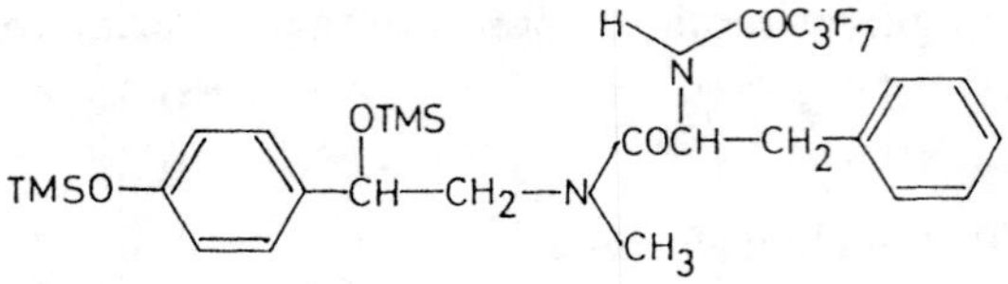

David G. Watson

Structure XIV. *p*-Synephrine derivative.

7.2 Chiral alkylating reagents

These derivatives may be applied to chiral carboxylic acids and the procedure for use of alkanols is described below.

(a) Heat together for 2 h at 80°C the following mixture:
- analyte   200 μg
- (−)- or (+)-2-butanol or 2-pentanol   80 μl
- acetyl chloride   20 μl

(b) Evaporate the excess reagents and dissolve the residue in EtOAc.

The procedure (*Protocol 42*) may also be used to trans-esterify amino acids which have been previously methylated to increase their solubility.

Protocol 42. Derivatization of amino acids with 2-butanol

1. Heat at 60°C for 30 min the following mixture:
   - amino acid   200 μg
   - MeOH containing 3% acetyl chloride   100 μl
2. Evaporate the reagents and treat the residue as above but instead of dissolving the residue in EtOAc dissolve it in 50 μl acetonitrile and add 50 μl trifluoroacetic anhydride.
3. Heat the solution for 15 min at 60°C.
4. Evaporate the reagents and dissolve the residue in EtOAc.

7.3 Formation of diastereomeric amides

This procedure (see *Protocol 43*) has been used to determine the enantiomeric composition of ibuprofren.

Protocol 43. Derivatization with *R*-(+)-α-phenylethylamine

1. Leave the following solution at room temperature for 1 h.
   - carboxylic acid   200 μg
   - freshly distilled thionyl chloride   100 μl

2. Evaporate the thionyl chloride and add 50 μl toluene containing 10 mg/ml of R-(+)-α-phenylethylamine to the residue and leave at room temperature for 10 min.

3. Dilute to an appropriate volume for analysis with EtOAc.

References

1. Knapp, D. (1979). *Handbook of analytical derivatisation reactions*. Wiley Inter-science, New York.
2. Blau, K. and King, G. (ed.) (1977). *Handbook of derivatives for chromatography*. Heyden and Son, London.
3. Jaeger, H. (ed.) (1985). *Glass capillary chromatography in clinical medicine and pharmacology*. Marcel Dekker, New York.
4. Pierce, A. E. (1968). *Silylation of organic compounds*. Pierce Chemical Co., Rockford, Il.
5. Tanaka, K., Hine, D. G., West-Dull, A., and Lynn, T. B. (1980). *Clin. Chem.*, **261**, 1839.
6. Woolard, P. M. (1983). *Biomed. Mass Spectrom.*, **10**, 143.
7. Lombardo, L. (1984). *Tetrahedron Lett.*, **25**, 227.
8. Aizpurua J. M. and Palomo C. (1985). *Tetrahedron Lett.*, **26**, 475.
9. Corbett, M. E., Scrimgeour, C., and Watt, P. W. (1987). *J. Chromatogr. Biomed. Appl.*, **419**, 263.
10. Miyazaki, H., Ishibashi, M., Itoh, M., and Nambara, T. (1977). *Biomed. Mass Spectrom.*, **4**, 23.
11. Bagghi, S. P. (1987). *J. Chromatogr. Biomed. Appl.*, **421**, 227.
12. Hakomori, S. (1964). *J. Biochem.*, **55**, 205.
13. Bates, C. D., Watson, D. G., Willmott, N., Logan, H., and Goldberg, J. (1991). *J. Pharm. Biomed. Anal.*, **9**, 19.
14. Wu, H.-L., Chen, S.-H., Funazo, K., Tanaka, M., and Shono, T. (1984). *J. Chromatogr.*, **291**, 409.
15. Baker, M. H., Howe, I., Jarman, M., and McCague, R. (1988). *Biomed. Environ. Mass Spectrom.*, **16**, 211.
16. Midgley, J. M., Watson, D. G., Healey, T., and McGhee, C. N. J. (1989). *Biomed. Environ. Mass Spectrom.*, **18**, 657.
17. Hartvig, P. and Vessman, J. (1974). *J. Chromatogr. Sci.*, **12**, 722.
18. Jenden, D. J., Hann, I. and Lamb, S. I. (1968). *Anal. Chem.*, **40**, 125.
19. Midgley, J. M., MacLachlan, J. M., and Watson, D. G. (1988). *Biomed. Environ. Mass Spectrom.*, **15**, 535.
20. Macfarlane, R. G., Macleod, S. C., Midgley, J. M., and Watson, D. G. (1989). *J. Neurochem.*, **53**, 1731.
21. Macfarlane, R. G., Watson, D. G., Midgley, J. M., and Evans, P. D. (1990). *J. Chromatogr. Biomed. Appl.*, **532**, 1.
22. Midgley, J. M., Watson, D. G., Macfarlane, R. G., Macfarlane, S. C., and McGhee, C. N. J. (1990). *J. Neurochem.*, **55**, 842.
23. Thenot, J.-P. and Horning, E. C. (1972). *Anal. Lett.*, **5**, 905.

24. Midgley, J. M., Watson, D. G., Healey, T. M., and Noble, M. (1988). *Biomed. Environ. Mass Spectrom.*, **15,** 479.
25. Poole, C. F. and Zlatkis, A. (1980). *J. Chromatogr.*, **184,** 99.
26. Brooks, C. J. W., Cole, W. J., and Barrett, G. M. (1984). *J. Chromatogr.*, **315,** 119.
27. Miyazaki, H., Ishibashi, M., Itoh, M., and Yamashita, K. (1984). *Biomed. Mass Spectrom.*, **11,** 377.
28. Husek, P. (1974). *J. Chromatogr.*, **91,** 475.
29. Macfarlane, R. G., Watson, D. G., and Midgley, J. M. (1990). *Rapid Commun. Mass Spectrom.*, **4,** 34.
30. Midgley, J. M., Andrew, R., Watson, D. G., Macdonald, N., Reid, J. L., and Williams, D. A. (1990). *J. Chromatogr.*, **399,** 207.
31. Watson, D. G., Rycroft, D. S., Freer, I. M., and Brooks, C. J. W. (1985). *Phytochemistry*, **24,** 2195.
32. Brooks, C. J. W. and Cole, W. J. (1987). *J. Chromatogr.*, **399,** 207.
33. Brooks, C. J. W., Gilbert, M. T., and Gilbert, J. D. (1973). *Anal. Chem.*, **45,** 896.
34. Midgley, J. M., Watson, D. G., Macfarlane, R. G., Shafi, N., and Brooks, C. J. W. (1988). *J. Pharm. Pharmacol.*, **40,** 86P.

6

Gas chromatography in analytical toxicology: principles and practice

ROBERT J. FLANAGAN

1. Introduction

Analytical toxicology is the detection, identification and, if appropriate, measurement of drugs and other poisons in biological and other relevant specimens to aid the diagnosis, treatment, prognosis, and in some cases the prevention of poisoning. The specimens encountered may range from the relatively simple, such as expired air, to amongst the most complex of samples, the residue from a decomposed body. This being said, most work is performed on commonly available samples such as blood and urine. In analytical toxicology, gas chromatography (GC) has three principal advantages over other widely-used techniques such as high-performance liquid chromatography (HPLC) and immunoassay. First, GC has a range of sensitive detectors ['universal': flame ionization detector (FID); selective: nitrogen/phosphorus detector (NPD), electron capture detector (ECD)] which can be used in parallel. Second, high efficiency (capillary) GC columns are now widely available. Third, GC is easy to interface with techniques giving direct information about compound identity such as mass spectrometry (MS) and Fourier-transform infrared spectrometry (FTIR). With GC, as with HPLC, qualitative and quantitative information can often be obtained simultaneously. Temperature programming in GC is analogous to gradient elution in HPLC and permits the analysis of compounds of different volatilities in one run. Moreover, the return to starting conditions is easy and the interdependence of molecular weight, retention time, and temperature is valuable in aiding peak assignment when screening for unknowns. In addition, GC retention data are reproducible between different centres/operators/instruments/ columns. Disadvantages of GC are that the analyte or a derivative should be stable at the temperature required for the analysis. In addition, some form of sample preparation is normally needed. This chapter reviews general aspects of the GC of drugs and other poisons in biological specimens and gives suggestions for further reading.

Robert J. Flanagan

2. Use of GC in analytical toxicology

GC remains the method of choice for gases and other volatiles such as ethanol and inhalational anaesthetics. GC is also widely used in the analysis of other compounds, in screening for unknowns, and as an interface to MS. However, sample volume and detector sensitivity may be limiting and potential interferences must always be considered. Moreover, many compounds, especially volatiles and metabolites such as N-oxides, are unstable in biological samples. Thus, in addition to the choice of the sample preparation procedure, the column, and the chromatographic and detection conditions, due consideration must be given to other factors including sample collection and storage, choice of an internal standard, and quality assurance.

2.1 Sample collection and storage

If feasible, analysis of a 'blank' sample collected into an appropriate container before real samples are collected is a wise precaution. In live patients, venous blood is normally collected from the antecubital vein remote from any infusion site. The femoral vein is often used in post-mortem cases. Care should be taken to avoid contamination with propan-2-ol or other alcohols used as a topical disinfectant prior to venepuncture (see Section 3.4.3). Glass or polystyrene blood-collection tubes coated with lithium heparin or EDTA are suitable for most purposes. Evacuated tubes with soft rubber stoppers such as Vacutainers (Becton Dickinson) should be used with caution since solvents and other volatiles may be lost. In addition, plasticizers and other compounds may be released and interfere not only in the analysis (see Section 2.5.1) but also in extraction procedures (1). Blood tubes containing soft gel separators should similarly be avoided. For urine or other specimens, plain bottles should be used and no preservative should be added except in special cases.

Some analytes such as tricyclic antidepressants (Section 3.2.13) bind to glass; plastic tubes are thus often preferable for sample collection and are less likely to break if frozen. Use of plain tubes (no anticoagulant) and separation of serum may be better if storage in a deep-freeze is required, since less precipitate will result than if plasma is similarly treated. Alternatively, thawed plasma should be centrifuged before analysis. Excessive haemolysis should be prevented since this may invalidate plasma or serum assays for highly protein-bound analytes (Section 2.2). Some compounds such as clonazepam, cocaine, nifedipine, nitrazepam, thiol drugs, and many phenothiazines and their metabolites are unstable in biological samples at room temperature on the bench; exposure to sunlight can cause up to 99% loss of clonazepam in serum after 1 h, for example (2). Covering the outside of the sample tube in aluminium foil is a simple precaution in this case. Special considerations in sample collection for volatiles are discussed in Section 3.4.4.

N-Glucuronides, such as nomifensine N-glucuronide, are also unstable and may be present in plasma at high concentration: on decomposition the parent compound is reformed (3). With ethanol and some other analytes, solid sodium fluoride (1% w/v) may be added to inhibit microbial and some other degradative enzymes (Section 3.4.3) while an excess of neostigmine is needed to inhibit cholinesterase if a substrate such as physostigmine is to be measured (4). Storage at -5 to $-20°C$ (or below) is recommended if the analysis cannot be performed immediately. However, even this may not be ideal since N-oxides and sulphoxides may be reduced to the parent compounds (5). Quinols such as 4-hydroxypropranolol, on the other hand, are readily oxidized and stabilization by addition of a reducing agent such as ascorbate or sodium metabisulphite is necessary.

2.2 Sample preparation

With GC some form of sample pretreatment is normally needed, even if this only consists of adding an internal standard. Further aims may be removal of water and/or other interfering compounds, and sometimes analyte concentration or even dilution to increase/decrease sensitivity. In plasma, many acidic and neutral organic compounds bind to albumin, while some bases bind to $α_1$-acid glycoprotein. Bound analytes must be released if 'total' analyte is to be measured; with plasma/serum analysis of samples containing added drug is a simple way of checking this. Alternatively, equilibrium dialysis or ultrafiltration may be used to separate 'free' and bound analyte, thus allowing the 'free fraction' to be measured directly. The plasma 'free fraction' is often in equilibrium with the cerebrospinal fluid (CSF) concentration. For strongly ($>90\%$) bound compounds 'free' plasma concentrations, hence CSF concentrations, are often very low; thus 'free' and/or CSF measurements can be difficult, especially as sample volume is often limited.

Different sample preparation procedures may be required for the same analyte depending on the number and type(s) of samples to be analysed, the sensitivity required, and whether metabolites or other compounds are also to be measured. Urine and bile contain fewer insoluble residues and may contain higher concentrations of compounds of interest than whole blood, plasma, or serum, and thus sample preparation may sometimes be simplified. It is important to prevent or, failing this, to understand any reactions occurring during sample preparation. The cholinesterase inhibitor physostigmine, for example, is rapidly hydrolysed if extracted at pH 9.5 or above (4), while nordextropropoxyphene, the major plasma metabolite of dextropropoxyphene (Section 3.2.10), rearranges to an amide at pH 11 or above (6). HPLC does have an advantage over GC in such studies in that the possibility of thermal decomposition during the analysis is not a factor.

Aspects of sample preparation for biomedical analysis have been reviewed (7). In GC headspace and 'purge and trap' analysis, liquid–liquid solvent extraction and solid-phase extraction (SPE), also known as sorbent extraction

Robert J. Flanagan

(SE), are widely used in sample preparation, as discussed below. Direct injection after addition of an internal standard and centrifugation to remove particulates is also sometimes used in the analysis of compounds such as alcohols and glycols (Section 3.4.3) in blood or urine.

2.2.1 Headspace and 'purge and trap' analysis

The principle underlying headspace analysis is that in a sealed vial at constant temperature equilibrium is established between volatile components of a liquid sample in the vial and the gas phase above it (the headspace). After allowing due time for equilibration (normally 15 min or so) a portion of the headspace may be withdrawn using a gas-tight syringe and injected into the GC column. This method has the advantage that the risk of contamination of the column with non-volatile residues is virtually eliminated. An internal standard may be added prior to the incubation, and quantitative analyses may be performed after constructing a calibration graph (see Section 2.5.2). This technique is widely used in the analysis of ethanol and other volatiles in biological samples (see Sections 3.4.3 and 3.4.4). A headspace procedure used in screening for a range of solvents and other volatile compounds in blood by GC (8) is given in *Protocol 1*.

Protocol 1. Headspace procedure for screening for volatile compounds

1. Add internal standard solution (25 mg/l ethylbenzene and 10 mg/l 1,1,2-trichloroethane in expired blood bank whole blood:deionized water, 1 + 24) (200 µl) to a 7 ml glass septum vial (Schubert, Portsmouth, UK) using a semi-automatic pipette.

2. Seal the vial using a crimped-on PTFE-lined silicone disc (Kontron, St Albans, UK).

3. Incubate the vial at 65°C in a heating block and, after 15 min, withdraw a portion (100–300 µl) of headspace using a warmed gas-tight glass syringe and inject on to the column.

4. Subsequently add the sample (whole blood, plasma or serum) (200 µl) to the same vial using a 1.0 ml plastic disposable syringe and, after at least 15 min, take a second portion (100–300 µl) of headspace for analysis.

5. Remove the plunger from the gas-tight syringe and place the assembly on the heating block until the next injection to ensure evaporation of any remaining analyte(s).

In the 'purge and trap' method, volatile compounds are liberated from the sample by bubbling with an inert carrier gas and subsequently either con-

174

densed in a receiver cooled usually with solid carbon dioxide or liquid nitrogen, or adsorbed on a cartridge filled with a material such as Tenax-GC (see Section 2.3.2). After a time trapped volatiles are flash-vaporized into a stream of carrier gas and carried onto the column. Alternatively cartridges filled with activated charcoal can be used to 'trap' the volatiles, which are then extracted into a small volume of carbon disulphide prior to the analysis. This technique is widely used in the analysis of volatile compounds in potable water samples, for example, but has not found wide application in the analysis of biological samples, mainly because of the difficulty in interpreting the results at concentrations below those which can be measured using ordinary headspace methods. Thermal desorption and other preconcentration techniques have been reviewed (9).

2.2.2 Liquid–liquid solvent extraction

Traditionally sample preparation prior to GC has been performed by extracting the sample with an excess of an inert, water-immiscible organic solvent at an appropriate pH (10). Phase separation is normally by centrifugation; filtration may introduce plasticizers and other contaminants, although such problems can be minimized by pre-washing. The solvent is usually removed by evaporation under a stream of compressed air or nitrogen before reconstituting the dried extract in a small volume of an appropriate solvent. Alternatively a SPE cartridge may be used to concentrate the analytes from the extraction solvent (see Section 2.2.3). Back-extraction of acids or bases into aqueous base or acid, respectively, can be used to remove neutral interferences. In addition, analyte concentration to improve sensitivity can be achieved at the solvent evaporation stage and solvents such as dichloromethane, which would interfere with both the NPD and ECD, can be used. 'Salting-out' by adding excess sodium chloride, for example, to the aqueous phase may improve extraction efficiency. On the other hand, solvent evaporation is tedious, interferences may be concentrated, and volatile analytes such as amphetamine (Section 3.2.1) may be lost.

Some commonly used extraction solvents are listed in *Table 1*. Mixtures of solvents may be used for specific purposes. Chloroform:propan-2-ol (9 + 1) (relative density (RD), >1), for example, has long been used to extract morphine and other opiates (see Section 3.2.10) while mixtures such as dichloromethane:hexane (1 + 1) (RD < 1) are useful if a chlorinated solvent is thought necessary but an upper layer is formed which simplifies removal of the extract. The inhalational toxicity and other hazards associated with use of some solvents should not be ignored. Benzene, for example, is a proven human carcinogen while occupational exposure to *n*-hexane or to *n*-butyl methyl ketone is associated with the development of peripheral neuropathy (*Table 2*). *iso*-Hexane (Fisons), a mixture of hexane isomers containing less than 5% *n*-hexane, has recently been introduced as a safer alternative to *n*-hexane.

Robert J. Flanagan

Table 1. Some widely used extraction solvents

Solvent	RD	BPt (°C)
n-Butyl acetate	0.88	125
Chloroform	1.49	61
Cyclohexane	0.78	81
1,2-Dichloroethane	1.25	83
Dichloromethane	1.32	40
Ethyl acetate	0.90	77
n-Heptane	0.68	98
Methyl tert-butyl ether	0.74	55
Petroleum ether[a]	0.65	40–60
Toluene	0.87	111
2,2,4-Trimethylpentane	0.69	99

RD, relative density; BPt, boiling point.
[a] A mixture of pentanes, hexanes, etc.—other boiling ranges available.

Table 2. Especial hazards associated with the use of some common solvents

Solvent	Hazard
Benzene	Human carcinogen
Carbon disulphide	Neurotoxicity
Carbon tetrachloride	Marked hepatorenal toxicity; may be human carcinogens
Chloroform	
1,2-Dichloropropane	
1,1,2,2-Tetrachloroethane	
Dichloromethane	Carbon monoxide poisoning (Section 3.4.1)
Diethyl ether	Highly flammable; may form explosive peroxides
Di-isopropyl ether	May from explosive peroxides
n-Hexane	Peripheral neurotoxicity
Hexan-2-one	
Trichloroethylene	Cardiotoxicity

Simple liquid–liquid extraction methods with direct analysis of the extract have been used for many years prior to GC (11). Hamilton repeating dispensers fitted with gas-tight glass syringes and Luer-fitting stainless steel needles are used for solvent and reagent additions whenever possible. Use of disposable glass test-tubes (60 × 5 mm internal diameter (i.d.), Dreyer tubes; Samco, Old Woking, UK) as extraction vessels minimizes the risk of contamination with the aqueous phase when obtaining a portion of the extract for analysis. The Eppendorf 5412 or an equivalent high-speed centrifuge gives rapid phase separation (30 sec) and also breaks up any emulsions which may have formed during the extraction.

176

This simple extraction/direct injection approach is quick and cheap but is not readily amenable to batch processing. There is also the risk of glass tubes breaking in the centrifuge. 'Salting-out' can also be used although an emulsion may ensue if excess salt is added. n-Butyl acetate and methyl *tert*-butyl ether (MTBE) give efficient extraction of many drugs and metabolites from plasma at an appropriate pH and form the upper layer, thus simplifying extract removal for analysis. These solvents do not interfere on NPD or ECD and the extracts are generally free from endogenous interferences. Unlike other ethers such as diethyl and di-isopropyl, MTBE does not form explosive peroxides at ambient temperature and thus antioxidants such as quinones are unnecessary. A simple method for the liquid–liquid extraction of basic drugs from urine prior to capillary GC–NPD (12) is given in *Protocol 2* (see also Section 3.1).

Protocol 2. Liquid–liquid extraction of basic drugs from urine prior to GC

1. Add 0.25 ml aqueous sodium hydroxide (1 mol/l) and 0.5 ml internal standard solution (5 mg/l prazepam in n-butyl acetate) to 1 ml urine in a 4.5 ml polypropylene tube (Sarstedt, Nottingham, UK).

2. Vortex-mix (30 s) and centrifuge (3000 r.p.m., 10 min; bench-top centrifuge).

3. Inject 2 µl of the n-butyl acetate extract on to the GC column.

A liquid–liquid micro-extraction procedure used to analyse nicotine and its metabolite, cotinine, in urine, plasma, or saliva by capillary GC–NPD (13) is outlined in *Protocol 3*. Antifoam and phenol red are added to inhibit emulsion formation and to aid visualization of the organic layer, respectively (see also Section 3.4.5).

Protocol 3. Liquid–liquid micro-extraction of nicotine and cotinine prior to GC

1. Add sample or standard (100 µl) to 60 × 5 mm i.d. glass tube (Dreyer tube).

2. Add 100 µl internal standard solution (117 µg/l 5-methylcotinine), 300 µl aqueous sodium hydroxide (5 mol/l), 20 µl antifoam/phenol red mixture [5% (v/v) Dow Corning antifoam RD emulsion (BDH), 200 mg/l phenol red (Sigma)] and 50 µl 1,2-dichloroethane.

3. Vortex-mix (1 min) and centrifuge (9950 g, 2 min; Eppendorf 5412 or equivalent).

4. Inject 2 µl of the 1,2-dichloroethane extract on to the GC column.

2.2.3 Solid-phase extraction

Extraction of drugs by adsorption on to solid materials such as Florisil (a synthetic magnesium silicate), ion-exchange resins, or activated charcoal, followed by washing with water and elution of compounds of interest using methanol, for example, is not a new idea. However, use of siliceous or other materials with relatively close particle size distribution (15 to 100 μm) in disposable polypropylene syringe barrels permits sequential extraction, washing, and finally reproducible elution of drugs and other analytes at relatively low pressures (14, 15). In addition to unmodified silica, a range of bonded-phase materials analogous to those used as HPLC packings is available (*Figure 1*). In all cases, the bonded phase is linked to a surface silanol moiety via a silyloxy bond (Si—O—Si). However, different manufacturers use different base silicas and different bonding chemistries, and it is thus not

Bonded Phase	Name
$-(CH_2)_{17}CH_3$	n-Octadecyl (C18, ODS)
$-(CH_2)_7CH_3$	n-Octyl (C8)
$-(CH_2)_5CH_3$	n-Hexyl (C6)
$-C_2H_5$	Ethyl (C2)
$-CH_3$	Methyl (C1)
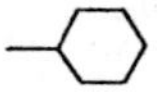	Cyclohexyl
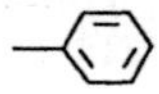	Phenyl
$-(CH_2)_3CN$	Cyanopropyl (Nitrile, CN)
$-(CH_2)_3OCH_2CHOHCH_2OH$	2,3-Dihydroxypropoxypropyl (Diol)
$-(CH_2)_3NH_2$	Aminopropyl (Amino, APS)
$-(CH_2)_3NH(CH_2)_2NH_2$	N-(2-Aminoethyl)aminopropyl
$-CH_2COOH$	Carboxymethyl (WCX)
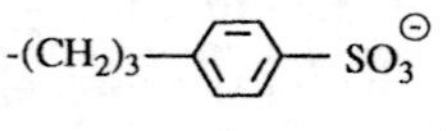 $-(CH_2)_3-\langle\rangle-SO_3^{\ominus}$	4-Sulphophenylpropyl (SCX)
$-(CH_2)_3SO_3^{\ominus}$	Sulphopropyl
$-(CH_2)_3N(CH_2CH_3)_2$	Diethylaminopropyl (DEA, WAX)
$-(CH_2)_3N(CH_3)_3^{\oplus}$	Trimethylaminopropyl (SAX)

Figure 1. Some bonded stationary phases used in silica-based, solid-phase extraction (SPE) columns.

surprising that their products may give very different results when used in a particular analysis, even though the bonded phase is ostensibly the same.

The development of SPE protocols remains largely empirical and most work has been done in conjunction with thin-layer chromatography (TLC) or HPLC. Different protocols may be required for urine, plasma, and whole blood; little work has been done on SPE for tissue digests (see Section 2.2.4). The steps in developing a method for physostigmine have been described in detail (5). It is helpful if the volume used in the final elution step is as small as possible. A major advantage of SPE is that batch processing can be simplified. A further feature when screening for unknowns is that a range of analytes can be extracted simultaneously, although this can be a problem if analysis of a single compound is required. Moreover, SPE columns are relatively expensive and it may not be possible to retain very water-soluble analytes. Indeed, in many cases simple liquid–liquid extraction with an appropriate solvent can be used to purify a lipophilic compound more quickly than with SPE. The risk of extracting interfering compounds such as plasticizers from the SPE column is a further consideration. On the other hand, analyte concentration may often be achieved more easily with SPE than with liquid–liquid extraction, while use of SPE columns to concentrate an analyte from a solvent extract may provide a quicker and possibly safer alternative to solvent evaporation (see Section 2.2.2).

A simple method for the SPE of acidic, basic, and neutral drugs from urine prior to TLC using pre-buffered Tox-Elut columns (Varian) was developed by Widdop (16). Urine (20 ml) was added to the columns and allowed to stand (2 min). Chloroform:propanol-2-ol $(9 + 1)$ $(3 \times 10$ ml) was then added, the combined eluates were evaporated to dryness under a stream of compressed air (60°C), and the residue reconstituted in methanol (100 µl). However, this method may give interferences on GC. A procedure for extracting acidic, basic, and neutral drugs from urine designed for use with GC is given in *Protocol 4* (15)—further protocols for other analytes are given in this same report.

Protocol 4. Solid-phase extraction of drugs from urine prior to GC

1. Add 5 ml urine to 2 ml phosphate buffer (0.1 mol/l, pH 6.0) in a glass test-tube and adjust to pH 5.5–6.5 using 0.1 mol/l aqueous sodium hydroxide or 1 mol/l aqueous acetic acid.

2. Insert SPE column (Clean-Screen, Worldwide Monitoring Corporation, Morrisville, USA) into vacuum manifold and wash with 1 ml methanol and 1 ml phosphate buffer (0.1 mol/l, pH 6.0).

3. Attach an 8 ml fritted reservoir to the top of the extraction column and add urine. Gently dry column under vacuum.

Protocol 4. *Continued*

4. Wash with 1 ml phosphate buffer (0.1 mol/l, pH 6.0) followed by 0.5 ml aqueous acetic acid (1 mol/l).

5. Dry column under vacuum (5 min) and wash with 1 ml hexane.

6. Elute acidic and neutral drugs with 4 × 1 ml dichloromethane.

7. Evaporate eluate to dryness under a stream of nitrogen at 30–40°C. Reconstitute extract in 0.1 ml ethyl acetate and inject 1–2 μl on to the GC column.

8. Wash columns with methanol (1 ml) and elute basic drugs with 2 ml methanolic ammonium hydroxide (2%, v/v).

9. Add 3 ml deionized water and 0.2–0.3 ml chloroform to eluate. Vortex-mix (15 s) and inject 1–2 μl of the chloroform layer on to the GC column.

2.2.4 Analysis of solid tissues

Some information as to the distribution of drugs between different tissues in man has come from studies performed post-mortem (17). There is little information on the distribution of drugs or other poisons within solid tissues in man; collection of *ca.* 5 g specimens from several sites from organs such as the brain is recommended if the whole organ is available. Conventionally, measurements in portions of organs such as liver and brain have been performed by mechanical homogenization and/or acid digestion on, say, *ca.* 5 g tissue prior to solvent extraction at an appropriate pH. However, digestion with proteolytic enzymes often gives much improved recovery when compared to conventional procedures (18) and has the advantage that, once the digest has been prepared, analogous methodology and calibration standards to those used with plasma can often be employed. It is obviously important to ensure that use of the enzyme preparation does not introduce interferences. A further potential problem is that conjugates and other metabolites may not survive. Various enzyme-based digestion procedures have been reviewed (19). A procedure developed to measure lignocaine in tissue specimens after digestion using subtilisin a (20) is given in *Protocol 5*.

Protocol 5. Tissue digestion using subtilisin a

1. Prepare solution (2 g/l) of lyophilized subtilisin a (Novo Nordisk, Windsor, UK) in sodium dihydrogen orthophosphate/disodium hydrogen orthophosphate buffer (7 mmol/l, pH 7.4).

2. Dissect *ca.* 100 mg wet weight portions of tissue, remove excess fluid on filter paper, add tissue to preweighed 10 ml tapered glass tubes, and record the exact weights.

3. Add subtilisin a solution (1.0 ml), seal the tubes with ground-glass stoppers, and incubate in a water bath (50°C, approx. 16 h).

4. Cool the tubes, mix the contents on a vortex-mixer, take 0.2 ml portions, and extract as for plasma/serum.

2.3 Columns and column packings

Conventionally, GC has been split into gas–solid chromatography (GSC) in which the stationary phase is an active solid and gas–liquid chromatography (GLC) in which the stationary phase is coated on to an inert support. However, these distinctions are becoming less useful with the widespread use of fused silica capillary columns in which the stationary phase is chemically bonded on to the inner surface of the column. As the mobile phase (carrier gas) has little effect on selectivity and only a small effect on efficiency, nitrogen is normally used with packed columns (flow rates *ca.* 30–60 ml/min) and helium with capillaries (flow rates *ca.* 1–10 ml/min). It is vital to use oxygen-free carrier gas, as the presence of even very small amounts of oxygen can result in oxidation of certain stationary phases. The ECD (Section 2.4.3) is especially sensitive to impurities, notably oxygen, in carrier or purge gas supplies. Although commercially supplied cylinders are a convenient source of the gases needed in GC operation, alternatives are available for air (simple compressor), hydrogen (electrolytic hydrogen generator), and nitrogen (generator). However, regular monitoring and maintenance, and the use of appropriate filters (oil, oxygen, moisture), are mandatory—indeed filters should be used even with gas from cylinders.

Most columns, especially packed columns, require conditioning to remove volatile impurities before use. Proprietary columns should be supplied with full instructions. Otherwise, the column (not connected to the detector) should be purged with carrier-gas to remove oxygen before heat is applied. The column should then be slowly brought to the required temperature, either using a programme or in a series of steps and held for *ca.* 12 h. The temperature used for conditioning is normally the maximum or slightly above the maximum at which it is planned to use the column, taking into account the maximum temperature for the stationary phase recommended for isothermal operation. Capillary columns should be conditioned according to the manufacturer's instructions and tested by injecting an appropriate mixture before use for sample analyses. If temperature programming is to be used, it is important that the column is not left at the maximum run temperature for a prolonged period. When conditioning columns, it is obviously important to ensure that the gas chromatograph oven temperature controller is working properly and that the carrier-gas flow is adequate as there is no detector output to help indicate if anything is wrong.

2.3.1 Injectors and injection technique

Analytical toxicology is essentially trace analysis and thus sensitivity is often limiting. Generally, then, it is important that as much of a sample extract as possible is injected on to the column, other factors (the 'signal-to-noise ratio') being equal. In most GC methods the difficulties inherent in reproducibly injecting relatively small volumes of an extract necessitate the use of internal standardization (see Section 2.5.2). With packed columns sample injection is usually via a syringe through a silicone rubber septum in the injection port—it is important to use 'low-bleed' septa, especially with sensitive detectors such as the ECD (Section 2.4.3). Use of a glass injection port liner which can be removed and cleaned minimizes the accumulation of non-volatile residues on the column, but on-column injection may be preferred if labile substances are to be analysed. In such cases contaminated packing at the top of the column should be replaced with fresh material if efficiency is affected. Normally, a minimal amount of solvent will be injected to reduce solvent effects at the detector. However, for gases and vapours much larger volumes can be injected via a gas tight syringe (see Section 3.4.4) or a gas-sampling valve.

A variety of injection devices can be used with capillary columns and the terminology employed can be very confusing. Nevertheless, the importance of using an appropriate injection technique when working with such columns cannot be overemphasized and the subject is worthy of specialized study if the full capabilities of modern columns are to be realized. With relatively narrow-bore capillaries (<0.32 mm i.d.) some form of inlet splitter is normally needed to prevent overloading with injection solvent. When using an inlet splitter, as with other GC injection devices, it is important that the portion of the injection solution passed to the column has the same composition as the rest of the solution. With wider bore capillaries (0.32 mm i.d. or greater) then (splitless) syringe injection either through a septum into a glass liner or directly into the column via a fused silica capillary is relatively simple and ensures deposition of the analytes on the column. If efficiency deteriorates then removal of the first few centimetres or even an entire coil from a well-used column often restores performance. However, insertion of a short (0.5–5 m) length of deactivated fused silica tubing ('guard column'), which can easily be discarded, between the injection port and the capillary column serves to minimize contamination of the analytical column.

2.3.2 Packed columns

Conventional borosilicate glass, stainless-steel or glass-lined stainless-steel packed columns (0.5 to 4 m × 2 to 4 mm i.d.) still find application in some areas. Although less robust, glass is less likely to adsorb polar analytes. In addition, settling of the packing, stationary phase oxidation and column contamination can be checked visually. Treatment of glass columns for *ca.* 8 h with a silanizing reagent such as trimethylchlorosilane (TMCS) (10% v/v, in toluene) and use of quartz or silanized glass wool to retain the packing may help minimize adsorptive effects. A good source of information on these and other aspects of packed column GC remains the book by Supina (21).

GSC packings are mainly used in the analysis of gases and solvents. Molecular sieve (synthetic zeolite) or silica gel packings are useful for the analysis of permanent gases and carbon monoxide (Section 3.4.1). The Chromosorb and Porapak series are cross-linked divinylbenzene polystyrene copolymers with maximum operating temperatures of *ca.* 250 °C. Alcohols from methanol to *n*-pentanol can be separated on Porapak Q or Chromosorb 102 (see Section 3.4.3). Tenax-GC is a porous polymer of 2,6-diphenyl-*p*-phenylene oxide and is used both as a stationary phase material and as a trap for volatiles prior to GC (Section 2.2.1). Carbopaks B and C are graphitized carbon blacks having surface areas of 12 and 100 m^2/g, respectively. The Carbopaks are usually used after modification with a light coating of a polar stationary phase such as Carbowax 20M, and can give good peak shapes and separations for alcohols and other volatiles. However, these materials are very friable and batch-to-batch variations in the peak shapes attained are common (see Section 3.4.4).

In packed column GLC the support should play little or no part in the separation. Calcined diatomaceous earth graded into appropriate size ranges (80–100 or 100–120 mesh) is widely used. Commercial brands include Chromosorb W and Chromosorb G (both Johns Manville) and Supelcoport (Supelco). Various deactivation procedures are employed, including acid washing to remove metallic impurities (denoted AW) and deactivation of surface silanols with hexamethyldisilazane (denoted HMDS) or other silaniz-ing reagents. A light initial coating of a polar stationary phase applied before the primary phase may increase the apparent deactivation of the support but may also influence the separations attained. Pre-coating with potassium hydroxide (2–5%, w/w) has been widely used to improve the peak shapes given by strong bases such as the amphetamines (Section 3.2.1). Unfortunately not all stationary phases are stable under strongly alkaline conditions.

Stationary phase loadings are normally expressed as percentage weight (or mass) of phase/weight (or mass) of support (% w/w or % m/m). The phase is normally applied to the support dissolved in an appropriate solvent. After standing the solvent is then removed either by gentle evaporation under vacuum in a rotary evaporator (taking care not to cause mechanical damage to the particles) or by filtration on a Buchner funnel followed by air drying. With the evaporation method the stationary phase may not be coated evenly. The filtration method gives a more uniform coating especially at lower phase loadings, but some stationary phase is lost at the filtration stage and experi-ence is needed to help assess the initial proportions of stationary phase and support to use. The stationary phase lost can be weighed by collecting the filtrate and evaporating to dryness.

There are several hundred stationary phases available but only a few have found wide application in analytical toxicology. As a general rule, retention

Robert J. Flanagan

on non-polar phases is influenced mainly by molecular weight. However, polar compounds such as many drugs and pesticides show increased separation and give better peak shapes on polar as compared to non-polar phases. McReynolds (22) used the retention of benzene, *n*-butanol, pentan-2-one, nitropropane, and pyridine to assess the 'polarity' of stationary phases. The sum of the difference of the retention indices (see Section 2.5.3) of each of these compounds on the stationary phase under investigation as compared to the retention index on a standard non-polar phase (squalane) is used to derive the McReynolds constant for the phase. This gives a measure of the polarity of the phase which can be used for classification purposes. The McReynolds constants for some stationary phases used in analytical toxicology are given in *Table 3*.

Apolane-87 (24,24-diethyl-19,29-dioctadecylheptatetracontane) was introduced some years ago with the aim of replacing squalane as the standard non-polar phase for GLC. This material had a much higher maximum operating temperature (260°C) than squalane, but proved difficult to manufacture and

Table 3. Some GC stationary phases used in analytical toxicology

Phase	Type	Maximum temperature (°C)	McReynolds constant
Squalane	Hydrocarbon	150	0
Apolane-87	C_{87} hydrocarbon	280	81
Apiezon L	Hydrocarbon grease	300	143
Silicone SE-30	100% Dimethyl	300	217
Silicone OV-1	100% Dimethyl	350	222
Silicone OV-101	100% Dimethyl	350	229
Silicone SP-2100	100% Dimethyl	350	229
Apiezon L/KOH	–	225	301
Silicone SE-54	94% Methyl, 5% phenyl, 1% vinyl	300	337
Silicone OV-73	94.5% Methyl, 5.5% phenyl	325	401
Silicone OV-7	80% Methyl, 20% phenyl	350	592
Silicone OV-17	50% Methyl, 50% phenyl	350	886
Silicone SP-2250	50% Methyl, 50% phenyl	350	886
Poly-A 103	Polyamide	275	1072
Carbowax 20M/KOH	–	225	1296
Silicone OV-210	50% Methyl, 50% trifluoropropyl	275	1550
Silicone OV-225	50% Methyl, 25% cyanopropyl, 25% phenyl	250	1813
CHDMS[a]	Cyclohexanedimethanol succinate	250	2017
Carbowax 20M	Polyethylene glycol	225	2318
FFAP	Substituted terephthalic acid	250	2546
SP-1000	Substituted terephthalic acid	250	2546
DEGS[b]	Diethyleneglycol succinate	200	3543

[a] Also known as Hi-EFF-8BP.
[b] Also known as Hi-EFF-1BP.

store because of auto-oxidation, and has now been abandoned. Apiezon L is a hydrocarbon grease and has the advantage that, unlike the polymeric silicone phases, it is stable if coated on to alkali-treated packing, although the maximum operating temperature is then only 225°C. SE-30, OV-1, and OV-101 are dimethylpolysiloxane stationary phases which are regarded as broadly equivalent. Polar interactions are minimal with these phases and thus separations occur largely on the basis of molecular weight. However, a disadvantage is that peak tailing may occur with polar compounds. In general peak tailing is reduced on more polar phases; SE-54, OV-7, OV-17, and OV-225 are amongst the more polar polysiloxane-based phases available (*Table 3*). Carbowax 20M (polyethylene glycol of average formula weight 20 000) is a high-polarity phase but has a relatively low maximum operating temperature. However, as with Apiezon L, precoating the support with KOH effectively minimizes peak-tailing of strong bases and the maximum operating temperature is unaffected. Polyester phases such as cyclohexanedimethanolsuccinate (CHDMS) and also polyamides such as Poly A 103 have been advocated for specific separations of polar compounds such as barbiturates (see Section 3.2.5), Mixtures of stationary phases have also been advocated for specific problems. For example, a commercially available mixture of SP-2110 and SP-2510-DA (Supelco) has been widely used in the analysis of common anticonvulsants (see Section 3.2.3).

2.3.3 Capillary columns

Capillary columns were originally developed with the aim of resolving complex mixtures and can have great separating power. Glass and stainless steel have both been used to prepare capillary columns but nowadays most applications in analytical toxicology use fused silica capillaries. An outer coating of polyimide protects the surface of the silica, thus maintaining mechanical strength and flexibility—the column may often be threaded through complex pipework to emerge, for example, at a detector jet. Typical dimensions are 0.1–0.53 mm i.d. and 10–50 m length. Column bleed, hence detector noise, are reduced compared to packed columns because: (i) the amount of stationary phase present is low, (ii) the phase is chemically bonded to the wall of the capillary. Thermal mass is low and thus temperature programming is facilitated. More importantly interfacing to MS is simplified because carrier-gas flow rates are also low. A further advantage with respect to packed columns is that when using polysiloxane stationary phases accumulation of siliceous deposits on NPD/ECD is minimized. On the other hand, one complication is that GC detectors other than the MS require higher flow rates than normally used with capillaries and thus a 'make-up' carrier gas supply is needed. PLOT (porous layer open tubular), SCOT (support-coated open tubular), and WCOT (wall-coated open tubular) have been used to describe different types of capillaries, but these terms are now virtually obsolete.

Although capillary columns can be prepared in-house, it is much more convenient to buy pre-tested columns. The stationary phases used are in general those developed for packed columns (*Table 3*), but better column deactivation and more uniform phase coating mean that good peak shapes are often obtained even when using low-polarity phases. In general, efficiency is increased with narrower column diameters/thinner film thicknesses, but sample capacity is greater with wider bores/thicker films and thus the risk of column overloading, usually manifest as 'fronting' (front-tailing), is minimized. Columns of 0.32 mm i.d. and 30–50 m length with a moderate phase loading (0.2–0.3 μm film thickness) and used with an appropriate injection system (Section 2.3.1) are suitable for many purposes in analytical toxicology such as screening for unknowns (see Section 3.1). Columns of 0.53 mm i.d. and 15–30 m length with relatively high phase loading (1–5 μm film thickness) have similar capacity and efficiency to packed columns at carrier-gas flow rates of 8–20 ml/min but with the advantages of better peak shapes, reproducibility, applicability in temperature programming, and ease of interface to MS. Such columns may often be substituted directly for packed columns in a given method. Several manufacturers supply conversion kits to enable these wide-bore capillaries to be used in instruments configured for use with conventional packed columns. Some companies also produce capillary GC method optimization/simulation software for use with microcomputers. GCOPS (Phase Separations) is a recent example and is designed to assist in the development of isothermal and temperature programmed separations; it can also be used as a teaching aid.

2.4 Detectors

Of the three commonly used GC detectors, the ECD is in essence non-destructive and thus can in theory be used in series with the FID or NPD. However, post-column splitter systems are usually used if multiple detection is desired. In such instances, the relative detector response can provide an additional parameter to aid compound identification although this is rarely used in practice. The first GC detector, the katharometer or hot-wire detector, also known as the thermal conductivity detector (TCD), is relatively insensitive but remains useful in analytical toxicology for permanent gases and for carbon monoxide, compounds which do not respond on more sensitive detectors (see Section 3.4.1). The flame photometric detector (FPD) can give a selective response for phosphorus- or sulphur-containing compounds, while the photoionization detector (PID) gives a higher signal-to-noise ratio than the FID for compounds such as barbiturates. However, both the FPD and the PID have found little application. Similarly the atomic emission detector (Hewlett-Packard), which can selectively detect any element except helium, has found little application thus far in analytical toxicology.

The following sections summarize the use of detectors in analytical toxicology.

2.4.1 Flame ionization detection

The high sensitivity (10^{-11} to 10^{-12} g/s methane), good stability, wide linear range (up to 10^6), and large number of compounds (essentially all organic molecules containing C-H bonds) responding to this detector have ensured that it remains widely used. The magnitude of the response is roughly proportional to the number of carbon atoms present, although this is reduced if oxygen or nitrogen are present in the molecule. The lack of response to water enables aqueous injections to be performed when measuring ethanol and similar compounds (Section 3.4.3). On the other hand, the risk of interference is high especially when analysing samples obtained post-mortem and thus selective detectors, especially the NPD, are favoured. The FID should normally be used in parallel with an NPD or ECD if retention measurements are to be based on Kóvats' retention indices (Section 2.5.3).

2.4.2 Nitrogen/phosphorus detection

Since many drugs and other poisons contain C-N bonds and many injection solvents and potential interferences do not, this detector (also known as the thermionic detector or the alkali flame ionization detector, AFID) is nowadays widely used in the N-selective mode (nitrogen:carbon response ratios *ca.* 5000:1). In such cases the P response is a disadvantage since phosphorus-containing plasticizers also show a good response (phosphorus:nitrogen response ratios *ca.* 10:1; ref. 23). On the other hand the P-selective mode (phosphorus:carbon response ratios *ca.* 50 000:1) can be exploited in a few instances, notably in the analysis of organophosphate pesticides (Section 3.3.2). Modern versions of this detector use an electrically heated rubidium silicate source and are somewhat more stable than early versions when a rubidium-containing bead was in essence 'balanced' over an FID. Heavily chlorinated solvents such as chloroform are not recommended for use with the NPD since they tend to volatilize rubidium from the source. *n*-Butyl acetate has proved a valuable extraction/injection solvent when used both with this detector and with the ECD (24). Use of nitrile-containing stationary phases such as OV-225 (*Table 3*) may cause increased baseline 'noise' with the NPD in the 'N' mode.

2.4.3 Electron capture detection

This detector shows an enhanced and selective response (*ca.* 10^3 g/s greater than the FID) to compounds containing a halogen, a nitro moiety and, to a lesser extent, to ketones. It is thus important in the analysis of halogenated solvents, pesticides, and some halogen- or nitro-containing drugs (notably the benzodiazepines—Section 3.2.6). Derivatization using reagents such as heptafluorobutyric anhydride is widely used not only to improve the volatility of suitable analytes but also to exploit the high sensitivity/selectivity attainable with the ECD. Indeed, for some compounds the sensitivity of this detector easily exceeds that of the mass spectrometer—detection limits of a few fg are possible for organochlorine pesticides (Section 3.3.1), for example. Modern ECDs can be operated in the constant-current mode giving linearity of response typically of the order of 10^4. Flow-rates of 30–60 ml/min are necessary for the efficient operation of the ECD and thus a 'make-up' or 'purge' gas supply at the end of the column is often needed. Oxygen-free nitrogen normally provides a suitable make-up gas. Addition of 'quench' gases such as methane is not normally necessary nowadays.

2.4.4 Mass spectrometry

The capillary GC provides almost the ideal sample delivery system for the MS since most vacuum systems can cope with the relatively low flow rates used. With packed columns an outlet splitter interface is needed to reduce the amount of carrier gas reaching the spectrometer (25). Nowadays quadrupole mass spectrometers are widely used, while the advent of the 'bench-top' MS in the form of the mass-selective detector (MSD; Hewlett-Packard), the ion-trap detector (ITD; Finnigan), and more recently the MD 800 (Fisons) has brought GC–MS within the reach of most analytical toxicology laboratories. The MS can be used in the electron impact (EI) mode to produce definitive information as to peak identity either by providing complete spectra or from the five principal peaks in the mass spectrum. Alternatively, the chemical ionization (CI) mode can be used to ascertain the molecular weight of the analyte. Quantitative information can also be obtained if the instrument is used in the selective ion monitoring (SIM) mode if appropriate calibration standards are available. The use of bench-top GC–MS, largely as applied to testing for drugs of abuse, one of the major applications of these instruments, has been reviewed by Deutsch (26). General aspects of the use of MS in the identification of drugs of abuse are discussed by Webb (25).

2.4.5 Fourier transform infrared

The idea of using infrared spectrometry as a GC detector is not new but only became practicable with the development of FTIR instruments. GC–FTIR has not been widely applied in analytical toxicology largely because of cost and the fact that, with the exception of isomers and of simpler molecules such as volatile solvents, GC–MS often gives more information. Some biomedical applications of GC–FTIR have been reviewed (27).

2.5 General considerations

2.5.1 Common interferences

Phthalate and other plasticizers may originate from plastic bags used to store transfusion blood, infusion tubing, and from soft plastic closures for blood tubes (28). Polyvinyl chloride (PVC), for example, can contain up to 40% (w/w) di-2-ethylhexyl phthalate and concentrations of this latter compound of up to *ca.* 0.5 g/l have been reported after storage of plasma in PVC bags for 14 days (29). A further consideration is that specimens obtained post-mortem may contain putrefactive bases, such as phenylethylamines and indole, which may interfere in the analysis of amphetamines and other stimulants (Section 3.2.1). The retention indices of a number of plasticizers and other substances which may be encountered in toxicological analyses on SE-30/OV-1/OV-101 were reported by Ramsey *et al.* (23).

Drugs (hence metabolites) may arise from unexpected sources. For example, quinine may originate from tonic water, caffeine from caffeinated beverages (tea, coffee, cola) and proprietary stimulants, chloroquine and related compounds from malaria prophylaxis, and phenylpropanolamine from cold cures. Lignocaine is commonly used in dental anaesthesia and lignocaine-containing gel is often used as a lubricant during procedures such as bladder catheterization or bronchoscopy; measurable plasma concentrations of lignocaine and some metabolites may be attained even after bladder catheterization. Blood-bank blood and commercially available horse serum often contain lignocaine, and sometimes metabolites from lignocaine are used topically during vene-puncture. The alkaloid emetine has been detected in plasma after giving syrup of ipecacuanha (ipecac) to suspected overdose patients to induce vomiting. Drugs such as diazepam, pethidine, or a phenothiazine may be given prior to computerized tomography (CT) scans, lumbar puncture, or other investigations. Iodinated hippuric acids are used as X-ray contrast media. Such compounds, and also drugs given in emergencies, e.g. anticonvulsants, may not be recorded on prescription sheets. Some compounds have very long elimination half-lives and/or undergo enterohepatic recirculation: e.g. chlor-promazine metabolites may be found in urine 18 months after stopping therapy (17).

2.5.2 Assay calibration and quality assurance

Several methods of assay calibration are available in GC (internal and external standardization and standard addition methods). Use of an internal standard can compensate for extraction losses as well as variations in injection volume, hence this method is much the most widely used. Internal standardization is also valuable in qualitative analyses if identification is to be based on relative retention times. The internal standard should show similar extraction, deriva-tization, chromatographic, and detection characteristics to the analyte(s) and should not interfere in the analysis. External standardization is useful when checking absolute recoveries and also in certain assays employing solid-phase extraction (Section 2.2.3) when it can be especially difficult to find a suitable internal standard. Other considerations in the choice of internal standard are discussed by Peng and Chiou (30) and by Huizer (31).

Construction of the calibration graph is normally by analysis of standard solutions containing each analyte at a range of concentrations prepared in analyte-free plasma, urine, or other appropriate fluid; an analysis before

Robert J. Flanagan

addition of the compound(s) of interest to ensure the absence of interferences is a prerequisite. If the internal standard method is being used then the calibration graph is of peak height or area relative to that of the internal standard against analyte concentration. Blood anticoagulated with citrate should not be used for standards since: (i) the blood is diluted; (ii) citrate has a high buffering capacity; and (iii) interference from plasticizers and other components is frequent. A common error when preparing standards is failure to allow for the contribution of salts of drugs such as hydrochlorides to the total weight. Reporting results of drug assays in molar units has no advantage in clinical practice. Generally SI mass units (mg/l, g/l, etc.) are preferred for liquid matrices, since this gives easier conversion from traditional units in legislation or in the literature:

0.1 mg% = 0.1 mg/100 ml = 0.1 mg/dl = 1 mg/l = 1 μg/ml = 1 ppm

Assessing the limit of accurate measurement (limit of sensitivity) is always difficult. A good practice is to prepare and analyse a synthetic standard at the claimed limit and measure the relative standard deviation (RSD = coefficient of variation, CV). Quality assurance (QA) is best performed by analysing appropriate specimens prepared by someone else. Ideally such specimens should be prepared independently using different stocks of pure compound, balances, etc. An interested pharmaceutical or other manufacturer may arrange for this. If the specimens have to be prepared internally, an experienced colleague should prepare them autonomously. QA materials for common drugs are available commercially but many are designed for immunoassay QA and may contain far too many compounds for use with GC or HPLC methods.

2.5.3 Retention indices

Although absolute retention times (volumes) and retention times relative to the retention time of a given compound (internal standard) can be useful ways of recording retention data, the Kóvats' retention index (32) provides a method of recording retention data which is independent of column length or operating temperature. Straight-chain (normal) hydrocarbons are assigned an index of 100 times the number of carbon atoms in the molecule (e.g. *n*-decane = 1000). The retention index of a given analyte at a given column temperature is then calculated by difference from the retention indices of the normal alkanes eluting before and after the analyte (see Chapters 1, 2, and 3). Retention indices can also be calculated from data generated on a temperature programme by applying the relevant formula applicable to individual ramps of the programme (33, 34). A detailed discussion of the use of relative retention time and retention index data in analytical toxicology is given by Huizer (31), and reference can be made to Chapters 1 and 3.

2.5.4 Derivatization

In GC, derivative formation is performed to achieve satisfactory chromatography, to improve the detection of characteristics of an analyte, or sometimes to provide additional evidence of compound identity. Derivatization may be used to shorten or lengthen the retention time as required, and also to permit the separation of enantiomers (Section 2.5.5). An obvious factor is that the analyte must be amenable to derivatization in the first place. It is also important that any internal standard undergoes the same derivatization reaction as the analyte. Use of a stable isotope in MS work is ideal. The reaction may be carried out during the extraction or other sample preparation procedure, on a dried residue, or after injection on to the GC column (on-column). If derivatization is to be performed prior to the injection then the reaction needs to be rapid and quantitative or problems may arise in assessing optimum reaction time. Removal of excess derivatizing reagent may be required and there are also the risks of decomposition of the derivatizing and other reagents, and of interference in the assay by contaminants, breakdown products, or by-products. On-column derivatization may not suffer from these drawbacks and has the advantage that the analyte can be injected directly. There are, however, other constraints.

Aspects of derivatization have been reviewed (35). Commonly used reactions include acetylation of primary amines, secondary amines, and phenolic hydroxyls using anhydrides such as acetic, propionic, trifluoroacetic, heptafluorobutyric, or pentafluorobenzoic. With paracetamol, extractive acetylation of paracetamol and N-butyryl-p-aminophenol (internal standard) can be performed using acetic anhydride and N-methylimidazole as catalyst (see Section 3.2.11). In addition to the acetylation reactions discussed above, imine formation follows treatment of primary amines with aldehydes and ketones such as acetone, benzaldehyde, or cyclohexanone. Most aliphatic and aromatic hydroxyls will form silyl ethers when treated with trimethylsilyl reagents at room temperature or, in the case of sterically-hindered compounds, with heating. Hexamethyldisilazane has been widely used. N,O-Bis(trimethylsilyl)trifluoroacetamide containing 1% (v/v) trimethylchlorosilane is a more powerful reagent. Since the silanizing reagents and the reaction products are easily hydrolysed, water must be rigorously excluded. However, tert-butyldimethylsilylimidazole forms ethers which are relatively stable to hydrolysis. One disadvantage to the use of silylated derivatives is accumulation of decomposition products at the detector, necessitating regular cleaning/servicing.

The reaction of boronic acids such as phenylboronic acid with 1,2- and 1,3-diols to form cyclic boronic acid derivatives is valuable in permitting the analysis of ethylene and propylene glycols on a packed column by GLC–FID (36; see Section 3.4.3). This reaction has also been utilized in the analysis of meprobamate after hydrolysis to a 1,3-diol (37). On-column methylation of barbiturates, hydantoins, and some carboxylic acids can be achieved by injecting the sample mixed with 0.2 mol/l trimethylanilinium hydroxide in methanol. However, 'ghost' peaks may arise on subsequent injection of derivatizing reagent if incompletely methylated material has been allowed to accumulate on the column. For carboxylic acids, methylation can be achieved by heating (60 to 100°C, 30 min) with 14% (w/v) boron trifluoride in methanol. After evaporating most of the solvent and adding water, the methylated derivative can be extracted into an organic solvent such as hexane. Diazomethane may also be used as a methylating reagent and reacts rapidly *in vitro* at room temperature; excess reagent may be simply removed by evaporation. However, care is needed since diazomethane is very toxic and potentially explosive.

2.5.5 Chiral separations

There is now much interest in the chromatographic separation of enantiomers of chiral drugs, pesticides, and other compounds. If the analyte contains a reactive functional group, then derivatization using an optically pure reagent and analysis of the resulting diastereoisomers on an achiral chromatographic system may prove adequate. Alternatively, analysis using a chiral stationary phase either directly or after diastereoisomer formation may be needed. HPLC has been widely applied in this area (38) but GC has also been used. A range of derivatizing reagents have been used prior to GC including, for amines, *N*-trifluoroacetylprolyl chloride (39) and α-chloroisovaleryl chloride (40). A number of chiral GC phases have also been described including amides, diamides, dipeptides, and polysiloxanes with chiral substituents (41, 42), but the low thermal stability of these phases precluded their widespread use. However, modification of polyphenylmethylsiloxanes by introduction of L-valine-*tert*-butylamide or L-valine-*S*-α-phenylethylamide moieties (43, 44) can give chiral phases with temperature stability as high as 230°C. Some of these phases are commercially available (Chirasil-Val) and have been widely used. More recently modified cyclodextrins have proved useful as chiral GC phases. Enantiomers of halogenated anaesthetics (enflurane, halothane, isoflurane), for example, can be resolved using cyclodextrin-coated glass capillary columns (45).

3. Applications of GC in analytical toxicology

In responding to a given analytical problem, many factors must be considered (30). Knowledge of the molecular weight and structure of the compound of interest is important even if a published method is to be followed. Information on any co-formulated or co-administered drugs may also be valuable. In addition to the primary literature, GC methods for specific analytes can sometimes be found in the compilations of Moffat (46) and of Baselt and Cravey (17). Information on dose, structure, metabolism, etc. can often be obtained from such sources or from more general works (47–51). Manufacturers may sometimes be able to give details of assay methods, current literature, and potential internal standards for particular compounds.

3.1 Screening for unknowns

The requirement for 'unknown screening' or 'drug screening' is to have the
ability to detect as wide a range of compounds as possible in as little sample
(plasma/serum/whole blood, urine/vitreous humour, stomach contents/
vomit, or tissues) as possible at high sensitivity but with no false positive(s).
Ideally some sample should be left to permit confirmation of the results using
another technique and quantification of any poison(s) present to aid clinical
interpretation of the results.

The concept of discriminating power was introduced by Moffat *et al.* (52)
with the aim of quantifying the ability of paper chromatography, TLC, and
GLC to give unequivocal identification of unknowns. In the case of GC, it
was found that retention data on a range of columns of differing polarities
were highly correlated (53). It was thus concluded that there was little to be
gained from the use of more than one column when screening for unknowns
and that a low-polarity phase such as SE-30/OV-1/OV-101 should be used for
this purpose since most of the compounds studied were eluted. In retrospect,
it seems clear that the high correlation between the retention times on
different phases should have been expected since the inter-relationship between molecular weight, volatility, and retention time is dominant in GC.
Moreover, a number of applications where combinations of polar and non-
polar packed columns gave useful discrimination of well-defined groups of
compounds have been published. In addition, the concept of discriminating
power took no account of selectivity introduced by either the sample preparation procedure or by the detection system. However, subsequent experience
with modern high-efficiency GC capillary columns has borne out the conclusion that a single low- or moderate-polarity column with temperature programming is satisfactory for most purposes if used with a selective sample
preparation procedure and some form of selective detection.

Nowadays, toxicology screening is normally performed using immunoassays and/or TLC and temperature-programmed capillary GC, with GC–MS
for difficult or important/medico-legal cases. The concept of systematic
toxicological analysis (54) brings together many of these strands but sometimes overstates the case for absolute reproducibility of retention data. In real
life, many factors (clinical and circumstantial evidence, availability of a particular poison, past medical history, occupation, number of peaks present on
the chromatogram, selective detector responses, etc.) are considered before
reporting results. Capillary GC–NPD has proved especially useful in screening for basic drugs (12, 55–58) although acidic drugs have also been analysed
on such systems (59).

In drug screening, low polarity phases such as SE-30/OV-1/OV-101 (see
Table 3) have the advantage that retention data for a large number of drugs
and other compounds of interest generated on packed columns are available
(60)—such data are generally directly transferable to capillary columns

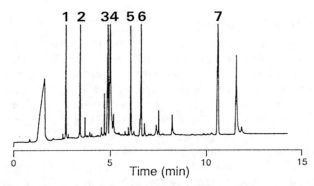

Figure 2. Capillary GC of drugs of abuse in urine: analysis of a sample from a patient prescribed methadone but who was found to also be taking pethidine and phentermine. Column: 25 m × 0.32 mm i.d. HP-5 (0.52 μm film). Column temperature: 90°C (0.5 min), then to 250°C at 40°C/min, then to 310°C at 5°C/min. Carrier gas: helium (flow rate *ca.* 3.5 ml/min). Detector: NPD. Extraction procedure: see *Protocol 2*. Injection: 2 μl sample extract. Peaks: 1, phentermine; 2, nicotine; 3, pethidine; 4, norpethidine; 5, EDDP (methadone metabolite); 6, methadone; 7, prazepam (internal standard). (Redrawn from ref. 12 with kind permission © Ann. Clin. Biochem.)

(31). Caldwell and Challenger (12), however, found that use of a 25 m × 0.32 mm i.d. fused silica capillary coated with cross-linked HP-5 (0.52 μm film) (Hewlett-Packard) was advantageous when screening for basic drugs. The analysis of an extract of a urine specimen performed using *Protocol 2* by GC–NPD using an HP-5 capillary column is illustrated in *Figure 2*. HP-5 is a 5% phenyl, 95% methylsilicone phase with a McReynolds constant similar to those of OV-73 and SE-54 (see *Table 3*) and has been used in several applications in our laboratory such as the analysis of organochlorine pesticides (Section 3.3.1). DB-5 (J&W) is said to be directly equivalent to HP-5.

3.2 Drugs

It is impossible to give full details of the analysis of even a few compounds of interest in the space available. Thus general remarks on the analysis of important groups of compounds have been provided, together with some key references. Immunoassays for some drugs and metabolites in plasma/serum are available in kit form, notably from Abbott and Sigma (both fluorescence polarization immunoassay) and Syva (EMIT). These products do have advantages in that factors such as selectivity, sensitivity, and reproducibility have been investigated beforehand, but may appear expensive and may not be readily applicable to specimens other than plasma or urine. The analysis of alcohols such as ethanol and other volatiles such as chlorinated hydrocarbons is discussed in Section 3.4. One important area, the GC of drugs of abuse, has been reviewed recently (31).

3.2.1 Amphetamines and related compounds

Dexamphetamine ((S)-α-methylphenylethylamine) and methylamphetamine ((S)-N,α-dimethylphenylethylamine) are commonly-abused central nervous system (CNS) stimulants. Many structurally similar compounds such as phenylethylamines (Section 2.5.1), N-ethylbenzenamine (61) and other drugs such as chlorphentermine, diethylpropion, ephedrine, fenfluramine, phentermine, and phenylpropanolamine (norephedrine) may be encountered, and thus any methods or combinations of methods used must normally be able to differentiate between these compounds and also between enantiomers such as R-amphetamine. Capillary GC–NPD is one commonly used achiral technique (12). The enantiomers of amphetamine and of methylamphetamine can be resolved after reaction with (S)-N-trifluoroacetylprolyl chloride (31). If solvent evaporation is used to concentrate an extract, then 0.5 ml 0.2% (v/v) methanolic hydrochloric acid should be added prior to the evaporation stage to prevent loss of analyte (Section 2.2.2). Pemoline is one stimulant which cannot be easily analysed by GC and HPLC is preferred (62).

3.2.2 Anticholinergics

Atropine is used as a preanaesthetic medication, in ophthalmic procedures, and as an antidote in poisoning due to cholinesterase inhibitors. The related compound hyoscine (scopolamine) is used as a premedication and to prevent travel sickness. Atropine and hyoscine are thermally labile but doses are low and thus capillary GC–NPD, and failing that GC–MS, are sometimes employed in assay methods. A further compound, orphenadrine, is used primarily as an anti-Parkinsonian agent and capillary GC–NPD is commonly used in measurement (17).

3.2.3 Anticonvulsants

Commonly used anticonvulsants include carbamazepine, ethosuximide, phenobarbitone, phenytoin (diphenylhydantoin), primidone, and sodium valproate. Most are weak acids. Primidone is metabolized to phenobarbitone (Section 3.2.5) and other compounds. Carbamazepine-10,11-epoxide is the principal active metabolite of carbamazepine. The benzodiazepines clobazam and clonazepam (Section 3.2.6) are used orally as anticonvulsants and high-dose intravenous diazepam is used to treat status epilepticus. Chlormethiazole and the barbiturate thiopentone are also used as intravenous anticonvulsants. In addition, oral chlormethiazole is used as a hypnotic in the elderly and to treat alcoholism and drug withdrawal. Chlormethiazole has many metabolites and less than 5% of a dose is excreted unchanged in urine. Pentobarbitone (Section 3.2.5) is a metabolite of thiopentone and may contribute to toxicity after prolonged thiopentone dosage. There are a number of newer anticonvulsants undergoing therapeutic trial including progabide, gabapentin, vigibatrin, and lamotrigine.

Much has been written about the measurement of anticonvulsants in plasma/serum and the subject has been thoroughly reviewed by Kapetanovic (63). In therapeutic drug monitoring (TDM), capillary GC and HPLC are both used. GC–NPD has greater selectivity/sensitivity for many compounds while GC–FID is necessary for ethosuximide and valproate. On the other hand, using HPLC, hydrophilic and other metabolites of carbamazepine can be analysed as well as the parent compound. In addition, HPLC is perhaps more amenable to automation. However, phenytoin, primidone, and phenobarbitone all require low-wavelength ultraviolet (UV) detection unless derivatized. A mixed SP-2110/SP-2510-DA (Supelco) phase permits the isothermal analysis of 11 common anticonvulsants and associated internal standards by GC on a packed column. Chlormethiazole is most easily analysed by capillary GC–NPD.

3.2.4 Antihistamines

This is a heterogeneous group and includes chlorpheniramine, cyclizine, diphenhydramine, doxylamine, pheniramine, terfenadine, and tripelennamine; many phenothiazines (Section 3.2.12) are used as antihistamines. There are many *N*-demethylated and other metabolites and, in general, capillary GC–NPD is the method of choice although HPLC assays have also been described for some compounds (17).

3.2.5 Barbiturates and related hypnotics

Barbiturates are 5,5'-disubstituted derivatives of barbituric acid. Substitution of sulphur for oxygen at position 2 gives thiobarbiturates such as thiopentone. This latter compound and phenobarbitone are used as anticonvulsants (Section 3.2.3); other barbiturates are used for euthanasia in veterinary medicine and barbitone/sodium barbitone is sometimes employed as a laboratory reagent. Barbiturates with relatively short half-lives such as amylobarbitone, butobarbitone, pentobarbitone, and quinalbarbitone are now rarely used as hypnosedatives in clinical practice in the UK. Nevertheless these compounds are widely abused either directly or when used to 'cut' other substances, and are subject to the UK Misuse of Drugs Regulations. As with many anticonvulsants, capillary GC is widely used to assay these compounds since normally only relatively low (200–220 nm) wavelengths give adequate sensitivity in HPLC. On-column methylation by injection with 0.2 mol/l trimethylanilinium hydroxide in methanol has been employed in packed column GC or GC–MS of barbiturates (25, 31).

3.2.6 Benzodiazepines

Benzodiazepines are used as tranquillizers and hypnotics, and in some cases as anticonvulsants (Section 3.2.3) and short-acting anaesthetics; some 60 compounds have been marketed. Temazepam (3-hydroxydiazepam) especially has been abused, often together with other drugs. Most benzodiazepines are extensively metabolized. Indeed, many are metabolites of other compounds.

Thus, diazepam gives nordiazepam, temazepam, and oxazepam (3-hydroxy-nordiazepam); the latter compounds are excreted as glucuronides and sulphates. The chromatographic analysis of the benzodiazepines, and the metabolism and analysis of 'newer' benzodiazepines (alprazolam, clobazam, flunitrazepam, ketazolam, lorazepam, midazolam, and triazolam) have been comprehensively reviewed (64, 65). GC–ECD is commonly used in the analysis of this group of compounds (24).

3.2.7 Cannabinoids
Cannabis preparations in their various forms are widely abused. Over 60 active constituents (cannabinoids) are known. In general, immunoassays are used for urine screening and GC or capillary GC–MS if confirmation of identity or quantification are required (25, 31). HPLC generally offers no advantage (66).

3.2.8 Local anaesthetics
Lignocaine (lidocaine) is employed as a local anaesthetic and is commonly found in lubricant gels used with urinary and other catheters (Section 2.5.1). Lignocaine is also used as an anti-arrhythmic but is only effective if given intravenously because of extensive presystemic metabolism. Cocaine hydrochloride is an effective local anaesthetic when used at concentrations of 1–20% (w/v), but is normally only applied topically because of the risk of systemic toxicity if given by other routes. Cocaine is also a major drug of abuse, cocaine free-base ('crack') being very rapidly absorbed when inhaled into the nasal passages or smoked. Other important local anaesthetics include benzocaine, bupivacaine, mepivacaine, procaine, and prilocaine. Both GC and HPLC have been applied widely in the analysis of this group of compounds (20, 67). However, non-isotopic immunoassays are usually preferred for screening for cocaine metabolites in urine.

3.2.9 Monamine oxidase inhibitors (MAOIs)
This group includes phenelzine and tranylcypromine. Plasma concentrations are low, since binding to monoamine oxidase occurs rapidly *in vivo* and unbound drug is excreted. Capillary GC–NPD is often used to assay these compounds (17).

3.2.10 Narcotic analgesics
This is a complex group and includes opiates such as morphine and codeine, heroin (diamorphine, diacetylmorphine) produced by treating morphine (or opium in the case of illicit preparations) with acetic anhydride, and synthetic analogues such as buprenorphine and dihydrocodeine. All are strictly controlled, but compounds with very similar structures such as the antitussive pholcodine (morpholinylethylmorphine) have virtually no opioid agonist activity and are available without prescription. Other compounds with similar

structures (nalorphine, naloxone) are potent opioid antagonists. Metabolism is complex, morphine being a metabolite of both heroin and codeine, for example. The GC and GC–MS of morphine and related compounds have been reviewed (25, 31). HPLC is important in the analysis of this group of compounds since many analytes are hydrophilic, thermally labile, or otherwise unsuited to GC unless derivatized (68, 69).

Many other synthetic compounds such as dextropropoxyphene (D-propoxyphene) and methadone, for example, are potent narcotic analgesics

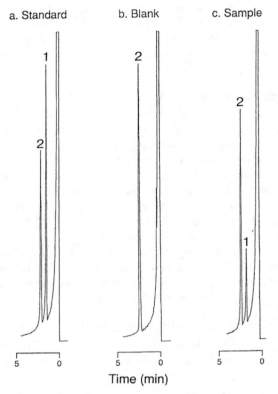

Figure 3. GLC of paracetamol in plasma (70). Column: 1.5 m × 4 mm i.d. glass packed with 3% SP-2250 on Chromosorb W HP (80–100 mesh). Column temperature: 235°C. Carrier gas: nitrogen (flow rate 40 ml/min). Detector: FID. Extraction procedure: (i) add sample or standard (100 μl), 200 mg/l *N*-butyryl-*p*-aminophenol (internal standard) in chloroform (50 μl), pH 7.4 phosphate buffer (50 μl) and acetylating reagent (acetic anhydride: *N*-methylimidazole:chloroform, 5 + 1 + 30) (50 μl) to Dreyer tube (see Section 2.2.2); (ii) vortex-mix (1 min) and centrifuge (9950 *g*, 3 min). Injections: 3–5 μl of chloroform extracts. Samples: (a) 200 mg/l paracetamol in water; (b) paracetamol-free human plasma; (c) plasma from a paracetamol overdose patient (paracetamol concentration 48 mg/l). Peaks: 1, acetylated paracetamol; 2, acetylated *N*-butyryl-*p*-aminophenol.

(opioid agonists) and their use is controlled in most countries. Dextropropoxyphene is extensively metabolized by N-demethylation and by other routes. Methadone is metabolized largely by N-demethylation and hydroxylation. Dextropropoxyphene is thermally labile but may be measured together with nordextropropoxyphene by HPLC with electrochemical oxidation detection (6); methadone and some metabolites may also be measured using this same system.

3.2.11 Paracetamol
Paracetamol (acetaminophen) is a widely used analgesic and sometimes occurs together with other drugs such as dextropropoxyphene (Section 3.2.10). The detection, identification, and measurement of paracetamol is of great importance in clinical toxicology, since clinical management of cases of acute poisoning is guided by the result. Spectrophotometric methods are unreliable and GLC–FID after derivatization, HPLC, enzyme-based assay, or some other selective method must be used. In the extractive acetylation method of Huggett *et al.* (70) (*Figure 3*) the analysis may be repeated without adding the acetylating reagent (acetic anhydride:N-methylimidazole: chloroform, 5 + 1 + 30) in order to confirm the identity of the peaks observed. Alternatively, different anhydrides can be used to give derivatives with different retention times (see *Table 4*).

3.2.12 Phenothiazines and haloperidol
Compounds such as chlorpromazine, perphenazine, prochlorperazine, promethazine, thioridazine, and triflupromazine are used as antihistamines, as tranquillizers, and in various psychiatric disorders. They are often extensively metabolized by N-dealkylation, sulphoxidation, and other pathways. Over 20 metabolites of chlorpromazine, for example, have been isolated; in all 168 are possible (17). Aspects of the analysis of this group of compounds, including

---

Table 4. Retention of derivatized paracetamol and N-butyryl-p-aminophenol on the SP-2250 column system (Section 3.2.11)

Derivative	Acetylating reagent [a]		Kóvats' retention index	
	Anhydride	Catalyst	Paracetamol	N-Butyryl-p-aminophenol
Acetyl	250	50	2145	2270
Propionyl	250	50	2235	2370
n-Butyryl	250	100	2330	2465
n-Heptanoyl	250	100	2660	2805
Benzoyl	250 [b]	200	2915	3055

[a] μl added to 1.5 ml chloroform.
[b] 20% (v/v) benzoic anhydride in chloroform.

---

analyte stability and the occurrence of diastereoisomers of thioridazine metabolites, have been discussed (71). Haloperidol is a butyrophenone derivative
and is widely used as an antipsychotic. Plasma concentrations are low (*ca.*
3–5 μg/l) after single doses and analysis is difficult by any method (17).

3.2.13 Tricyclic and other antidepressants

The tricyclic antidepressants include amitriptyline, clomipramine, dothiepin,
doxepin, imipramine, protriptyline, and trimipramine. *N*-Dealkylated metabolites are common and some, e.g. nortriptyline and desipramine, are also
used as drugs. Related compounds include the tetracyclic antidepressants
maprotiline and mianserin, and the isoquinoline derivative nomifensine.
Aspects of the analysis of these compounds, including problems of adsorption
on to glassware, have been discussed (71). The use of GC–NPD and GC–MS in
the analysis of amitriptyline, doxepin, imipramine, and their *N*-demethylated
metabolites has been discussed by Poklis *et al.* (72). Comparative data on the
performance of chromatographic assays (GC and HPLC) for amitriptyline,
nortriptyline, imipramine, and desipramine in an external QA scheme have
been provided by Wilson *et al.* (73).

3.3 Pesticides

3.3.1 Organochlorines

Organochlorine pesticides include aldrin, chlordane, dicophane (DDT), dieldrin (also a metabolite of aldrin), endrin, heptachlor, hexachlorobenzene
(HCB), 1,2,3,4,5,6-hexachlorocyclohexane (HCH = benzene hexachloride,
BHC), and lindane (gamma-HCH). These compounds were widely used but
persist in the environment, and lindane, which has a relatively short half-life
in vivo, is now the only member of this group to remain in common use in
Europe and North America. Solvent extraction followed by GC–ECD remains the method of choice for the analysis of these compounds in biological
specimens (74–77; see *Figure 4*). The phenolic metabolites of lindane may be
measured by GC-ECD after acid hydrolysis of conjugates followed by acetylation (78). A related compound pentachlorophenol (PCP) is still widely used
in wood preservatives and disinfectants, and as a contact herbicide. Solvent
extraction followed by GLC–ECD of an acetyl derivative again provides a
convenient method of analysis (79, 80). Metabolism is by conjugation, thus
hydrolysis of urine by boiling with strong acid yields results up to 17-fold
higher than if the sample is extracted directly (81).

3.3.2 Organophosphates and carbamates

Some organophosphates (OPs) and carbamates are used as herbicides and
fungicides and are relatively non-toxic to man. OP and carbamate insecticides,
however, inhibit acetylcholinesterase and some are extremely toxic. Measure-

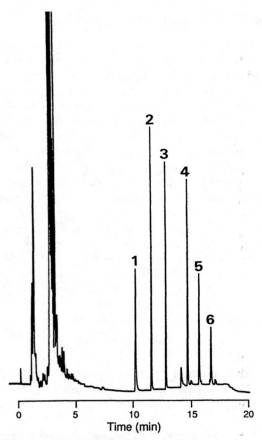

Figure 4. Capillary GC of chlorinated pesticides. Column: 30 m × 0.32 mm i.d. HP-5 (0.52 μm film). Column temperature: 60°C to 170°C at 20°C/min, then to 290°C at 10°C/min and hold (2 min). Carrier gas: helium (flow rate *ca.* 2.5 ml/min). Detector: ECD (sensitivity: 10 kHz f.s.d., Hewlett-Packard 5890). Make-up gas: nitrogen (flow rate 35 ml/min). Injection: 1 μl mixture [each compound 0.1 ng/l except 2,2-bis(4-chlorophenyl-1,1,1-trichloroethane) (p,p′-DDT), 0.2 ng/l] in hexane (mixed isomers). Peaks: 1, lindane; 2, heptachlor; 3, isodrin; 4, endrin; 5, p,p′-DDT; 6, methoxychlor.

ment of plasma and erythrocyte cholinesterase activity can be used to assess the severity of exposure to these agents. OPs themselves can be measured after solvent extraction by GC–NPD (82) or by GC–FPD (83, 84). Chlorinated OPs such as chlorpyriphos and its major metabolite have also been measured by GC–ECD after gel permeation chromatography (85). OPs such as parathion which contain a nitro moiety have also been measured by GC–ECD (86). Carbamates such as carbaryl have been measured by GC–ECD after derivatization with heptafluorobutyric anhydride (87).

3.4 Gases, solvents, and other poisons

3.4.1 Carbon monoxide

Carbon monoxide (CO) poisoning remains the single most common cause of fatal poisoning in developed, Western countries and most probably in the rest of the world as well. Motor vehicle exhausts, defective heating systems, and smoke from all types of fires are common sources. Some 40% of an absorbed dose of dichloromethane is also metabolized to CO although the significance of this finding is unclear as regards acute dichloromethane poisoning. Spectrophotometric measurement of carboxyhaemoglobin is commonly used to assess the severity of CO poisoning but can be unreliable in the presence of other pigments. GC is the most reliable method for measuring blood CO itself which can then be related to haemoglobin measured spectrophotometrically to give the percentage of carboxyhaemoglobin (%HbCO) in the sample. Liberation of CO by mixing with ferricyanide in a headspace vial followed by analysis on a column packed with 60/80 mesh 0.3 nm molecular sieve is a convenient method (88). CO itself does not respond on the FID, but reduction to methane by reaction with hydrogen in the presence of a nickel catalyst allows an FID to be used which is more convenient and gives greater sensitivity than the TCD (88). Some of the factors affecting the stability of CO in stored blood samples have been discussed (89).

3.4.2 Cyanide

Cyanide (CN^-) poisoning may be encountered after the inhalation of hydrogen cyanide (HCN), or after the ingestion of hydrocyanic acid, or of potassium or sodium cyanides. Cyanide solutions used in electroplating may release HCN if acidified, while a number of naturally occurring nitriles are metabolized to cyanide ion *in vivo*. Thiocyanate insecticides also give rise to cyanide *in vivo*. In addition, cyanide is often present in the blood of fire victims due to the inhalation of HCN from the partial combustion of wool, silk, and synthetic polymers such as polyurethanes and polyacrylonitriles; carbon monoxide is usually also present. Although colorimetric methods based on microdiffusion are available for measuring blood cyanide, a headspace GC method based on NPD after liberation of HCN using acetonitrile as the internal standard has also been described (90). Factors affecting the stability of cyanide in stored blood samples have been discussed (91).

3.4.3 Alcohols and glycols

Despite the introduction of evidential breath ethanol instruments, the measurement of blood ethanol in road traffic cases remains an important analysis. In addition, ethanol from alcoholic drinks is very frequently encountered in hospital admissions due to acute poisoning. However, poisoning with industrial alcohol (methylated spirit) containing various denaturants, notably methanol, still occurs. Ethanol may also be given to treat poisoning with methanol or with ethylene glycol. Methanol itself is used as a general and laboratory solvent, in car radiator antifreeze, and in windscreen washer additives. Serious outbreaks of acute methanol poisoning have occurred in several countries in recent years due to the sale of illicit 'alcoholic' drinks based on methanol. Isopropanol (propan-2-ol), used in lotions for topical administration, window and windscreen washers, and as a solvent for toiletries, is metabolized to acetone and is much less toxic than methanol.

Although enzymatic methods and simple instruments based on fuel cells are available for the measurement of blood ethanol, GC–FID remains the method of choice if a range of compounds of interest (primarily methanol, isopropanol, and acetone) are to be measured. Direct injection (92, 93) or headspace (94) methods have been employed for ethanol and are often applicable to the other compounds. Since ethanol may be both formed and destroyed in biological specimens *in vitro* by microbial action, correct preservation and storage of specimens is especially vital. Addition of 1% (w/v) sodium fluoride seems adequate for most purposes. The subject has been thoroughly reviewed (95).

Ethylene glycol is used mainly in car radiator antifreeze as a concentrated (20–50%, v/v) aqueous solution, sometimes together with methanol. A rise in plasma osmolality may be a useful although non-specific indicator of poisoning with this compound, but measurement of ethylene glycol itself is required to confirm the diagnosis and to monitor treatment. Propylene glycol (propan-1,2-diol) is used as a solvent in the pharmaceutical and food industries and is relatively non-toxic. Although direct analysis of these compounds after dilution with internal standard solution on a column packed with a porous polymer such as Chromosorb 101 is feasible, a better method is to form the corresponding phenyl boronates. Thus in the case of plasma/serum protein precipitation with acetonitrile containing the internal standard (propan-1,3-diol) and derivatization reagent (15 g/l phenylboronic acid in 2,2-dimethoxypropane) can be followed by analysis of a portion (0.5–2 μl) of the supernatant on an OV-101 packed column (36; see *Figure 5*). Note that the phenylboronates of ethylene and of propylene glycols are resolved on this system but not on more polar phases such as OV-17.

3.4.4 Solvents and other volatiles

Today, if anaesthesia is excluded, acute poisoning with solvents and other volatile substances usually follows deliberate inhalation of vapour in order to become intoxicated [volatile substance abuse (VSA)]. Patients who ingest solvents or solvent-containing products, either by accident or deliberately, and the victims of domestic and industrial mishaps, provide further groups which may suffer acute poisoning by these compounds (96). Toxicity due to fumigants such as bromomethane or to compounds used primarily as chemical intermediates, on the other hand, is more commonly associated with

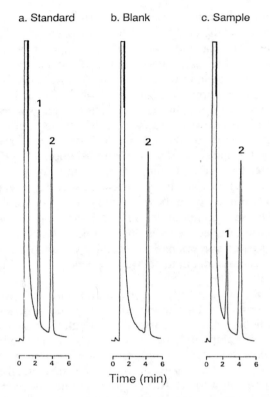

Figure 5. GLC of ethylene glycol in plasma. Column: 2.1 m × 4 mm i.d. glass packed with 3% OV-101 on Chromosorb W HP (80–100 mesh). Column temperature: 150°C. Carrier gas: nitrogen (flow rate 50 ml/min). Detector: FID. Sample preparation: (i) add sample or standard (100 µl), 0.50 g/l propan-1,3-diol (internal standard) in acetonitrile (200 µl) and derivatizing reagent [15 g/l phenylboronic acid in 2,2-dimethoxypropane containing 0.2% (v/v) hydrochloric acid] (100 µl) to Dreyer tube (see Section 2.2.2); (ii) vortex-mix (30 s) and centrifuge (9950 *g*, 1 min). Injections: 2 µl supernatants. Samples: (a) standard solution prepared in heparinized bovine plasma containing 1.0 g/l ethylene glycol; (b) ethylene-glycol-free human plasma; (c) plasma from a patient poisoned with ethylene glycol (ethylene glycol concentration 0.4 g/l). Peaks: 1, ethylene glycol phenylboronate; 2, propan-1,3-diol phenylboronate. (Redrawn from ref. 36 with kind permission © Ann. Clin. Biochem.)

occupational exposure. Since the mid 1970s, concern as to the consequences of the release of massive quantities of organochlorine and organobromine compounds such as chlorofluorocarbons (CFCs) into the atmosphere has led to the planned phased withdrawal of many of the volatile halogenated compounds in current use (97). Commercial 'butane' (liquefied petroleum gas, LPG) or dimethyl ether (DME) have already largely replaced CFCs as propellants in aerosols, for example, in many countries. Only time will tell which halogenated volatiles will be in widespread use by the year 2000, although in the case of refrigerants the trend is to polyfluorinated compounds such as 1,1,1,2-tetrafluoroethane (FC 134a).

Headspace GC using either packed (98) or capillary (8) columns together with temperature-programmed GC and split-flame ionization/electron capture detection (FID/ECD) often provides a convenient mode of analysis for many volatiles in blood or in tissue digests. Ramsey and Flanagan (98) used a packed column [2 m × 2 mm i.d. 0.3% (w/w) Carbowax 20M on Carbopack C] programmed from 35 to 175°C in the headspace GC of volatiles. On-column septum injections of up to 400 μl of headspace could be performed and thus good sensitivity (of the order of 0.1 mg/l or better using 200 μl of sample) could be obtained. Moreover, most analytes could be retained without resort to subambient operation. Disadvantages included the poor resolution of some very volatile substances, the long total analysis time (40 min), and variation in the peak shape given by alcohols between different batches of column packing.

A 60 m × 0.53 mm i.d. fused silica capillary coated with the dimethylpolysiloxane phase SPB-1 (5 μm film thickness) offers many advantages over the packed column system described above (8). Good peak shapes are obtained for polar analytes such as ethanol and on-column injections of up to 300 μl of headspace can be performed with no discernable loss of efficiency. Sensitivity is thus at least as good as that attainable with packed columns. Most commonly abused compounds, including many with very low boiling points such as bromochlorodifluoromethane (BCF), n-butane, dichlorodifluoromethane (FC 12), DME, fluorotrichloromethane (FC 11), isobutane, and propane, can be retained and differentiated at an initial temperature of 40°C followed by programming to 200°C. The total analysis time is 26 min. The reductions in costs and in the time taken in recycling which arise directly from the use of this relatively high starting temperature are considerable. Quantitative measurements can be performed either isothermally or on a temperature programme. The analysis of a qualitative standard mixture (*Table 5*) on the temperature programme on the SPB-1 capillary column is illustrated in *Figure 6*. In all, retention and detector response data for 244 compounds are available (8).

3.4.5 Nicotine

Nicotine is commonly encountered in tobacco, although usually in concentrations insufficient to cause acute poisoning except when ingested by young children. However, nicotine occurs in higher concentrations in some herbal medicines and may also be used as a fumigant in horticulture. Nicotine is metabolized by N-demethylation to give cotinine. The possibility of acute poisoning notwithstanding, the major requirement for the measurement of nicotine and/or cotinine in biological specimens is in studies of smoking behaviour. Capillary GC–NPD remains the method of choice for the measurement of nicotine and cotinine in biological fluids (13; see *Figure 7*).

Table 5. Headspace capillary GC of volatiles: composition of the qualitative standard mixture

(a) Qualitative standard mixture [prepared in 125 ml gas sampling bulb (Supelco 2-2146)]

Compound	Amount added (ml) [a]
Bromochlorodifluoromethane	0.005
n-Butane	[b]
Dimethyl ether	1.0
Isobutane	[b]
Fluorotrichloromethane	0.02
Dichlorodifluoromethane	0.3
1,1,2-Trichlorotrifluoroethane	0.5
Propane	[b]

[a] Volume of vapour phase in headspace vial.
[b] 2.0 ml commercial 'butane' (mixture of n-butane, isobutane, and propane) added.

(b) Liquid components mixture (add 10 μl to mixture of gaseous components in gas sampling bulb (see above)

Compound	Amount added (ml)
Acetone	7.5
Butanone	5.0
Carbon tetrachloride	0.05
Chloroform	0.5
Ethanol	5.0
Ethylbenzene	2.5
Halothane	0.1
n-Hexane	5.0
Methyl isobutyl ketone	2.5
Propan-2-ol	5.0
Tetrachloroethylene	0.025
Toluene	2.5
1,1,1-Trichloroethane	0.25
1,1,2-Trichloroethane	1.0
Trichloroethylene	0.25
2,2,2-Trichloroethanol	0.015

4. Conclusions

The refinement of capillary column technology in the last 10 years or so and the introduction of cheaper 'bench-top' mass spectrometers has led to a resurgence in the use of GC in analytical toxicology. This trend looks set to continue. Many sources of information may be helpful when up-dating old methods or setting up new ones. A compilation of artefacts which may be

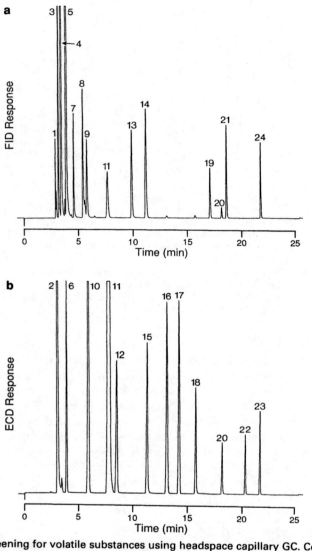

Figure 6. Screening for volatile substances using headspace capillary GC. Column: 60 m × 0.53 mm i.d. SPB-1 (5 μm film). Column temperature: 40°C (6 min), then to 80°C at 5°C/min, then to 200°C at 10°C/min. Carrier gas: helium (8.6 ml/min). Detector sensitivities (f.s.d., 5:1 split): FID 3.2 nA, ECD 64 kHz (Hewlett-Packard 5890). Injection: *ca*. 10 μl qualitative standard mixture (see *Table 5*). Peaks: 1, propane; 2, dichlorodifluoromethane; 3, dimethyl ether; 4, isobutane; 5, *n*-butane; 6, bromochlorodifluoromethane; 7, ethanol; 8, acetone; 9, isopropanol; 10, fluorotrichloromethane; 11, 1,1,2-trichlorotrifluoroethane; 12, halothane; 13, butanone; 14, *n*-hexane; 15, chloroform; 16, 1,1,1-trichloroethane; 17, carbon tetrachloride; 18, trichloroethylene; 19, methyl isobutyl ketone; 20, 1,1,2-trichloroethane (internal standard); 21, toluene; 22, tetrachloroethylene; 23, 2,2,2-trichloroethanol; 24, ethylbenzene (internal standard). (Reproduced from ref. 8 with kind permission.)

Robert J. Flanagan

a. Plasma   b. Urine   c. Saliva

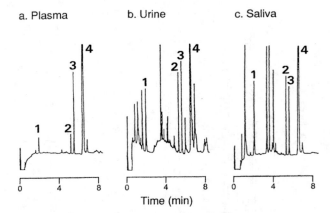

Time (min)

Figure 7. GC of nicotine and cotinine in human plasma, urine and saliva. Column: 7 m × 0.32 mm i.d. HP-FFAP (0.52 μm film). Column temperature: 70°C to 115°C at 40°C/min and hold (1.5 min), then to 200°C at 40°C/min and hold (2.25 min), then to 210°C at 40°C/min and hold (2.75 min); total analysis time 10 min. Carrier gas: helium (head pressure 105 kPa). Detector: NPD. Extraction procedure: see *Protocol 3*. Injections: 2 μl sample extracts. Nicotine and cotinine concentrations (μg/l): plasma 23 and 27, urine 73 and 100, saliva 115 and 130, respectively. Peaks: 1, nicotine; 2, cotinine; 3, 3-methylcotinine (internal standard); 4, caffeine. (Redrawn from ref. 13 with kind permission.)

encountered has been published (99). Manufacturers' catalogues often contain up-to-date information, although important experimental details may be lacking. Chromatograms, for example, may have been obtained using pure compounds and/or concentrated solutions may have been injected; in some cases the actual amount injected may not be given. As well as the formal scientific literature, the trend to publications sponsored by manufacturers or funded from advertising has produced some useful magazines; LC-GC International (Aster) is probably the best of these. Such publications also help keep in touch with developments not only in capillary GC, but also in HPLC, capillary (zone) electrophoresis [C(Z)E], and supercritical fluid chromatography (SFC, SCFC) (100).

Acknowledgements

I thank the colleagues who have assisted in preparing this review, especially Mr J. Ramsey (St George's Hospital Medical School), and Dr M. Ruprah and Mr P. J. Streete (Poisons Unit).

References

1. Shang-Qiang, J. and Evenson, M. A. (1983). *Clin. Chem.*, **29**, 456.
2. Wad, N. (1986). *Ther. Drug Monit.*, **8**, 368.
3. Dawling, S. and Braithwaite, R. (1980). *J. Pharm. Pharmacol.*, **32**, 304.

4. Hurst, P. R. and Whelpton, R. (1989). *Biomed. Chromatogr.*, **3**, 226.
5. Whelpton, R. (1978). *Acta Pharmacol. Suecica*, **15**, 458.
6. Flanagan, R. J., Ramsey, J. D., and Jane, I. (1984). *Human Toxicol.*, **3**, 103S.
7. McDowall, R. D. (1989). *J. Chromatogr.*, **492**, 3.
8. Streete, P. J., Ruprah, M., Ramsey, J. D., and Flanagan, R. J. (1992). *Analyst*, **117**, 1111.
9. Thomas, C. L. P. (1991). *Chromatogr. Anal.*, April, 5.
10. Whelpton, R. (1989). *Trends Pharmacol. Sci.*, **10**, 182.
11. Flanagan, R. J. and Withers, G. (1972). *J. Clin. Pathol.*, **25**, 899.
12. Caldwell, R. and Challenger, H. (1989). *Ann. Clin. Biochem.*, **26**, 430.
13. Feyerabend, C. and Russell, M. A. H. (1990). *J. Pharm. Pharmacol.*, **42**, 450.
14. Van Horne, K. C. (ed.) (1985). *Sorbent extraction technology*. Analytichem International, Harbor City, California.
15. Harkey, M. R. (1989). In *Analytical aspects of drug testing*, (ed. D. G. Deutsch). Wiley, New York, p. 59.
16. Widdop, B. (1980). In *Toxicological aspects. Proceedings of the IXth Congress of the European Association of Poison Control Centres and the European Meeting of the International Association of Forensic Toxicologists*, (ed. A. Kovatsis). TIAFT, Thessaloniki, p. 231.
17. Baselt, R. C. and Cravey, R. H. (1989). *Disposition of toxic drugs and chemicals in man*, (3rd edn). Year Book Medical, Chicago.
18. Osselton, M. D. (1977). *J. Forensic Sci.*, **17**, 189.
19. Shankar, V., Damodaran, C., and Sekharan, P. C. (1987). *J. Anal. Toxicol.*, **11**, 164.
20. Monkman, S. C., Armstrong, R., Flanagan, R. J., Holt, D. W., and Rosevear, S. (1989). *Biomed. Chromatogr.*, **3**, 88.
21. Supina, W. R. (1974). *The packed column in gas chromatography*. Supelco, Bellefonte, Pennsylvania.
22. McReynolds, W. O. (1970). *J. Chromatogr. Sci.*, **8**, 685.
23. Ramsey, J. D., Lee, T. D., Osselton, M. D., and Moffat, A. C. (1980). *J. Chromatogr.*, **184**, 185.
24. Rutherford, D. M. (1977). *J. Chromatogr.*, **137**, 439.
25. Webb, K. S. (1991). In *The analysis of drugs of abuse*, (ed. T. A. Gough). Wiley, Chichester, p. 175.
26. Deutsch, D. G. (1989). In *Analytical aspects of drug testing*, (ed. D. G. Deutsch). Wiley, New York, p. 87.
27. Lacroix, B., Huvenne, J. P., and Deveaux, M. (1989). *J. Chromatogr.*, **492**, 109.
28. Ching, N. P. H., Jham, G. N., Subbarayan, C., Grossi, C., Hicks, R., and Nealon, T. F. (1981). *J. Chromatogr.*, **225**, 196.
29. Dine, T., Luyckx, M., Cazin, M., Brunet, Cl., Cazin, J. C., and Goudaliez, F. (1991). *Biomed. Chromatogr.*, **5**, 94.
30. Peng, G. W. and Chiou, W. L. (1990). *J. Chromatogr.*, **531**, 3.
31. Huizer, H. (1991). In *The analysis of drugs of abuse*, (ed. T. A. Gough). Wiley, Chichester, p. 24.
32. Kovats, E. (1961). *Z. Anal. Chem.*, **181**, 351.
33. Van den Dool, H. and Kratz P. D. (1963). *J. Chromatogr.*, **11**, 463.
34. Lee, J. and Taylor, D. R. (1983). *Chromatographia*, **16**, 286.

Robert J. Flanagan

35. Blau, K. and King, G. S. (1977). *Handbook of derivatives for chromatography.* Heyden, Chichester.
36. Flanagan, R. J., Dawling, S., and Buckley, B. M. (1987). *Ann. Clin. Biochem.,* **24,** 80.
37. Flanagan, R. J. and Chan, M. W. J. (1989). *Analyst,* **114,** 703.
38. Zief, M. and Crane, L. J. (1988). *Chromatographic chiral separations.* Marcel Dekker Inc., New York.
39. Hoopes, E. A., Pelzer, E. T., and Bada, J. L. (1978). *J. Chromatogr. Sci.,* **16,** 556.
40. König, W. A., Stoelting, K., and Kruse, K. (1977). *Chromatographia,* **10,** 444.
41. König, W. A. (1982). *J. High Resolut. Chromatogr. Chromatogr. Commun.,* **5,** 588.
42. Liu, R. H. and Ku, W. W. (1983). *J. Chromatogr.,* **271,** 309.
43. König, W. A., Benecke, I., and Sievers, S. (1981). *J. Chromatogr.,* **217,** 71.
44. König, W. A. and Benecke, I. (1983). *J. Chromatogr.,* **269,** 19.
45. Meinwald, J., Thompson, W. R., Pearson, D. L., König, W. A., Runge, T., and Francke, W. (1991). *Science,* **251,** 560.
46. Moffat, A. C. (ed). (1986). *Clarke's isolation and identification of drugs.* (2nd edn). Pharmaceutical Press, London.
47. Budavari, S., ed. (1989). *The Merck index,* (11th edn). Merck & Co, Rahway, New Jersey.
48. Hayes, W. J. and Laws, E. R. (eds). (1991). *Handbook of pesticides.* Academic Press, San Diego.
49. McEvoy, G. K. (ed). (1991). *AHFS drug information.* American Society of Hospital Pharmacists, Bethesda.
50. Proctor, N. H., Hughes, J. P., and Fischman, M. L. (1988). *Chemical hazards of the workplace.* J. B. Lippincott Co., Philadelphia.
51. Reynolds, J. E. F. (ed). (1989). *Martindale, the extra pharmacopoeia,* (29th edn). Pharmaceutical Press, London.
52. Moffat, A. C., Smalldon, K. W., and Brown, C. (1974). *J. Chromatogr.,* **90,** 1.
53. Moffat, A. C., Stead, A. H., and Smalldon, K. W. (1974). *J. Chromatogr.,* **90,** 19.
54. De Zeeuw, R. A. (1989). *J. Chromatogr.,* **488,** 199.
55. Hime, G. W. and Bednarczyk, L. R. (1982). *J. Anal. Toxicol.,* **6,** 247.
56. Fretthold, D., Jones, P., Sebrosky, G., and Sunshine, I. (1986). *J. Anal. Toxicol.,* **10,** 10.
57. Taylor, R. W., Greutink, C., and Jain, N. C. (1986). *J. Anal. Toxicol.,* **10,** 205.
58. Watts, V. W. and Simonick, T. F. (1986). *J. Anal. Toxicol.,* **10,** 198.
59. Bogusz, M., Bialka, J., Gierz, J., and Klys, M. (1986). *J. Anal. Toxicol.,* **10,** 135.
60. DFG/TIAFT (1985). *Gas-chromatographic retention indices of toxicologically relevant substances on SE-30 or OV-1,* (2nd edn.). VCH Publishers, Weinheim.
61. Christopherson, A. S., Bugge, A., Dahlin, E., Møorland, J., and Wethe, G. (1988). *J. Anal. Toxicol.,* **12,** 147.
62. Tomkins, C. P., Soldin, S. J., MacLeod, S. M., Rochefort, J. G., and Swanson, J. M. (1980). *Ther. Drug Monit.,* **2,** 255.
63. Kapetanovic, I. M. (1990). *J. Chromatogr.,* **531,** 421.
64. Jones, G. R. and Singer, P. P. (1989). In *Advances in analytical toxicology,* (ed. R. C. Baselt), vol 2. Year Book Medical, Chicago, p. 1.
65. Sioufi, A. and Dubois, J. P. (1990). *J. Chromatogr.,* **531,** 459.

66. Harvey, D. J. (1984). In *Analytical methods in human toxicology*, Part 1, (ed. A. S. Curry). Macmillan, London, p. 257.
67. Tucker, G. T. and Lennard, M. S. (1984). In *Analytical methods in human toxicology*, Part I, (ed. A. S. Curry). Macmillan, London, p. 159.
68. Daldrup, T., Michalke, P., and Szathmary, S. (1985). In *Practice of high performance liquid chromatography*, (ed. H. Engelhardt). Springer, New York, p. 241.
69. Tagliaro, F., Franchi, D., Dorizzi, R., and Marigo, M. (1989). *J. Chromatogr.*, **488**, 215.
70. Huggett, A., Andrews, P., and Flanagan, R. J. (1981). *J. Chromatogr.*, **209**, 67.
71. Whelpton, R. (1984). In *Analytical methods in human toxicology*, Part 1, (ed. A. S. Curry). Macmillan, London, p. 139.
72. Poklis, A., Soghoian, D., Crooks, C. R., and Saady, J. J. (1990). *Clin. Toxicol.*, **28**, 235.
73. Wilson, J. F., Tsanaclis, L. M., Williams, J., Tedstone, J. E., and Richens, A. (1989). *Ther. Drug Monit.*, **11**, 196.
74. Dale, W. E., Miles, J. W., and Gaines, T. B. (1971). *J. Assoc. Off. Anal. Chem.*, **53**, 1287.
75. Radomski, J. L., Deichmann, W. B., Rey, A. A., and Merkin, T. (1971). *Toxicol. Appl. Pharmacol.*, **20**, 175.
76. Barquet, A., Morgade, C., and Pfaffenberger, C. D. (1981). *J. Toxicol. Environ. Health*, **7**, 469.
77. Saito, I., Kawamura, N., Uno, K., and Takeuchi, Y. (1985). *Analyst*, **110**, 263.
78. Angerer, J., Heinrich, R., and Laudehr, H. (1981). *Int. Arch. Occup. Environ. Health*, **48**, 319.
79. Woiwode, W., Wodarz, R., Drysch, K., and Weichardt, H. (1980). *Int. Arch. Occup. Environ. Health*, **45**, 153.
80. Siqueira, M. E. P. B. and Fernicola, N. A. G. G. (1981). *Bull. Environ. Contam. Toxicol.*, **27**, 380.
81. Egerton, T. R. and Moseman, R. F. (1979). *J. Agric. Food Chem.*, **27**, 197.
82. Osterloh, J., Lotti, M., and Pond, S. M. (1983). *J. Anal. Toxicol.*, **7**, 125.
83. Gabica, J., Wyllie, J., Watson, M., and Benson, W. W. (1971). *Anal. Chem.*, **43**, 1102.
84. Heyndrickx, A., Van Hoff, F., De Wolff, L., and Van Peteghem, C. (1974). *J. Forensic Sci. Soc.*, **14**, 131.
85. Guinivan, R. A., Thompson, N. P., and Bardalaye, P. C. (1981). *J. Assoc. Off. Anal. Chem.*, **64**, 1201.
86. Kadoum, A. M. (1968). *Bull. Environ. Contam. Toxicol.*, **3**, 247.
87. Mount, M. E. and Oehme, F. W. (1980). *J. Anal. Toxicol.*, **4**, 286.
88. Guillot, J. G., Weber, J. P., and Davoie, J. Y. (1981). *J. Anal. Toxicol.*, **5**, 264.
89. Chace, D. H., Goldbaum, L. R., and Lappas, N. T. (1986). *J. Anal. Toxicol.*, **10**, 181.
90. Zamecnik, J. and Tam, J. (1987). *J. Anal. Toxicol.*, **11**, 47.
91. Bright, J. E., Inns, R. H., Tuckwell, N. J., and Marrs, T. C. (1990). *Human Exp. Toxicol.*, **9**, 125.
92. Curry, A. S., Walker, G. W., and Simpson G. S. (1966). *Analyst*, **91**, 742.
93. Manno, B. R. and Manno, J. E. (1978). *J. Anal. Toxicol.*, **2**, 257.
94. Penton, Z. (1985). *Clin. Chem.*, **31**, 439.
95. Cory, J. E. L. (1978). *J. Appl. Bacteriol.*, **44**, 1.

96. Flanagan, R. J., Ruprah, M., Meredith, T. J., and Ramsey, J. D. (1990). *Drug Safety*, **5,** 359.
97. Flanagan, R. J. (1993). In *Toxicology and Drug Analysis*, (ed. H. Brandenburger, and R. A. A. Maas). de Gruyter, Berlin, in press.
98. Ramsey, J. D. and Flanagan, R. J. (1982). *J. Chromatogr.*, **240,** 423.
99. Middleditch, B. S. (1989). *Analytical Artifacts. GC, MS, HPLC, TLC and PC. J. Chromatogr. Libr.*, Vol. 44. Elsevier, Amsterdam.
100. Novotny, M. (1989). *J. Pharmaceut. Biomed. Anal.* **7,** 239.

7

Gas chromatography in clinical biochemistry

JAGADISH CHAKRABORTY

1. Introduction

During the past two decades advances in the detection systems, column technology and automation in sampling and data processing resulted in virtually universal acceptance of gas chromatography (GC), particularly in combination with mass spectrometry (MS), as an indispensable analytical tool in biomedical research. For multicomponent analyses of complex biological specimens, GC uniquely combines high resolution of components with detectors that offer high sensitivity, and simultaneous characterization and quantification of analytes.

During the expansion phase of clinical biochemistry laboratory services in the 1960s and 1970s, the choice of equipment and methods was based on criteria set out for bulk-processing of specimens in hospital laboratories. Automated 'workhorse' systems, with increasing sophistication and flexibility, were brought into use to cope effectively with the large workload without a comparable increase in number of staff. The key features in the selection of such equipment systems are:

(a) capability of being used for a range of assays and assay formats required in the laboratory

(b) cost and cost-effectiveness

(c) operational characteristics, e.g. user-friendliness, robustness, maintenance

(d) efficient analytical performance, e.g. quality of analysis and its reliability, and high sample throughput

Gas chromatography instrumentation and procedures do not quite meet these above criteria, e.g. sample manipulation essential for GC analyses does not lend itself to the analytical routine of hospital biochemistry laboratories and is not compatible with the goal of rapid 'turnaround time'. This explains why GC still remains a specialist analytical tool in clinical biochemistry service with an important role mainly in three areas of application as follows:

(a) assay of drugs and other toxins

(b) provision of reference methods

(c) development of new methods specially those involving multiple components

Of these, the GC assay of drugs and related compounds is dealt with in Chapter 6. However, in the following sections a practical approach is given to other uses of GC that are of interest to clinical biochemists in the investigation of patients or medical research.

2. Applications of GC

2.1 Organic volatiles

Human breath, body fluid, plasma, urine, and saliva contain a vast number of 'organic volatiles' which consist of essential nutrients, metabolic intermediates and waste products, environmental contaminants, and low molecular weight substances involved with various metabolic processes. Knowledge of the composition of these complex mixtures offers a considerable potential for the recognition of biochemical fingerprints characteristic of the aetiology, pathogenesis, or the diagnosis of diseases. For such analyses, GC, especially in combination with mass spectrometry, provides the most effective analytical tool, although such systems are likely to be found only in specialist laboratories. In view of their large number and chemical diversity, the study of organic volatiles poses special problems with regard to sample preparation for chromatography, resolution, sensitivity of detection, and the processing of the vast quantity of data generated from such measurements. In this field, as in most other biochemical applications of GC, the use of capillary columns and MS detection systems have become essential features of standard methodology. The preparation of samples, involves procedures of varying complexity and efficiency, depending on the nature of the biological specimen and the analytes under consideration.

2.1.1 Sample preparation

The preparation of samples for GC analysis of biological specimens for organic volatiles has received a great deal of attention over the years, and the methods reported in the literature may be classified into three broad categories:

- solvent extraction
- head space techniques
- direct chromatography

Extraction of specimens with a suitable organic solvent and concentration of the extract prior to chromatography has been used for urine, plasma, blood, breast milk, amniotic fluid, and saliva. Limitations to this approach include:

- the need for pure solvents
- loss of analyte during evaporation
- solvent suitability for analytes

In direct headspace analysis, the sample, e.g. serum or urine, is equilibrated with the headspace in a suitable container. A portion of this headspace gas is then injected for analysis. More elaborate headspace trapping devices combine separation of the volatiles from the sample matrix with subsequent enrichment of the constituents. Such a system, suitable for small volumes of body fluids, is known as the transevaporator sampling technique. It contains a microcolumn packed with Porosil E (pore silica gel), into which the sample is injected. In one mode of use, helium is passed through the column to remove the volatiles which are then collected in a trap (Tenax-GC, a porous polymer, 2,6-diphenyl-p-phenylene oxide). Alternatively, the volatiles in the specimen may be extracted by the vapour of an appropriate solvent, e.g. 2-chloropropane and collected on a suitable adsorbent, such as Tenax-GC or glass beads. The substances in this adsorbent trap are thermally desorbed in a stream of helium, condensed in a cooled short pre-column and finally 'flashed' on to the main chromatographic column. Similarly, in the methods for breath analysis the sample is passed through a device in which organic compounds are retained, while nitrogen and oxygen present pass through unhindered. This process can be accomplished by cold trapping, chemical interaction, or adsorptive binding. Suitably diluted samples of body fluids have been used directly for the chromatography of, for example, ethanol, acetone, and other metabolites. If needed, volatiles can also be trapped and concentrated on a small, cooled pre-column of an adsorbent or a stainless steel capillary tube immersed in liquid nitrogen.

2.1.2 Alkanes

Although oxidative stress is now thought to be involved in the pathophysiology of several diseases and aging, laboratory tests for its assessment in the human is not yet established. One of the proposed markers of oxidant stress status that looks attractive is the analysis of lipid peroxidation products, pentane and ethane, in breath. In quantitative assay, there are many possible sources of error, particularly those arising from sample collection, storage, injection technique, and contaminants of similar volatility. Sampling usually involves collection of breath for a period in a gas-tight bag and then passing a known volume of the sample through a liquid-nitrogen-cooled adsorbent pre-column. The adsorbed alkanes are desorbed thermally at the time of analysis. Some examples of alkane concentrations in normal human breath are given in *Table 1* and the procedure used for such measurements is outlined below (1).

(a) Collect samples of alveolar breath by use of Haldane–Bristley tube and store in 50 ml polyethylene syringes and analyse within 5 h. For total breath samples use gas-tight bags. Volatile alkanes are not lost from polyethylene syringes for at least 10 h, but after that their concentrations begins to decline. To minimize peak broadening from large sample volume, carry out 'cold trap' concentration of sample.

(b) Analyse breath using a chromatograph fitted with a gas sampling valve and a FID with column and conditions as follows:

- column   2 m stainless steel packing Chromosorb 102
- oven temperature   50°C for 1 min
  50°C/min to 100°C
  15°C/min to 190°C
- injector temperature   150°C
- detector temperature   225°C
- t_R, alkane   ethane 2.7 min, propane 4.56 min
  butane 6.74 min, pentane 8.85 min

The measurement of pentane in breath, as shown in *Figure 1*, was proposed as a good index of lipid peroxidation, thus providing a sensitive test for assessing vitamin E status (2).

2.1.3 Acetone and other small molecules

Some examples of volatile substances that may accumulate in conditions such as diabetes and renal patients are shown in *Table 2* (3). In standard headspace analysis on packed column, retention indices were introduced to the screening of over 60 volatile organic compounds in blood (4). A blood sample (0.5 ml), mixed with 2 ml distilled water, is incubated in a sealed vial for 20 min at 55°C. An aliquot of the vapour phase is chromatographed (column: 2.1 m × 2.6 mm, glass tube with packing 100% OV-17 on Chromosorb W HP, 80–100 mesh or Porapak P), connected to a FID detector. The limit of detection of

Table 1. Alkane concentrations in normal human breath (1)

Alkane	nmol/l of air (mean ± SD)		
	Alveolar breath	Total breath	Room air
Ethane	0.88 ± 0.04	0.85 ± 0.04	0.80 ± 0.39
Propane	0.81 ± 0.20	0.87 ± 0.08	0.44 ± 0.57
Butane	0.64 ± 0.09	0.54 ± 0.12	1.45 ± 0.88
Pentane	3.7 ± 1.2	2.4 ± 0.71	0.03 ± 0.06

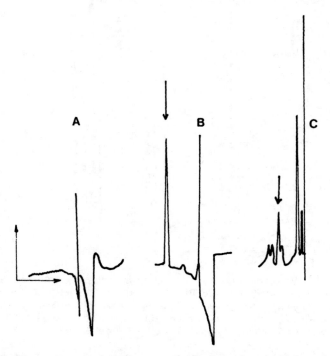

Figure 1. Breath pentane analysis by gas chromatography. A, hydrocarbon-free air; B, pure pentane; C, breath samples from patients. Pentane peaks are marked by arrows. Chromatography was carried out on a stainless steel coil column (1/8 in i.d. × 8 ft) packed with Porasil-D, held isothermally at 60°C and connected to a FID detector. (Reproduced with kind permission from ref. 2.)

Table 2. Some volatile metabolites which are raised in body fluids in certain diseases (3)

Condition	Body fluid	Metabolites
Uraemia	Plasma	Methyl mercaptan, acetone, 2-butanone chloroform, benzene, toluene, dipropylketone, 4-heptanone, cyclohexanone
Diabetes	Urine	1-ethanol, 1-propanol, 2-propanol, 2-methyl-1-propanol, 1-butanol, 2-methyl-1-butanol, 3-methyl-1-butanol, 1-pentanol, 1-octanol, 2-pentanone, acetone, 2-heptanone, 3-hepten-2-one, cyclohexanone

the volatiles tested ranges from 1×10^{-3} to 10×10^{-3} µg when direct injection is used, whereas the corresponding values for the headspace analysis varied between 5×10^{-2} to 20 µg and higher for polar compounds. A similar procedure (*Figure 2*), where diluted plasma was exposed to 3-hydroxybutyrate dehydrogenase, acetoacetate decarboxylase, and lactate dehydrogenase

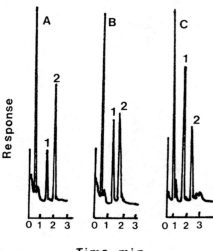

Figure 2. GC analysis of plasma ketone bodies of a diabetic patient. Chromatographic conditions were as described on p. 218. Peaks: 1, acetone; 2, methyl ethyl ketone (internal standard). Deproteinized plasma was (A) untreated, (B) treated with acetoacetate decarboxylase, and (C) treated with acetoacetate decarboxylase and 3-hydroxybutyrate dehydrogenase. (Reproduced with kind permission from ref. 5. 1985 © Amer. Assoc. Clin. Chem.)

prior to chromatography, improves the measurement of hydroxybutyrate, acetoacetate, and acetone (5) and this procedure is described below.

(a) Dilute plasma diluted threefold with phosphate buffer, pH 8, containing methylethyl ketone (3 mg/l) as internal standard. Incubate with or without enzyme reagents, in a 10 ml vial with rubber and aluminium seal at 50°C and 1 ml headspace gas injected into the gas chromatograph for acetone assay.

(b) Analyse using a chromatograph fitted with a FID using the following conditions:
- glass column   packed 2 m × 3 mm 10% PEG 600 on chromosorb WAW, 80/100 mesh
- injector temperature   140°C
- detector temperature   140°C
- column temperature   77°C
- carrier gas N_2
- flow rate   30 ml/min

The lowest concentration of 3-hydroxybutyrate detectable in plasma is 2 μmol/l and the CV of acetone, acetoacetate, and 3-hydroxybutyrate assay over the concentrations 75–334 μmol/l is under 4%.

Headspace gas chromatography on a packed PEG column is also suitable for the determination of blood acetate levels. Methylation of acetate, e.g. by BF$_3$-methanol reagent makes the method sufficiently reproducible and reliable (6).

For acetone, ethanol, and related endogenous compounds, methods are available for both breath and body fluid (7). As described previously, concentration of the relevant volatile organic compounds in breath beforehand, by cryogenic trapping, chemical interactions, or adsorptive binding methods, increases the sensitivity of the assay (*Protocol 1, Figure 3*). Such a procedure for breath ethanol involved on-column concentration of sample, followed by stepwise heating of column from 35°C to 190°C for chromatography (8). Breath ethanol levels between 2.23 and 6.51 nmol/l were found in normal subjects, and additional peaks of acetaldehyde, acetone, and methanol were also observed regularly.

Protocol 1. Assay of endogenous acetone in breath [a] (7)

1. Collect breath sample in Myler bags preloaded with internal standard, 13.2 nmol isopropyl alcohol. Heat the bag to 65°C and pump the sample to the chromatography column, maintained at 35°C. Heat the column to 120°C and assay the eluted analytes.

2. Analyse using a chromatograph fitted with FID and the following conditions:

   - column   Porapak Q, 80–100 mesh
   - detector temperature   225°C
   - ramped   35°C to 120°C (2 min),
   - isothermal   120°C (25 min),
   - ramped   190°C (2 min)
   - carrier gas   N$_2$
   - flow rate   70 ml/min

3. Plot the area under curve (AUC) ratio, AUC$_{acetone}$/AUC$_{int \cdot std}$ against acetone concentration.

[a] A quality control specimen (20 nmol/l) gives 25 ± 2.3 (mean ± SD, $n = 6$) and CV 9.2%. The acetone concentrations in the normal human breath range between 10.0 and 48.4 nmol/l with mean ± SD of 23.2 ± 12 nmol/l.

2.2 Organic acids

Short-chain fatty acids, excreted in faeces, are of interest in disorders of the large bowel because they are considered to be markers of fermentation in proximal part of the colon. GC is most suited for such measurements (9).

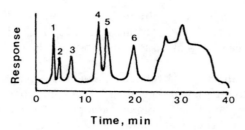

Figure 3. GC-FID analysis of breath from a normal subject. Peaks: 1, methanol; 2, acetaldehyde; 3, ethanol; 4, acetone; 5, isopropyl alcohol (internal standard); 6, isoprene. Chromatography was carried out on Porapak Q, 80–100 mesh column at 120°C and detection was by FID. (Reproduced with kind permission from ref. 7. 1987 © Elsevier Science Publishers.)

Body fluid content and composition of amino acids and fatty acids, saturated and unsaturated, on the other hand, are carried out in a wide range of investigations in health and disease. With the exception of the diagnostic uses in certain rare familial conditions, body fluid amino acid analysis is often a part of research on the pathophysiology of diseases concerning, e.g. nutrition, neurotransmitters, muscles, and endocrine systems. For these determinations, GC is one of the available options. As an analytical tool it is, however, more valuable for fatty acid analysis in body fluid components. Such acids, which would be of interest in lipid-related disorders, may be present as non-esterified acids or a major portion could be fatty acids derived, by chemical hydrolysis, from triacylglycerols, cholesterol esters, phospholipids, and other complex lipids. Another important application of GC and GC–MS, albeit mostly in pharmacological research, is in the assay of prostanoids, e.g. prostaglandins, thromboxanes, and their metabolites in biological specimens (10).

A special group of acidic metabolites found in serum and urine at concentrations between 10 μg and 100 mg/100 ml are referred to as 'organic acids'. These are derived from amino acids, carbohydrates, fatty acids, biogenic amines, and from ketogenesis. In one study, up to 143 of these organic acids were identified, with respect to their chemical structures, in urine from healthy subjects (*Table 3*; ref. 11). They were classified as dicarboxylic acids, oxocarboxylic acids, hydrocarboxylic acids, aromatic acids, furan carboxylic acids, nitrogen-containing acids, and acid conjugates. It was also observed that the qualitative pattern of excretion of the organic acids was constant and reproducible. Studies of the profiles of the organic acids have helped in the detection and characterization of the biochemical defects underlying organic acidaemias and acidurias found in some hereditary diseases, acquired metabolic disorders, such as diabetes mellitus, and other diseases, e.g. those of the liver and kidneys (11). Except in pathophysiological states where profound change in organic acids occurs, pre-fractionation of sample extract by thin-

Table 3. List of organic acids increased in urine of patients with ketoacidosis due to diabetes mellitus or other metabolic disorders (3)

Adipic acid, suberic acid, succinic acid,
lactic acid, 2-hydroxybutyric acid, 3-hydroxyisobutyric
acid, 2-methyl-3-hydroxybutyric acid, 2-hydroxyisovaleric
acid, 3-hydroxyvaleric acid, 3-hydroxyhexanoic acid,
3-hydroxyoctanoic acid, 5-hydroxyhexanoic acid,
2-hydroxy-2-methyllevulinic acid, 3-hydroxyoctanedioic acid,
3-hydroxyoctenedioic acid, 3-hydroxydecanedioic acid,
3-hydroxydecenedioic acid, 3-hydroxydodecanedioic acid,
3-hydroxydodecenedioic acid, 3-hydroxytetradecenedioic
acid, 3-hydroxytetradecadienedioic acid

layer or column chromatography may be necessary for complete GC analysis, even with a high-efficiency capillary column.

2.2.1 Analysis of organic acids

For the isolation of short-chain fatty acids from biological materials such as faeces, steam distillation has been used; vacuum transfer in conjunction with alkaline freeze-drying and the use of ultrafiltration may be more efficient (9). Body fluids, e.g. serum, need deproteinizing by precipitation or processing by dialysis, ultrafiltration, or precipitation before GC analysis may be undertaken. A biological sample of small size may be processed with a relatively larger volume of organic solvents, e.g. methanol or chloroform-methanol mixture, thereby combining precipitation of proteins with the isolation of acid analytes (see *Protocol 2, Figure 4*). Urine is generally acidified before organic acids are isolated by solvent extraction or by some form of column chromatography. For extraction, solvents are usually chosen from ethylacetate, ethanol, ether, or a mixture of them; it is simple and rapid, but efficiency for some acids, e.g. polyhydroxy acids, can be inadequate. Inorganic acids in the sample are removed by precipitation with barium hydroxide, but some organic acids may also be lost in the process. Some essential adjuncts to isolation and pre-fractionation of acidic substances in body fluids are TLC, ion-exchange resins, pre-packed cartridges (e.g. Extrelut, Sep-Pak) and immuno-affinity chromatography (13). Organic acids are normally made into alkyl or halogenated-alkyl, halogenated-arylalkyl, or trimethylsilyl (TMS) esters before chromatography. Methyl esters are easily formed by using diazomethane and produce predictable mass spectral patterns; its reaction, however, with carbonyl groups, amino groups, and double bonds complicates data interpretation. Hydroxyl groups present are changed to TMS-ethers, amino groups to *N*-acyl derivatives, and carbonyl groups are stabilized by conversion to *O*-methyl or ethyl oximes. The extent to which any combination of the pre-GC

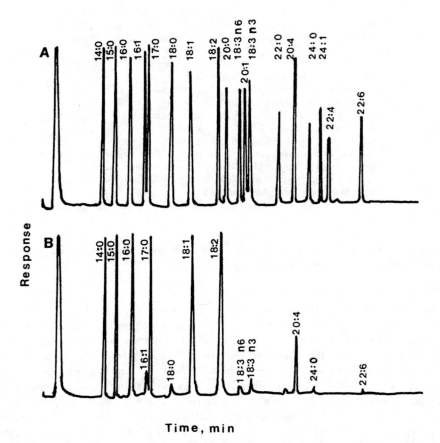

Figure 4. Chromatograms of standard fatty acid mixtures and fatty acyl moieties from cholesteryl ester fraction of male plasma. (A) Standard mixture of methyl esters of fatty acids with individual members indicated by the usual notation. (B) Fatty acyl moieties of the cholesteryl ester fraction from normal male plasma. Chromatographic conditions were as given in *Protocol 2*. (Reproduced with kind permission from ref. 12. 1990 © Elsevier Science Publishers.)

sample cleaning up/separation steps may be required and the choice of chemical manipulation depends on the analysis at hand. As for detection, the halogenated derivatives allow the use of the sensitive electron-capture detector; but a far superior alternative, especially for assaying prostaglandins, thromboxanes, and related compounds in body fluids, is the mass spectrometer in the negative ion–chemical ionization mode (NICI) making use of single ion monitoring (SIM) or, more recently, tandem MS (NICI–MS–MS). An example of such procedures applied to the measurement of two urinary products of thromboxane A_2 is illustrated in *Protocol 3*. The scheme in *Figure 5* gives the basis of these analyses.

Protocol 2. Measurement of plasma fatty acids (12)

1. Extract plasma (0.4 ml) + IS extracted with ice-cold methanol and chloroform.

2. Evaporate extract to dryness and separate lipid groups on thin-layer chromatography.

3. Elute lipid fractions and methylate by BF_3-methanol. Extract methyl esters of fatty acids into hexane and evaporate extract to dryness under N_2 at 30 °C. Take up residue in trimethylpentane for analysis.

4. Analyse on a chromatograph fitted with FID using the following conditions

   - column   50 m × 0.22 mm WCOT CP-Sil 88 d_f 0.2 μm
   - oven temperature   180 °C and 225 °C
   - carrier gas   helium
   - flow rate   30 cm/s

Protocol 3. Assay of two metabolites of thromboxane A_2 in urine[a] (10)

1. Acidify aliquots of urine mixed with the corresponding tetradeuterated standard to pH 3, incubate for 2 h (only for 11-dehydrothromboxane B_2), and pass through reverse-phase C_{18} cartridge.

2. Esterify to PBF esters using PFBBr and pass esters through silica normal-phase cartridge. After methoximation and silylation, remove excess reagents on a silical normal-phase cartridge.

3. Analyse[a] on a tandem mass spectrometer using negative ion chemical ionization and the following conditions:

   - column   30 m × 0.25 mm, DB-1
   - initial temperature 2 min at 100 °C
   - temperature rise   100 to 250 °C at 30 °C/min
                           250 to 300 °C at 5 °C/min
   - final temperature   300 °C for 10 min
   - carrier gas   helium
   - inlet pressure   750 Torr

[a] Recoveries of 2,3-dinorthromboxane B_2 and 11-dehydrothromboxane B_2 at 100–1500 pg/ml were between 93–95%.

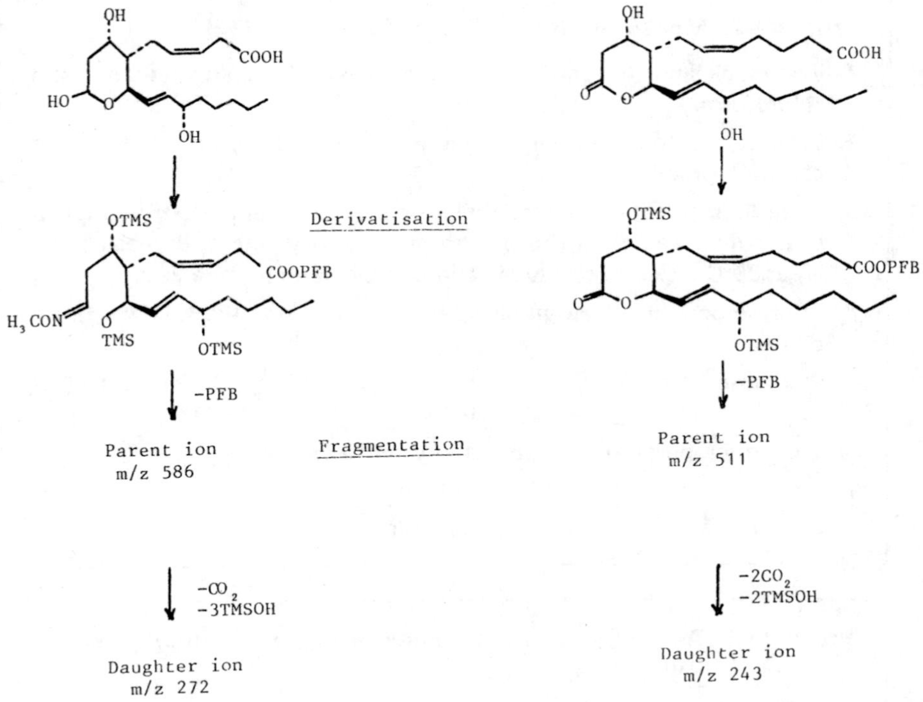

Figure 5. Scheme for derivatization and MS detection of 2,3-dinorthromboxane B_2 and 11-dehydrothromboxane B_2 (10). PFB, pentafluobenzyl ($CH_2C_6F_5$); TMS, trimethylsilyl [$Si(CH_3)_3$]; MO, methoxime (= $NOCH_3$).

2.2.2 Analysis of amino acids

The choice of technique for amino acid analysis is between HPLC and GC; the sensitivity with GC is claimed to be higher, 10–100 pg can be detected with electron-capture of suitable halogenated derivatives and a lower level still with MS. Amino acids have to be converted to suitable derivatives for GC analysis. Before carrying out reactions for preparing derivatives, samples require some form of clean-up, such as deproteinization or more commonly the use of a cation-exchange resin. Both the amino and carboxyl groups may take part in the preparation of derivatives and some examples of the acylation and esterification products are: *N*-fluoroacetyl butyl esters, *N*-heptafluoro-butyryl(HFB)-*n*-propyl esters, HFB-isoamyl esters, HFB-isobutyl esters, and pentafluoropropionyl hexafluoroisopropyl esters. Overall, the choice of the entire procedure—sample isolation and preparation of derivatives—has to be made and validated with respect to the individual amino acids at hand.

For the detection of amino acids, GC–MS with selected ion monitoring provides a marked improvement over FID and ECD (*Table 4*). Although sensitivity of chemical ionization (CI) and electron impact ionization (EI) are similar, CI produces higher mass ions which can improve the selectivity of amino acid analysis in body fluids (14). Use of MS also allows distinction between natural amino acids and ^{15}N-enriched analogues, which can be useful in monitoring the quality of assays and in the study of amino acid metabolism in various diseases. The conditions used for the GC–MS analysis of amino acids listed in *Table 4* are as follows:

- inlet temperature   180°C
- initial temperature   60°C for 3 min
- ramp rate   5°C/min to 180°C
- column   30 m × 0.25 mm, DB-5
- derivatives   pentafluoropropionyl (PFP) and hexafluoroisopropyl (HFIP) esters.

Table 4. Selection of ions for the GC–MS analysis of amino acids (14)

Amino acid	Ionization mode			
	EI		CI	
	Ion (m/z)	Abundance (%)	Ion (m/z)	Abundance (%)
GABA	176	100	232	100
			206	80
Tyrosine	253	100	478	100
Glycine	176	100	372	100
Alanine	190	100	238	100
Glutamic acid	202	100	230	100
Glutamine	202	100	230	100
	230	80	426	50
Aspartic acid	384	100	384	100
Ornithine	216	100	–	–
	176	80	–	–
Phenylalanine	298	100	426	100
Lysine	230	100	589	100
	176	80	–	–
Leucine	203	100	428	100
	232	70	232	80
	371	60		
Isoleucine	202	100	232	100
	232	70	428	20
	371	60		
Methionine	202	100	446	100
Proline	216	100	260	100
Tryptophan	245	100	481	100
Valine	203	100	414	100
	218	75		

2.3 Cholesterol and related compounds

Cholesterol assay in body fluids is at the centre of current research and laboratory investigations in cardiovascular diseases. Although GC methods for cholesterol in body fluids have been available for a long time, their use has been limited to research studies and for reference purposes. The enzymic assays of serum cholesterol have now almost completely superceded the spectrophotometric methods used previously. Similarly, for the routine analysis of steroids, immunoassays have largely replaced GC. The exceptions are certain special disorders, e.g. inborn errors of steroid metabolism especially in the newborn infant, polycystic ovarian syndrome, or virilizing tumours where body fluid steroid profile or some form of discriminant function may be of more diagnostic significance than any individual steroid (15, 16). GC, however, can still be a valuable analytical tool for analysing bile acids and their metabolites in bile, intestinal content, faeces, and serum in research on hepatic, hepatobiliary, and intestinal diseases.

2.3.1 Analysis of cholesterol, metabolites, and bile acids

Cholesterol in body fluids can to be extracted with organic solvents as shown in *Protocol 4* (17). Sample preparation generally begins with a hydrolysis step, chemical or enzymic, to convert cholesterol esters to cholesterol. Dilution of body fluids with, say, 0.1 M NaOH releases bile acids from albumin. Bile acids and their conjugates can be extracted by anion-exchange resins, e.g. Amberlite XAD-2. Cartridges packed with materials such as octadecylsilane-bonded silica provide an easy and effective way of retaining bile acids and steroids, which can then be eluted with organic solvent(s) (18). The conjugated metabolites of steroids and bile acids have to be hydrolysed first, by chemical or enzymic means, producing the appropriate alcohol or acid before the sample may be further processed for GC. The acids are converted to methyl, ethyl, or some fluoroalkyl esters. The hydroxyl groups in the steroid nucleus may be changed to acetate or fluoroacetate, or silyl or methyl ether; the carbonyl group can be made into an oxime. The processing of serum and urine samples and the preparation of derivatives for bile acid and 17-oxogenic steroid (metabolites of androgens and corticosteroids) analyses are illustrated in *Protocols 5* and *6*. Flame ionization detection is adequate for monitoring urinary steroid profiles as shown in *Figure 6*. For the measurement of the minor constituents of body fluids e.g. serum 3α-hydroxy-7-oxo-5-β-cholanoic acid, 0.16 ± 0.08 nmol/l, an effective approach would be to use GC in conjunction with a mass spectrometer detector using single ion monitoring (SIM) and a deuterium-labelled analyte as internal standard (*Figure 7*).

Protocol 4. Assay of cholesterol in body fluids[a] (17)

1. Mix urine or saliva (2 ml) with water and internal standard epicoprostanol, extract with chloroform/methanol (2:1, v/v), and evaporate organic extract to dryness.

2. Heat residue in methanolic KOH at 65°C for 1 h.

3. Extract hydrolysate with hexane and evaporate the extract to dryness. React residue with pentafluorobenzoyl chloride, 90°C for 1 h.

4. Remove excess reagent, dissolve residue in ethylacetate, and analyse using a GC-ECD with following conditions:
   - column   5 m × 0.53 mm, HP-1 methylsilicone megabore
   - isothermal temperature   253°C
   - carrier gas   methane/argon (5:95, v/v)
   - flow rate   6 ml/min

[a] Cholesterol in normal urine 2.29 µmol/l ± SD 0.9, saliva 3.8 µmol/l ± SD, 2.3, CV 4.2%, within-batch and 8.2% between batch; limit of detection, 100 pg cholesterol injected.

Protocol 5. Determination of serum bile acids (18)

1. Dilute serum (0.2–0.5 ml) + IS (multi-deuterium-labelled bile acids) with 0.1 M NaOH and incubate at 65°C for 15 min.

2. Pass sample through Bond-Elut C-18 cartridge, wash with water, and elute bile acids with 90% ethanol. Remove solvent under reduced pressure.

3. Hydrolyse extract with cholylglycine hydrolase at 37°C for 18 h. Clean-up hydrolysate on C_{18} cartridge as before and solvolyse residue at 37°C for 18 h.

4. Extract bile acids from the acidified mixture into ethylacetate and convert to ethyl ester-dimethylethylsilyl ether (DMES) and where appropriate to ethyl ester-methyloxime-DMES.
   Remove excess reagent using Sephadex LH-20 with $CHCl_3$/*n*-hexane/methanol (20:20:1, v/v/v) as eluent.

5. Analyse on GC with FID and MS using the following conditions:
   - column   20 m × 0.2 mm, HiCap-CBP1, FSOT
   - isothermal temperature   275°C
   - carrier gas   He
   - linear velocity   40 cm/s

Jagadish Chakraborty

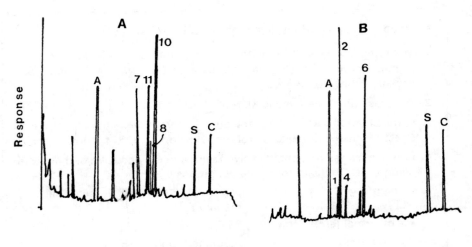

Figure 6. GC analysis of the 17-oxogenic steroids from urine of a patient with (a) congenital adrenal hyperplasia due to 17-hydroxylase defect and (b) a patient with 5-reductase deficiency. Chromatographic conditions were as given in *Protocol 6*. Peaks: 1, 5α-androstane-3α,17β diol; 2, 5β-androstane-3α,17β diol; 4, aetiocholanolone; 6, 11-hydroxyaetiocholanolone; 7, pregnanediol; 8, 17β-formyl-5β-androstane-3α,11β diol; 10, pregnane-3α,11β,20α-triol; 11, pregnane-3α,11β,20β-triol. Internal standards: A, 5α-androstane-3α,17α diol; S, stigmasterol; C, cholesteryl butyrate. (Reproduced with kind permission from ref. 16. 1990 © Amer. Assoc. Clin. Chem.)

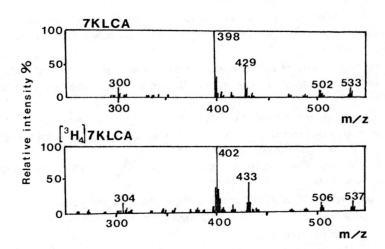

Figure 7. Mass spectra of the ET-MO-DMES ether derivatives of 3α-hydroxy-7-oxo-5β-cholanoic acid (7KLCA) and [³H₄]7KLCA. (Reproduced with kind permission from ref. 18. 1990 © Elsevier Science Publishers.)

228

Protocol 6. Assay of urinary 17-oxogenic steroids (16)

1. Adjust urine sample (5 ml) to pH 7 with 0.1 M NaOH.

2. Reduce with 10% w/v $NaBH_4$ in 0.1 M NaOH 45 min at 60–65°C. Remove excess $NaBH_4$ by heating with acetic acid.

3. Oxidize with aqueous sodium metaperiodate (10%, v/v), 10 min at 60–65°C.

4. Hydrolyse formates with 5 M NaOH and extract free steroids into 15 ml $CHCl_3$. Dry extract at 40°C. Reconstitute residue in 1 ml ethanol, mix with internal standards, and evaporate mixture to dryness.

5. Prepare ethoxime (60°C, 1 h) and trimethylsilyl ether (room temperature, 4 h) derivatives. Dry sample under N_2 and dissolve residue in 250 μl cyclohexane for analysis by GC.

6. Analyse by GC–FID and MS using the following conditions:
   - column — 20 m × 0.1 mm, OV-1, OTC
   - detector temperature — 250°C
   - injector temperature — 180°C
   - ramp rate — 180°C to 270°C at 2.5°C/min

2.4 Amines and related compounds

Important examples of compounds in this group are catecholamines, polyamines, 5-hydroxytryptamine, and their metabolites. Other than their limited use in diagnosis, e.g. catecholamines/metabolites in pheochromocytoma and 5-hydroxyindoleacetic acid in carcinoid tumours, the study of biogenic amines are mostly connected with research that includes psychiatric disorders, regulation of sympathetic activity, blood circulation, hormonal control, mental abnormalities, schizophrenia, anorexia nervosa, and diabetes mellitus. Altered patterns of erythrocyte polyamines were reported in chronic renal failure, sickle cell anaemia, cystic fibrosis, and liver diseases, but they appear more promising for estimating the extent of chemo- and radiotherapeutically-induced tumour cell death in cancer patients (19).

2.4.1 Analysis of catecholamines

Catecholamines in body fluids can first be extracted with diethyl ether or ethyl acetate and then back-extracted into an acidic aqueous phase. Solid phase matrices used for extraction include activated alumina, cation-exchange resins and boric acid-bound affinity gel. For acidic metabolites, samples are acidified before extraction with organic solvent (19). For polyamine, the sample pretreatment steps include extraction with alkaline butanol, cation-

229

exchange chromatography, and adsorption onto silica gel. Amines and their metabolites have to be converted to suitable derivatives for analysis by GC.

2.4.2 Analysis of catechols and indoles

Although in recent years HPLC methods have gained wide use for the analysis of endogenous amines and indole compounds, GC, with ECD and particularly MS, continues to be used as an accurate and precise alternative. The derivatives chosen for neutral catechols tend to be haloacyl (e.g. trifluoroacetyl, pentafluoropropionyl) and/or silyl ethers (e.g. trimethylsilyl, *tert*-butyl-dimethylsilyl or dimethyl-*n*-propylsilyl). In the absence of a MS detector, fluoro derivatives with ECD give adequate sensitivity for plasma catecholamine and 5-hydroxytryptamine assay, whereas FID may be sufficient for many urinary metabolites such as HVA, VMA, and 5-hydroxyindoleacetic acid. *Protocol 7* describes the analysis of the acidic and alcoholic products of catecholamines in CSF, plasma, and urine. These determinations can be conducted with good precision using acetylation or silylation of phenolic hydroxy groups, esterification of carboxyl group (pentafluorobenzyl ester), and acylation of aliphatic hydroxyl groups (20). Of the MS detection modes, NICI gives the highest sensitivity. *Protocol 8* provides another example of current practice in GC, i.e. the use of mass spectrometric detection in conjunction with deuterated analogues of the analytes, in this instance for the assay of 6-hydroxytryptamine (21).

Protocol 7. Chromatography of acidic metabolites of catecholamines (20)

1. Use samples of:
   - plasma   1 ml
   - CSF   0.5 ml
   - urine   0.1 ml

   as appropriate.

2. Extract with ethyl acetate from acid pH and back extract into buffer at pH 7.7.

3. Acetylate in aqueous medium, extract into ethyl acetate from acid pH and evaporate organic extract to dryness.

4. Esterify to yield the pentafluorobenzyl derivative, extract into ethyl acetate from acid pH, and evaporate the organic solvent.

5. Acetylate in anhydrous medium, extract derivatives into ethyl acetate from acid pH, and evaporate the organic solvent.

---

6. Analyse[a] using electron capture and mass spectrometer as detectors with the following conditions:

- column   20 m × 0.22 mm, CPSIL-19, FSOT
- initial temperature   260°C for 2 min
- initial rate   5°C/min
- final temperature   300°C
- carrier gas   He
- flow rate   0.55 ml/min

[a] Recoveries for homovanillic acid, vanillylmandelic acid and dihydroxyphenyacetic acid are between 86 and 94%.

Protocol 8. Assay of 5-hydroxytryptamine (HT)[a] in plasma (21)

1. Mix plasma (200 µl) with 100 ng deuterated HT, mix deproteinized supernatant with borate buffer pH 10.0 (200 µl) and n-butanol/diethyl ether (5 ml, 1:4, v/v).

2. Evaporate organic extract to dryness and react residue with pentafluoropropionic anhydride (100 µl, 140°C for 4 h). Evaporate mixture to dryness and reconstitute in ethyl acetate (20 µl).

3. Analyse on GC–MS using SIM for ions at m/z 451 and 454 and the following conditions:

- column   2 m × 3 mm, 1.5% OV-1, packed
- injector temperature   220°C
- column temperature   200°C
- carrier gas N_2
- flow rate   50 ml/min

[a] HT was detectable at picogram levels; plasma concentrations found in normal subjects were 295 ± 0.92 ng/ml.

2.4.3 Analysis of polyamines

Simultaneous measurement of polyamines such as 1,3-diaminopropane, putrescine, cadaverine, spermidine, and spermine in body fluids or erythrocytes is possible by GC using capillary columns and FID, ECD, NPD, or MS detector; the urinary metabolites have to be first hydrolysed in acid (see *Figure 8*). Most procedures involve adsorption of analytes on a solid phase, e.g. silica gel followed by the preparation of halo-acyl derivatives, e.g. heptafluorobutyryl amine.

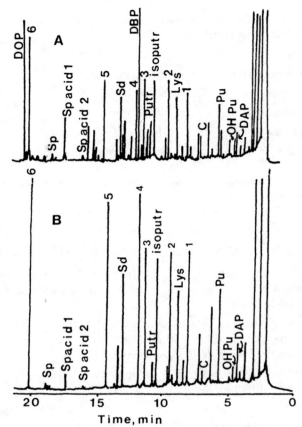

Figure 8. GC of methyl-heptafluorobutyryl derivatives of polyamines and metabolites from acid-hydrolysed normal urine. Detectors: A, flame ionization: B, nitrogen-phosphorous. Abbreviations: DAP, 1,3-diaminopropane; OHPu, 2-hydroxyputrescine; Pu, putrescine; C, cadaverine; 1, 1,6-diaminohexane; lys, lysine; 2, 1,7-diaminoheptane; isoputr, isoputreanine; Putr, putreanine; 3, N-methylisoputreanine; DBP, dibutylphthalate; 4, bis(3-aminopropyl)amine; Sd, spermidine; 5, N-(3-aminopropyl)-1,5-diaminopentane; Sp, acid, 2, N,N-bis(2-carboxyethyl)-1,4-diaminobutane; Sp, spermine; 6, N,N-bis(3-aminopropyl)-1,5-diaminopentane; DOP, dioctylphthalate. Compounds 1–6 are added internal standards. Column: 35 m × 0.2 mm i.d. fused silica capillary coated with cross-linked methyl silicone gum, 0.11 μm film thickness. (Reproduced with kind permission from ref. 19. 1984 © Amer. Assoc. Clin. Chem.)

2.5 Polyols and sugars

The reduction products of sugars, e.g. inositol, glycerol, and sorbitol, are known to play important roles in processes associated with human physiology and pathology. GC methods have been developed to analyse polyols in body fluids and tissues and applied to the study of brain carbohydrate metabolism in health and disease.

2.5.1 Analysis of polyols, plasma, and CSF

Plasma and CSF, with added internal standard, e.g. rhamnose, may be
processed by shaking with organic solvent such as methanol. After drying the
liquid phase *in vacuo*, the common practice is to convert polyols to silyl ethers
prior to GC (22). Recoveries of polyols from CSF are between 84 and 92%,
whereas plasma values tend to be lower, 45 to 75%. Use of capillary columns,
e.g. 25 m × 0.3 mm FSOT with 0.5 μm cross-linked methyl silicone stationary
phase produces the required resolution and both FID and MS have been used
for detection (see *Figure 9*). The CSF concentrations of aldoses, ketoses and
polyols reported in a group of healthy subjects are quoted in *Table 5*.

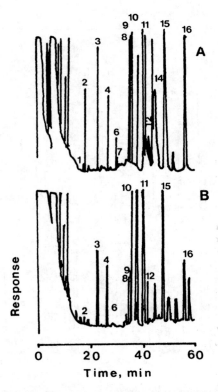

Figure 9. Chromatograms of aldo and keto sugars and polyols from cerebrospinal fluid
and plasma. (A) Cerebrospinal fluid; (B) plasma. Peaks: 1, threitol; 2, erythreitol; 3, 4,
internal standard; 6, arabitol; 7, ribitol; 8, 12, mannose; 9, fructose; 10, anhydroglucitol;
11, 15, glucose; 14, glucitol; 16, myoinositol. Column: 25 cm × 0.31 mm i.d., coated with
0.52 μm film of cross-linked methyl silicone. Temperature programming: 50 to 140°C at
10°C/min, then 140°C to 190°C at 1°C/min and finally to 260°C at 30°C/min. (Reproduced
with kind permission from ref. 22. 1989 © Elsevier Science Publishers.)

Table 5. GC assay of CSF aldoses, ketoses, and polyols in healthy subjects (22)

Carbohydrate	Concentration[a] mg/l, (mean ± SD)	Within-day precision[b] mean mg/l	CV (%)
Erythritol	2.4 ± 0.9	4.1	9
Arabitol	4.8 ± 0.9	4.9	8
Ribitol	1.61 ± 0.1	1.6	6
Mannose	10.1 ± 2.3	12.6	11
Fructose	25.5 ± 11.1	23.3	9
Anhydroglucitol	19.4 ± 5.3	20.2	8
Glucose	613 ± 8.83	543	3
Glucitol	7.7 ± 1.5	6.0	5
Myoinositol	33.0 ± 4.6	45.2	5

[a] Ten males and four females.
[b] Pooled CSF from five healthy subjects.

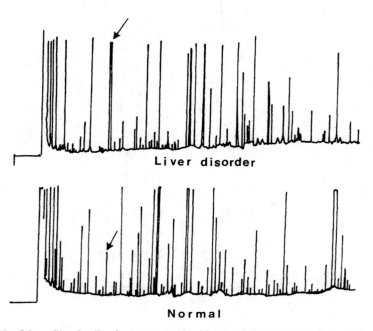

Figure 10. GC profile of saliva from a normal subject and a patient with liver disorder. The diagnostic peak has a retention time of 12.02 min, in the liver disease profile (arrow). Silyl derivatives were chromatographed on column, 50 m × 0.22 mm i.d., fused silica WCOT coated with CP sil 8 CB (8% phenylmethylphenyl silicone), 0.13 μm thick. Initial temperature 150°C, initial time 0 min., progress rate 4°C/min, final temperature 300°C Carrier gas helium 2 ml/min. Detector: Finnigan MAT GC/MS 1020/OWA. (Reproduced with kind permission from ref. 23. 1986 © Elsevier Science Publishers.)

3. Conclusions

With the availability of autosampling devices, good choice of capillary columns, sensitive and robust nitrogen-phosphorous detector, the operation of mass spectrometer in various modes, and versatile data processing systems, GC is bound to remain at the core of biomedical research for the future. It would be inappropriate to judge the role of GC in the clinical biochemistry laboratory in terms of the 'turnaround time' used for routine assays, because the potential capabilities of GC as a technique is far beyond that of even the most advanced of the automated analytical systems currently used in clinical biochemistry laboratories. GC complements the modern clinical biochemistry laboratory service by providing the most effective and, in many instances, the only available means of undertaking complex, multi-component analyses needed for pathobiological research and laboratory investigations of human disease with regard to its diagnosis, treatment, and prevention.

An example of the unique areas of application of GC that represent potentially major advances in diagnostic medicine, is the analysis of volatiles in breath and body fluids. As shown in *Figure 10* (23) for saliva, the composition of such materials is highly complex and only partly characterized. With the use of modern GC technology, in conjunction with a data library for identification of the constituents, it may become possible to identify what may be called 'metabolic profiles' or 'chromatographic fingerprints' for specific clinical conditions. The findings from such multi-component analyses, whether for environmental pollutants/toxins/drugs or endogenous products of body metabolism, can provide a wider window for viewing the biochemical processes that may be associated with the aetiology or are the consequences of pathological changes. Progress in this direction will not only enhance the quality of the clinical biochemistry laboratory service but also contribute to our understanding of the intricate biochemical mechanisms underlying human diseases.

References

1. Zarling, E. J. and Clapper, M. (1987). *Clin. Chem.*, **33**, 140.
2. Lemoyne, M., Gossum, A. Van, Kurian, R., Ostro, M., Axler, J., and Jeejeebhoy, K. N. (1987). *Am. J. Clin. Nutr.*, **46**, 267.
3. Niwa, T. (1986). *J. Chromatogr.*, **379**, 313.
4. Uehori, R., Nagata, T., Kimura, K., Kudo, K., and Noda, M. (1987). *J. Chromatogr.*, **411**, 251.
5. Kimura, M., Kabayashi, K., Matsuoka, A., Hyashi, K., and Kimura, Y. (1985). *Clin. Chem.*, **31**, 596.
6. Akane, A., Shoju, F., Kazuo, M., Setsunori, T., and Hiroshi, S. (1990). *J. Chromatogr.*, **529**, 155.
7. Phillips, M. and Greenberg, J. (1987). *J. Chromatogr.*, **422**, 235.

8. Phillips, M. and Greenberg, J. (1987). *Anal. Biochem.*, **163**, 165.
9. Scheppach, M., Fabian, C. E., and Kasper, H. W. (1987). *Am. J. Clin. Nutr.*, **46**, 641.
10. Uedelhoven, W. M., Messe, C. O., and Weber, P. (1989). *J. Chromatogr.*, **497**, 1.
11. Liebich, H. M. (1990). *J. Chromatogr.*, **525**, 1.
12. Roemen, T. H. M. and Keizer, H., and Van der Vusse, G. J. (1990). *J. Chromatogr.*, **528**, 447.
13. Mackert, G., Reinka, M., Schweer, H., and Seyberth, H. W. (1989). *J. Chromatogr.*, **494**, 13.
14. Singh, A. K. and Ashraf, M. (1988). *J. Chromatogr.*, **425**, 245.
15. Weykemp, C. W., Penders, T. J., Schmidt, N. A., Borburgh, A. J., Caiseyde, J. F. V., and Wolthers, B. J. (1989). *Clin. Chem.*, **35**, 2281.
16. Honour, J. W., Tsang, W. M., and Patel, H. (1990). *Ann. Clin. Biochem.*, **27**, 338.
17. Schwertner, H., Johnson, E. R., and Lane, T. E. (1990). *Clin. Chem.*, **36**, 519.
18. Eguchi, T., Miyazaki, H., and Nakayama, F. (1990). *J. Chromatogr.*, **525**, 25.
19. Muskiet, F. A. J., Berg, G. A. V., Kingma, A. W., and Halie, R. (1984). *Clin. Chem.*, **30**, 687.
20. De Jong, A. P. M. and Rok, R. M. (1986). *J. Chromatogr.*, **382**, 19.
21. Baba, S., Uton, M., and Horie, M. (1984). *J. Chromatogr.*, **307**, 1.
22. Kusmierz, J., DeGeorge, J. J., Sweeney, D., May, C., and Rapoport, S. I. (1989). *J. Chromatogr.*, **497**, 39.
23. Lochner, A., Weisner, S., Zlatkis, A., and Middleditch, N. S. (1986). *J. Chromatogr.*, **378**, 267.

8

Chiral separations by gas chromatography

DAVID R. TAYLOR

1. Introduction

The separation of one enantiomer (optical isomer) from another is difficult to achieve by any chromatographic procedure. This is because any two enantiomers are structurally related to one another in the same way as an object and its mirror image, and therefore have identical physical and chemical properties as long as they remain in an achiral (symmetrical) environment. Only when observed in plane-polarized light, as in a polarimeter, or when interacting with a chiral (asymmetrical) reagent or chiral surface, do two enantiomers reveal differences which can be used in analysis or to bring about their separation.

This inherent difficulty of separation, which greatly complicates the analysis of mixtures of enantiomers, would not be particularly important were it not for the asymmetry of living tissue, which follows from Nature's use of mainly single enantiomers of amino acids and carbohydrates. Its significance was revealed with particular brutality in the response of pregnant women to the two enantiomers of thalidomide, the active ingredient of which was the 50:50 (racemic) mixture of both enantiomers of *Structure 1*. Because of their quite different metabolic pathways in the asymmetric human biosystem, the (S)-enantiomer (*Structure Ia*) displayed previously unsuspected teratogenic effects, the desired therapeutic effect residing exclusively in the (R)-isomer (*Ib*).

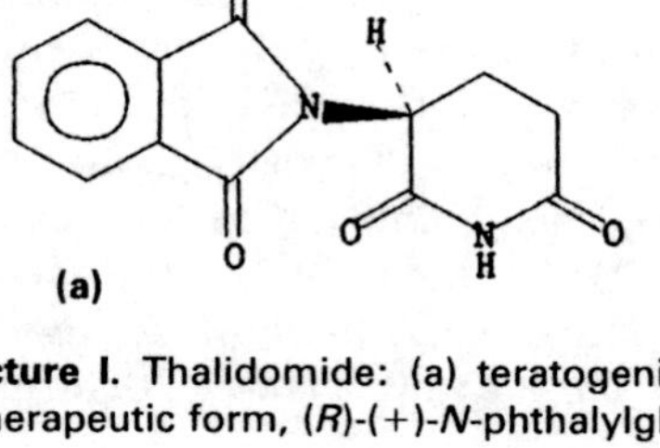

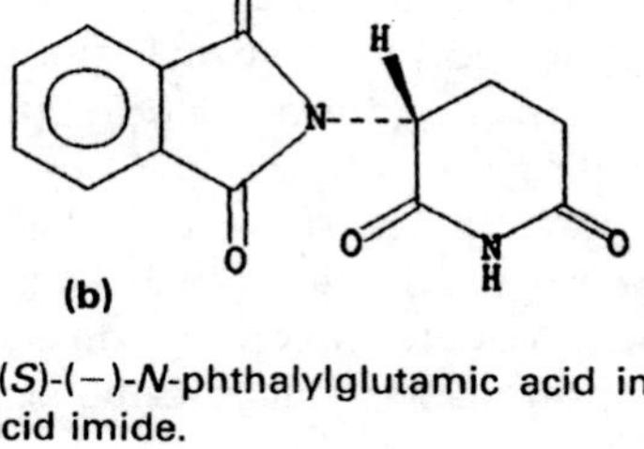

Structure I. Thalidomide: (a) teratogenic form, (S)-(−)-N-phthalylglutamic acid imide; (b) therapeutic form, (R)-(+)-N-phthalylglutamic acid imide.

Once the significance of this phenomenon was recognized, stricter control of the purity of enantiomers used in medicine, food, and agriculture has been expected. As a result, analysts now require reliable methods to determine the concentration of both enantiomers of chiral products, chiral intermediates in their synthesis, and their degradation products and metabolites. Chiral gas chromatography is just one of the methods available, and offers a viable alternative to chiral liquid chromatography and chiral spectroscopic techniques. It offers the typical advantages of any GC method, namely high peak capacity, simplicity, accuracy, and speed of analysis. Recent developments have also enabled chiral GC to gain the edge over chiral HPLC in achieving the enantioselective resolution of compounds lacking significant functionality.

Before describing the procedures for chiral gas chromatography in more detail, some useful terms and definitions will be dealt with. Reference to general reviews of chiral GC is also appropriate (1–3).

1.1 Terminology and definitions

Historically the study of optical isomers began with the work of Pasteur in the late nineteenth century. It is scarcely surprising that the terminology appropriate to those early days of science has had to be refined and elaborated. Particularly important is the unambiguous assignment to a particular optical isomer of its unique absolute configuration, and it is this aspect which normally causes the most confusion.

By polarimetry, a single enantiomer can be designated (+) or (−) according to whether it rotates plane-polarized light clockwise or anticlockwise. Mixtures in which the (+)-isomer predominates will display clockwise rotation, but a racemic mixture of exactly equal parts of the (+)-and (−)-isomer will fail to display this property. The absolute configuration of an enantiomer can be determined by various means, such as X-ray crystallography, but not yet by polarimetry, because we still have no means of predicting the sign or magnitude of rotation of polarized light from a knowledge of the structure. Once determined, configuration is described by assigning to each asymmetric centre in the molecule the label (*R*) or (*S*), determined by application of a set of rules originated by Cahn, Ingold and Prelog. These rules replace the older terminologies such as (D) and (L) which nevertheless persist in certain areas of chemistry, particularly those concerned with amino acids and sugars.

An atom bears the label (*R*) if it has attached to it four substituents such that, viewed along an axis pointing from the atom to the substituent with lowest priority, the remaining substituents have priority decreasing in a clockwise direction (*Figure 1*). Conversely, the atom will be designated (*S*) if, viewed along the same axis, the remaining substituents have priorities which decrease in an anticlockwise sense (*Figure 2*). For the purposes of these rules, priority is determined by considering the Relative Atomic Mass (RAM) of atoms directly attached to the atom whose chirality label is to be determined:

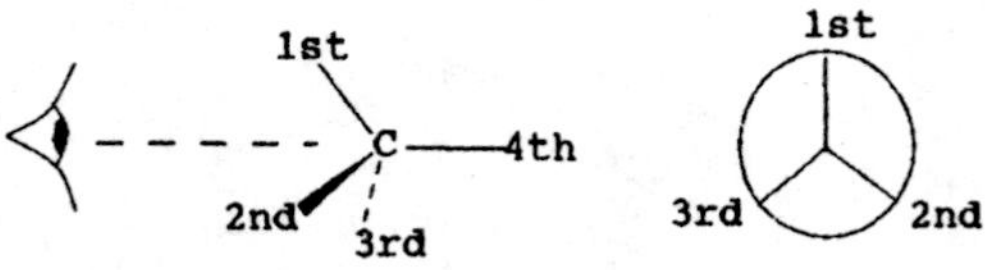

Figure 1. (*R*)-Enantiomer (1st-highest priority).

higher atomic mass implies higher priority. When two substituents have the same atom joined to the chiral centre, preference is determined by considering the atomic masses of atoms one bond further away, and so on until a priority is indicated. Atoms joined *via* double bonds are counted twice, but with the same mass. For further details of the application of these rules, reference should be made to a textbook of stereochemistry (4).

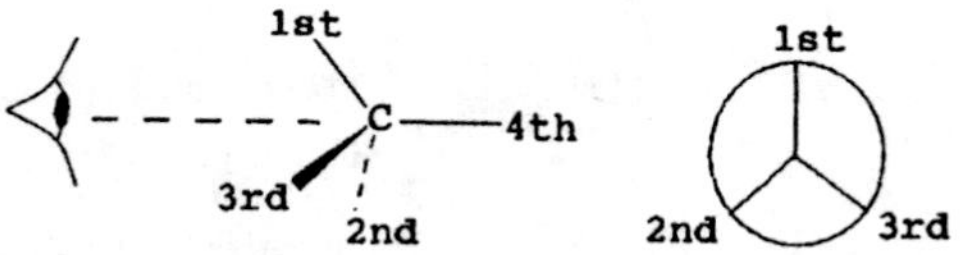

Figure 2. (*S*)-Enantiomer (1st-highest priority).

When a molecule contains two asymmetric atoms, as in tartaric acid, the number of possible isomers rises to four. However, in such a case one-half of the molecule may constitute a reflection of the remaining half, when the two enantiomers become equivalent leading to fewer possible diastereoisomers. Thus, in tartaric acid the (*R,R*)-isomer is not identical to the (*S*)-isomer (*Structure II*); these are two optically-active enantiomers. However, the (*R,S*)-enantiomer is identical to its mirror image, the (*S,R*)-isomer, and so there are only three diastereoisomers, not four. This would not be the case for monoethyl tartrate, in which there are four enantiomers, (2*R*, 3*S*; 2*R*, 3*R*; 2*S*, 3*R*; and 2*S*, 3*S*) (*Figure 3*). It should be noted that in any such set of isomers, the (2*R*, 3*R*)-isomer is not enantiomerically related to the (2*R*, 3*S*)-isomer; these two substances are diastereoisomers, and will normally have detectably different physical, spectroscopic, and chemical properties. Such diastereoisomers can therefore be separated by conventional procedures, at least in theory.

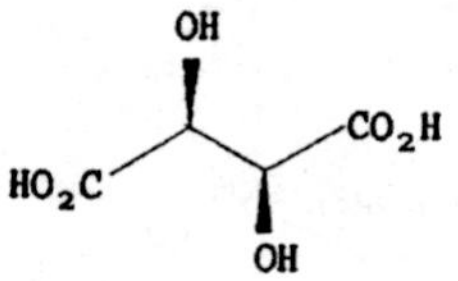

Structure II.

$PhOCHMeCO_2H$

Structure III.

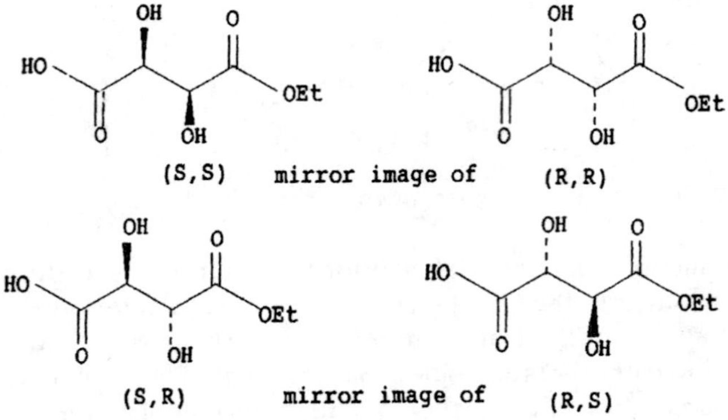

Figure 3. Enantiomers of monoethyl tartrate.

2. Role of derivatization in chiral separations by GC

Because diastereoisomers which are not enantiomers have distinctive properties, they may be separated by conventional techniques such as chromatography, distillation, or recrystallization. Hence, one of the standard procedures for determining the relative abundance of two enantiomers, let us say the (R)- and (S)-isomers of 2-phenoxypropionic acid (*Structure III*), is to convert them into separable diastereoisomers by reaction with an optically pure reagent. For example, if these two acids were treated with one molecular proportion of (R)-1-phenylethanol (*Structure IV*), the (R,R)- and (S,R)-esters (*Structures V* and *VI*) would be formed. These are potentially separable for analytical purposes by GC or HPLC using conventional achiral stationary phases.

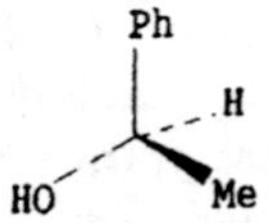

Structure IV.

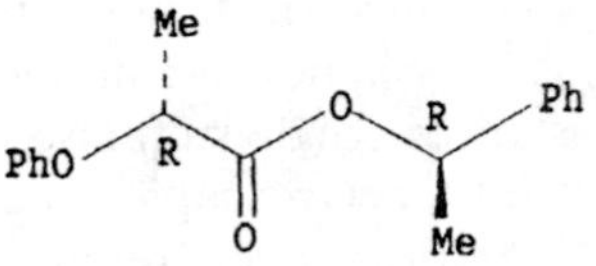

Structure V.

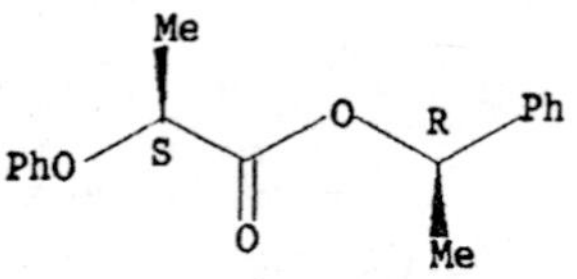

Structure VI.

Me$_2$CHCHClCOOH

Structure VII.

240

Potential disadvantages of chiral derivatization followed by GC analysis of the mixture of diastereoisomers are:

(a) cost of a suitable chiral reagent of high purity
(b) failure of the chiral reagent to react quantitatively with both substrate enantiomers
(c) different chromatographic response factors of diastereoisomers
(d) inaccuracies arising from the use of even slightly impure chiral reagent
(e) poor volatility of the diastereoisomers

In spite of these disadvantages, diastereoisomer formation prior to GC analysis has been widely used for many years, and should be considered a perfectly viable alternative to a more direct GC analytical procedure using a chiral stationary phase.

A number of generally useful chiral derivatizing reagents for the synthesis of separable diastereoisomers are listed in *Table 1*. A general principle in their selection is that they must display high reactivity towards available

Table 1. Chiral derivatizing reagents for gas chromatography[a]

Group/reagent	Derivative formed[b]	Type of analyte suitable
Amino group		
2-Chloroisovaleryl chloride	$NHCOCHClCHMe_2$	Amino acids Amines
Drimanoyl chloride[b]	$NHCOR^1$	Amino acids Amines
trans-Chrysanthemoyl chloride	$NHCOR^2$	Amino acids Amines
N-Trifluoroacetylprolyl chloride	$NHCOR^3$	Amino acids Amines
N-Pentafluorobenzoylprolyl chloride	$NHCOR^4$	Amines
N-Trifluoroacetylalanyl chloride	$NHCOCHMeCOCF_3$	Amines
(*S*)-2-Methoxy-2-trifluoromethyl- phenylacetyl chloride	$NHCOR^5$	Amines
(−)-Menthyl chloroformate	$NHCOR^6$	Amino acids Amines
Hydroxyl group		
2-Phenylpropionyl chloride	OCOCHMePh	Hydroxyacids Alcohols
2-Phenylbutyryl chloride	OCOCHEtPh	Alcohols
O-Acetyllactic acid chloride	OCOCHMeOAc	Alcohols
Drimanoyl chloride	$OCOR^1$	Hydroxyacids
trans-Chrysanthemyl chloride	$OCOR^2$	Hydroxyacids Alcohols
N-Trifluoroacetylalanyl chloride	$OCOCHMeNHCOCF_3$	Alcohols
(*S*)-2-Methoxy-2-trifluoromethyl- phenylacetyl chloride	$OCOR^5$	Amphetamines

David R. Taylor

Table 1. (*continued*)

Group/reagent	Derivative formed[b]	Type of analyte suitable
(−)-Menthyl chloroformate	OCOR^6	Hydroxyacids Alcohols
1-Phenylethyl isocyanate	OCONHCHMePh	Hydroxyacids Alcohols
1-(1-Naphthyl)ethyl isocyanate	OCONHCHMeNaph	Alcohols
Chlorodimethyl(menthyloxy)silane	OSiMe$_2$R^6	Alcohols
Chlorodimethyl(bornyloxy)silane	OSiMe$_2$R^7	Alcohols
Carboxylic acid (COOH)		
Butan-2-ol	CO$_2$CHMeEt	Amino acids Hydroxyacids Ketoacid
3-Methylbutan-2-ol	CO$_2$CHMeCHMe$_2$	Alkanoic acids Amino acids Hydroxyacids N-Methylamino acids
(−)-Menthol	COR^6	Alkanoic acids Amino acids Hydroxyacids
2-Amino-4-methylpentane	CONHCHMeBu-i	Amino acids
Ethyl esters of amino acids	CONHCHRCO$_2$Me	Amino acids
Carbonyl		
(+)-2,2,2-Trifluoro-phenylethyl hydrazine	>C=NNHCH(CF$_3$)Ph	Ketones
O-(−)-Menthylhydroxylamine	>C=NR^6	Carbohydrates
2,3-Butanediol	MeCHOCR$_2$OCHMe$_2$	Ketones Lactones
Butan-2-ol	But-2-yl glycoside	Carbohydrates

[a] The book by W. A. König (1) contains much useful information and key references regarding preparation of chiral reagents and derivatisation procedures.
[b] Key to groups R^1–R^6:

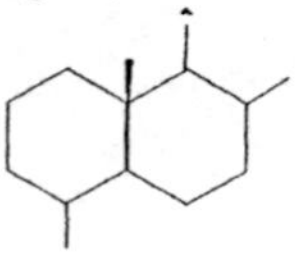

R^1 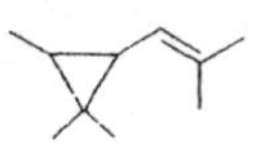   R^2 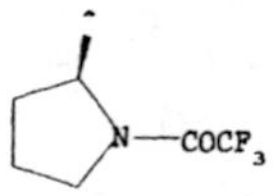   R^3 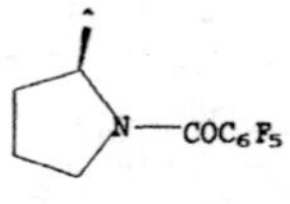   R^4 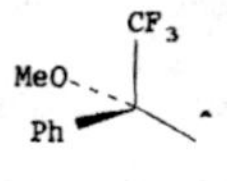   R^5 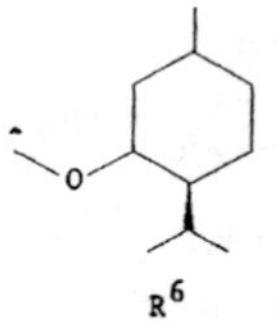   R^6 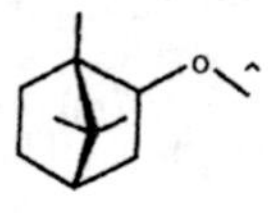   R^7

functionality in the analyte under conditions which avoid any racemization in the analyte or the chiral reagent. Particularly useful in GC is the transformation by this process of functional groups which lead to high boiling points, such as hydroxyl-, carboxyl-, and amino-groups, into functions with reduced intermolecular forces. Consequently, a high proportion of the reagents in *Table 1* are chiral acid chlorides and chloroformates. For GC analysis they should not be of high molecular mass, although *N*-trifluoroacetyl-amino acids have been used. The acid chlorides of 2-chloroisovaleric acid (*Structure VII*) and Mosher's acid (*Structure VIII*), and (−)-menthol (*Structure IX*) and its chloroformate and hydroxylamine derivatives (*Structures X and XI*) have been the most widely used reagents for this form of diastereoisomer analysis by GC after derivatization.

Disappointingly, there are few readily available optically active organosilanes of high purity; reagents such as the recently reported chloro(menthyloxy) dimethylsilane (*Structure XII*) (5), should prove to be extremely useful for the routine formation of volatile diastereoisomeric silyl ethers and esters prior to GC analysis.

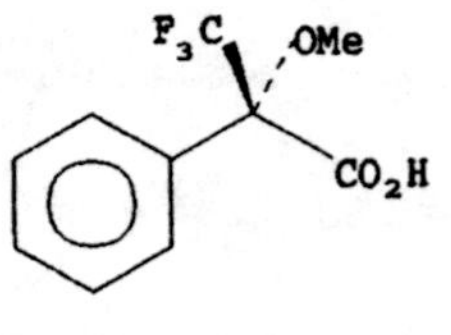

R = H

R = COCl

R = NH$_2$

R = SiMe$_2$Cl

Structure VIII.

Structures IX–XII. (IX) R = H; (X) R = COCl; (XI) R = NH$_2$; (XII) R = SiMe$_2$ Cl.

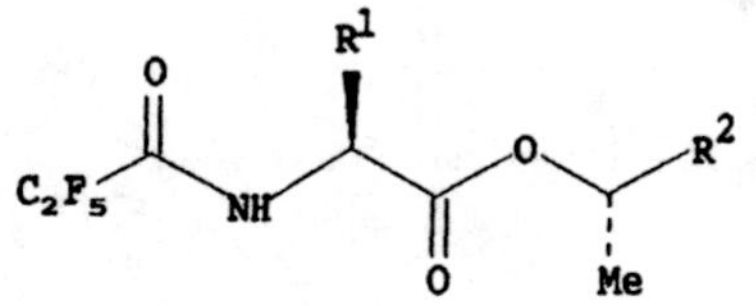

Structure XIII.

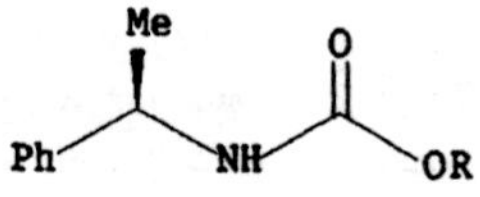

Structure XIV.

2.1 Typical protocols for diastereoisomeric derivatization

Protocols 1 and *2* below describe the procedure to be followed to prepare volatile diastereoisomers from racemic amino acids and alcohols. For a review of the reagents available and references to their use and optimal analysis conditions, see König (3).

Protocol 1. Synthesis and analysis of
N-pentafluoropropionylamino acid
O-(S)-2-alkyl esters (6)

1. Prepare N-pentafluoropropionylamino acid O-(S)-2-alkyl esters (*Structure XIII*) by heating for 1 h at 100°C with sonication in a 1 ml 'Reacti-Vial' the following:
   - the amino acid   0.6 mg
   - acetyl chloride (20%, v/v) in
   - the (S)-2-hydroxyalkane   0.1 ml

     e.g. butan-2-ol (Fluka)
           3-methylbutan-2-ol (Chemical Dynamics)
           hexan-2-ol (Chemical Dynamics)
           octan-2-ol (Fluka)

2. Evaporate to dryness in a stream of N_2 and heat for 20 min at 100°C
   - residual ester in DCM   100 μl
   - pentafluoropropionic anhydride (Pierce)   25 μl

3. Remove the reagents with a stream of N_2 and redissolve in 50 μl/ml DCM.

4. Analyse the N-TFA-amino acid ester in DCM on a suitable GC using the following conditions[a]:
   - CP-SIL 5 WCOT or FSOT capillary   25 m × 0.32 mm i.d.
   - temperature range   100–200°C
   - split mode of injection   1 μl
   - carrier gas   He at 100 kPa
   - detector   FID or ECD

[a] These conditions should give t_r of ca. 6–12 min and selectivities (α) of 1.02–1.06.

Protocol 2. Synthesis and analysis of diastereoisomeric carbamates of racemic alcohols (7)

1. Prepare the diastereoisomeric carbamates (*Structure XIV*) of racemic alcohols in a 1 ml 'Reacti-Vial' by heating for 7 h at 110°C the following:
   - racemic alcohol   1 μl
   - (R)-1-phenylethyl isocyanate   1.5 μl

   Add 0.5 ml methanol to destroy excess reagent.

2. Analyse the derivatives in a suitable GC using the following conditions:

- WCOT or FSOT capillary (SE54,
  DB210 or Carbowax 20M)   25 m × 0.32 mm i.d.
- temperature: isothermal or   170–180°C
                            ramped   from 170°C at 2°C min^{-1}
- carrier gas   He at 100 kPa
- detector   FID or NPD

3. GC on chiral stationary phases (CSP)

As an alternative to the formation of such volatile diastereoisomeric derivatives, mixtures of enantiomers may be analysed directly by gas chromatography using chiral stationary phases, generally termed CSP. Commercially available CSP of this type (*Table 2*), which for use in GC should ideally be both high-boiling and thermally stable liquids, include peptides, chiral polysiloxanes, certain asymmetric complexes of transition metals, and the recently developed liquid polyalkyl derivatives of cyclodextrins. Before considering these in turn, since they each have special features which define their scope and limitations, their underlying mechanism of operation will be discussed.

In gas chromatography there are no chiral mobile phases: the common gases such as helium and nitrogen have symmetrical molecules. Hence, the only way to obtain an asymmetrical retention environment is to employ chiral

Table 2. Commercially-available CSP for direct chiral GC analyses

Type	Selector	Temperature limit (°C)	Applications	Supplier
Lauroyl-proline-naphthylethylamide		170	Amino acid derivatives	Sumitomo
Lauroyl-valine-*tert*-butylamide		190	N-TFA-amino acid methyl esters	Supelco
Chirasil-VAL (L or D)		200–230	Many types	Chrompack Macherey–Nagel
XE-60-VAL.NHCHMePh		200–220	Many types	Chrompack
Chiraldex cyclodextrin phases (range of modified α, β, and γ-cyclodextrins)		300	Many types	Astec Technicol
Permethyl β-cyclodextrin dissolved in CP-Sil		250	Barbitals, free diols, dioxalanes, etc.	Chrompack
Lipodex cyclodextrin phases (range of modified α, β, and γ-cyclodextrins)		300	Many types	Macherey–Nagel

stationary phases. In molecular terms this means that the analyte enantiomers are in equilibrium between the mobile phase and a chiral stationary phase such that, if we are to observe enantioselective separation, the position of equilibrium in any theoretical plate is different for each enantiomer. In other words, since the phase ratio β is the same for the two enantiomers, if $k'(R) \neq k'(S)$ then $K(R) \neq K(S)$. The molecular implications of this are that the enantiomers must enter into reversible diastereoisomeric complexation with chiral structures in the CSP. Since K can be related to changes in enthalpy and entropy by the application of well-established thermodynamic laws, this means that the enthalpy and entropy changes for the two enantiomers as they form these transitory complexes must differ in an appropriate manner such that

$$-RT \log K = \Delta H - T\Delta S \qquad \text{and}$$
$$K(R) \neq K(S).$$

In general terms, we might expect these ΔH and ΔS values to be controlled by processes such as:

- hydrogen bonding
- dipole–dipole interactions
- π–π interactions
- Van der Waals interactions (steric effects)
- hydrophobic interactions (dispersion forces)

between the analyte molecule and asymmetric regions of the CSP. If we could understand these processes well, we would be in a position to design CSP with some precision. Unfortunately, it has to be admitted that such processes are not well understood; as a result, there remains a great deal of empiricism in the selection of CSP for a particular analysis. In the case of CSP for GC, the main limitation on such a column selection is that there are, to date, a much more limited range of commercially available CSP than is the case in HPLC. Whether this situation changes in the near future will depend to a large extent upon demand; in the recent past, it has been the HPLC analysts who have been demanding improved CSP, primarily because it has been the HPLC analysts who have faced the most stringent requirements for regulating the purity of pharmaceutical and agrochemical products.

Faced with a limited selection of chiral stationary phases for the GC analysis of enantiomers, the following general considerations appear to apply

- thermal limit to operation of CSP
- performance criteria such as column efficiency
- reproducibility of manufacture
- cost
- range of chiral analytes separable on the CSP

Although few GC analysts prepare their own capillary columns, some information regarding CSP formation will nevertheless be included in the sections which follow. A knowledge of the chemistry involved will lead to greater understanding of the limitations of the various CSP, and of the problems overcome in their manufacture.

3.1 Phases based on monomeric peptides

The credit for developing the first CSP for GC goes to Gil-Av and Feibush (8), a contribution which should not be underestimated since it predates enantioselective resolution by HPLC by several years. The Israeli group coated 50–100 m glass capillaries with trifluoroacetyl (TFA) derivatives of long-chain esters of amino acids; they found that N-TFA-(S)-isoleucine lauryl ester (*Structure XV*) combined a suitable liquid range with adequate to poor thermal stability, and were able to report modest separations of esters of racemic N-TFA-valine, N-TFA-leucine and N-TFA-alanine, with α-values in the range 1.02–1.06.

Having demonstrated the potential for such procedures, Gil-Av, Feibush, and others explored the performance and resolving power of related CSP (*Table 3*), and from this series of investigations over nearly two decades, the diamide CSP N-lauryl-(S)-valine *tert*-butylamide (*Structure XVI*) has emerged as one of the most powerful (9). It was accordingly taken to commercial development and emerged as the Supelco SP-300 GC packing on Supelcoport, with a recommended upper temperature limit of 140°C. Supelco's Application Notes Bulletin 765 commends its use for N-trifluoroacetylamino acid methyl esters and its inventors quote α-values for these analytes on a 150 ft glass capillary ranging from 1.06 to 1.28; similar values are obtainable with a 4 m packed column with a 10% w/w coating of the stationary phase (10).

The literature of this period contains many references to analogous mono-, di- and tripeptides, particularly suitable for the resolution of racemic volatile derivatives of amino acids, such as N-docosanoylvaline *tert*-butylamide (*Structure XVII*) (10), carbonylbis-(N-valine isopropyl ester) (*Structure XVIII*) (11), and cyclohexyl and isopropyl N-TFA-valinylvaline (*Structure XIX*) (12). Subsequently the separations achieved included N-TFA-β- and γ-amino acid esters (11), trifluoroacetyl derivatives of amino alcohols (13) and amines (14), and *tert*-butylamides of 2-halogenocarboxylic acids (15). Four main structural types of CSP were studied, namely, N-TFA-α-amino acid esters, N-TFA-

Structure XV.

Structure XVI.

dipeptide esters, carbonyl-bis(amino acid esters) (11) (so-called ureido phases), and *N*-lauroylamino acid amides, especially their *tert*-butylamides, *tert*-octylamides (16), and their arylalkylamides. The last-mentioned type (*Structure XX*) became commercially available from Sumitomo Chemical Company as Sumipax-CC OA 500, the amino acid in this case being (*S*)-proline and the amide that derived from (*S*)-1-(α-naphthyl)ethylamine. *N*-Lauroyl-α-(1-napthyl)ethylamide *Structure XXI* and its *N*-(1R,2R)-*trans*-chrysanthemoyl analogue (*Structure XXII*) (18) were also found to be useful as capillary phases.

The mechanism of enantioselective retention on such monomeric phases has been extensively discussed but is still not well understood. The usual

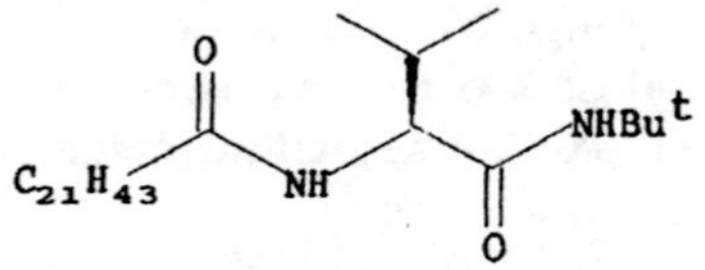

Structure XVII.

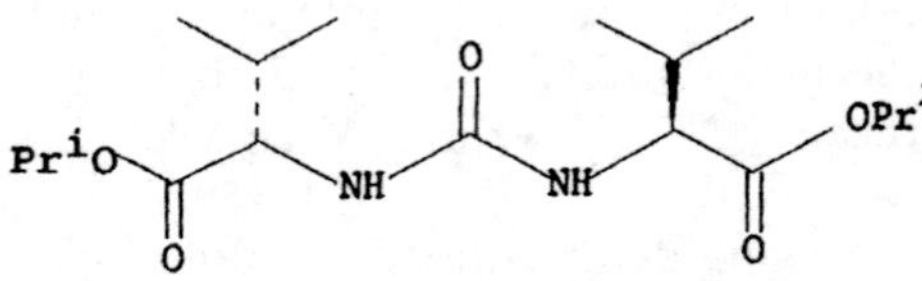

Structure XVIII.

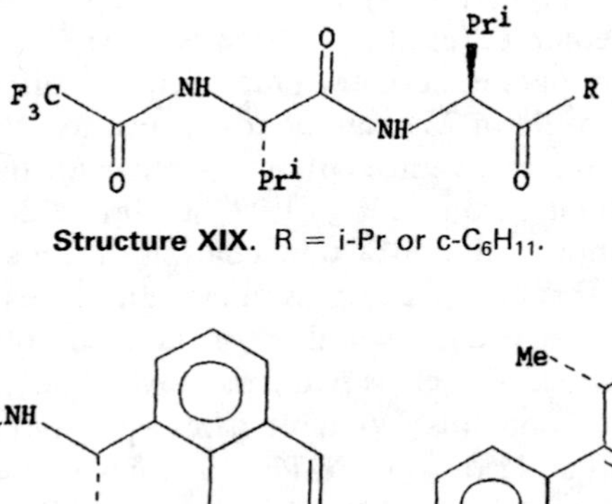

Structure XIX. R = i-Pr or c-C_6H_{11}.

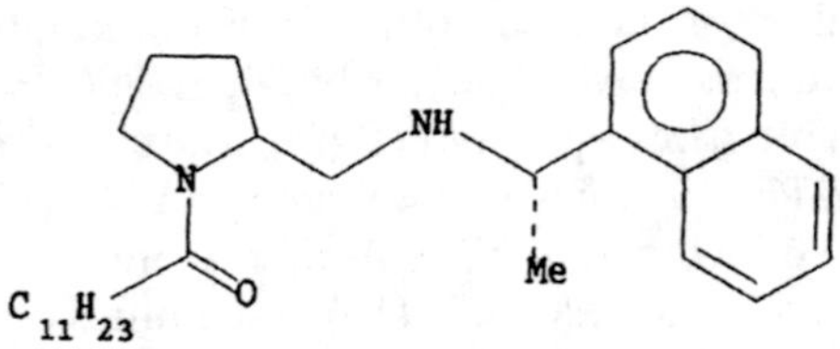

Structure XX.

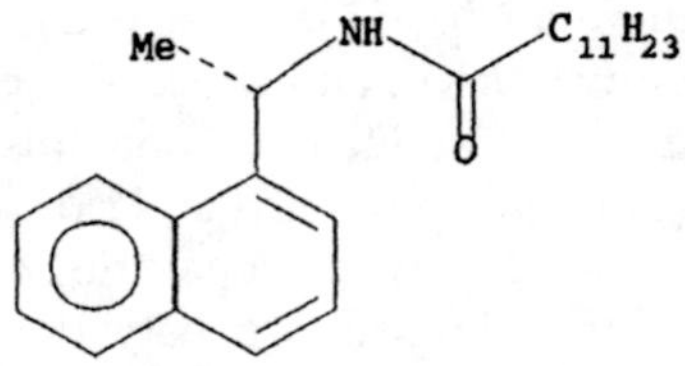

Structure XXI.

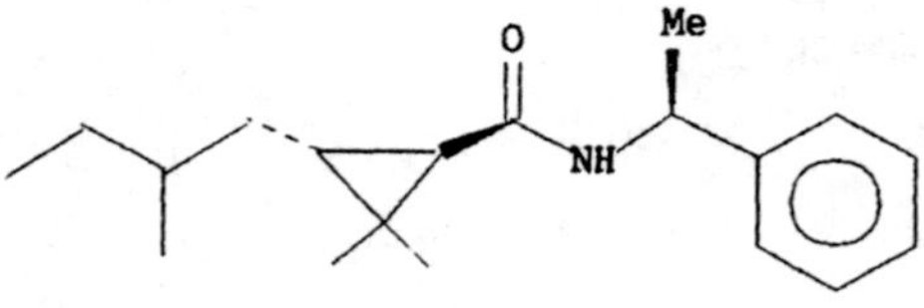

Structure XXII.

Table 3. Amide stationary phases for gas chromatography

Type of selector and examples	Temperature limit (°C)	Applications	Reference
Amides			
$C_{11}H_{23}CONHCHMePh$	120	N-TFA-amines	69
		Amides	
		N-TFA-amino acid esters	
$C_{11}H_{23}CONHCHMeNaph$	130	N-TFA-amines	69
		Amides	
		N-TFA-amino acid esters	
RCONHCHMeNaph [R = (1R,2R)-trans-chrysanthemoyl]	110	Chrysanthemamides	18
$RCO_2CHPhCONHCHMeNaph$ [R = (1R,2R)-trans-chrysanthemoyl]	110	Chrysanthemamides	18
$C_{11}H_{23}CO_2CHPhCONHCHMeNaph$	130	N-TFA-amines	3
Amino acid amides			
$C_{11}H_{23}CO.VAL.NHBu$-t	190	N-TFA-amino acid isopropyl esters	9
		N-TFA-amino acid methyl esters	10
$C_{11}H_{23}CO.VAL.NHCMe_2C_{15}H_{31}$	110	N-TFA-amino acid isopropyl esters	19
$C_{11}H_{23}CO.VAL.NHCH(Hex^n)C_5H_{11}$	140	N-TFA-O-acyl aminoalcohols	13
$C_{21}H_{43}CO.VAL.NHBu$-t	190	N-TFA-amino acid isopropyl esters	19
		N-TFA-amino acid methyl esters	10
$C_{21}H_{43}CO.VAL.NHCMe_2C_{15}H_{31}$	170	N-TFA-amino acid isopropyl esters	19
		N-TFA-amino acid methyl esters	10
tert-BuCO.VAL.NHC_{12}H_{25}	130	N-TFA-amino acid esters	19
$C_{11}H_{23}CO.PRO.NHCHMeNaph$	160	N-pentafluropropanoyl-arylamines	70
Dipeptide esters			
N-TFA-ALA.ALA.OC_6H_{11}	120	N-TFA-amino acid esters	71
N-TFA-LEU.LEU.OC_6H_{11}	110	N-TFA-amino acid ethyl esters	72
N-TFA-LEU.VAL.OC_6H_{11}	110	N-TFA-amino acid ethyl esters	72
N-TFA-MET.MET.OC_6H_{11}	150	N-TFA-amino acid isopropyl esters	73
N-TFA-NLEU.NLEU.OC_6H_{11}	130	N-TFA-amino acid isopropyl esters	74
N-TFA-NVAL.NVAL.OC_6H_{11}	130	N-TFA-amino acid esters	74

Table 3. (*continued*)

Type of selector and examples	Temperature limit (°C)	Applications	Reference
N-TFA-PHE.ASP.(OC$_6$H$_{11}$)$_2$	165	*N*-PFP-amino acid esters	75
		Chloroisovaleryl amino acid esters	76
N-TFA-PHE.LEU.OC$_6$H$_{11}$	140	*N*-TFA-amino acid isopropyl esters	77
N-TFA-PHE.PHE.OC$_6$H$_{11}$	165	*N*-(2-Cl-3-Me-pentanoyl)amines	75
		N-PFP-amino acid isopropyl esters	
		Chloroisovaleryl amines	76
N-TFA-PRO.PRO.OC$_6$H$_{11}$	110	*N*-TFA-proline esters	78
N-TFA-sarcosylproline cyclohexyl ester	110	Proline	78
N-PFP-VAL.LEU.OC$_6$H$_{11}$	110	*N*-TFA-amino acid ethyl esters	72
N-TFA-VAL.LEU.OC$_6$H$_{11}$	110	*N*-TFA-amino acid ethyl esters	72
N-Ac-VAL.VAL.OCHMe$_2$	140	*N*-TFA-amino acid esters	12
C$_5$H$_{11}$CO.VAL.VAL.OC$_6$H$_{11}$	160	*N*-TFA-amino acid esters	79
N-TFA-VAL.VAL.OCHMe$_2$	140	*N*-TFA-amino acid esters	12
N-TFA-VAL.VAL.OC$_6$H$_{11}$	110	*N*-TFA-amino acid esters	14

Tripeptide esters

N-TFA-LEU.LEU.LEU.OC$_6$H$_{11}$	110	*N*-PFP-amino acid methyl esters	72
N-TFA.VAL.VAL.VAL.OCHMe$_2$	140	*N*-TFA-amino acid *tert*-butyl esters	12

Ureides

[Pr^{i}O.LEU.NH]$_2$CO	150	*N*-TFA-amines	80
[Pr^{i}O.VAL.NH]$_2$CO	150	*N*-TFA-amines and amino acid esters	11
[EtO.VAL.NH]$_2$CO	150	*N*-PFP-amines	80
[Bu^{t}O.VAL.NH]$_2$CO	150	*N*-HFB-amines	80
[Pr^{i}O.VAL.VAL.NH]$_2$R (R = 6-ethoxy-2,5-triazinyl)	150	*N*-TFA-amines	81
[Pr^{i}O.VAL.VAL.VAL.NH]$_2$R (R = 6-ethoxy-2,5-triazinyl)	180	*N*-PFP-amines	81

Polysiloxane amides

Chirasil-VAL	200–230	Many types	73
OV225-VAL.NHBu^t	230	*N*-Perfluoroacyl amino acid esters	26
Z.LEU.OV225	130	Bis(TFA)aminols	27

Z.VAL.OV225	130	N-TFA-amines	27
Silar-10C-VAL.NHBu^t	150	N-TFA-amino acid isopropyl esters	82
XE-60-VAL.NHCHMePh	200–220	Many types	30
XE-60-VAL.NHBu^t	220	N-Perfluoroacyl amino acid esters	34
Me$_2$SiO-containing 7% (CH$_2$)$_3$OC$_6$H$_4$CONHCHMePh	280	N-PFP-amino acid isopropyl esters	36
Carbowax 20M crosslinked to CH$_2$ = CHCO.VAL.NHCHMePh	190	N-TFA-amino acid isopropyl esters	35

mechanistic theory involves the reversible formation of diastereoisomeric complexes *via* hydrogen bonding between the analyte and the amide functions of the CSP. Any theory of this kind must explain, and preferably predict, the elution order of the separated enantiomers; this is usually visualized in terms of ease of formation of diastereoisomeric complexes such as those shown in *Figure 4*, which are based on the associative interactions of pleated sheet proteins (13).

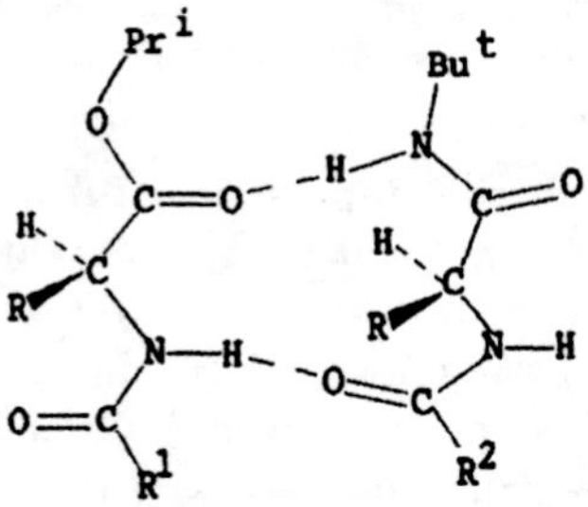

Figure 4. C$_5$–C$_7$ hydrogen-bonded association complex of two (*S*)-amino acid derivatives.

Rather curiously, the way in which such interactions are visualized in chiral GC theory is quite different from similar interactions in chiral HPLC. In chiral GC, hydrogen bonds are regarded as dominant and are drawn in coplanar arrangements between the docked analyte molecule and the chiral stationary phase unit beside which it is 'docked' (*Figure 4*). In chiral HPLC, the polar bonds are believed to lie above and below one another in a dipole stack (*Figure 5*), there being strong evidence for the correctness of this arrangement from X-ray analyses of typical complexes. The experimental observation which has to be explained is that, on both the early (*S*)-isoleucine monoamide CSP and the later (*S*)-valine diamide *tert*-butylamide, (*S*)-enantiomers of amino acids are more retained and (*R*)-enantiomers less retained and so eluted first. Until a substantial set of exceptions emerges, this can be used to determine the more abundant configuration, and hence the enantiomeric excess (*ee*), of volatile α-amino acid derivatives. It has been reported that N-trifluoroacetyl β- and γ-amino acid esters elute in the reverse order, i.e. (*R*)-isomers more, (*S*)-isomers less retained on (*S*)-CSP of the peptide type (19), while it is claimed that on the carbonyl-bis[(*S*)-valine] phases, the elution order of various types of analyte is predictable by considering the steric requirements of groups around the chiral carbon centre (11), so that analyte enantiomers with a clockwise arrangement of groups of decreasing size elute later, and those with an anti-clockwise sequence earlier.

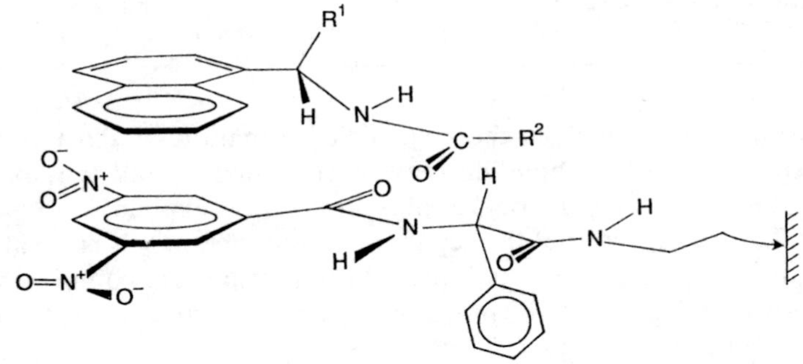

Figure 5. Dipole stack mechanism (after Pirkle), showing interactions complex of *N*-DNB-(*R*)-phenylglycine with acyl derivative of 1-(1-naphthyl)-alkylamine.

Such semi-empirical relationships are useful when a series of structurally related analytes is involved, the key functionality remaining constant: the theory is not sufficiently well-developed to permit the prediction of elution orders, or conversely to assign absolute configuration from elution order, for diverse analytes with a range of functionality. I have long advocated the use of computer-assisted molecular graphics to explore these relationships, but as yet the technique has not yielded a firm exploitable framework. Indeed, this approach has seldom been used in chiral GC on this type of phase.

3.2 Phases based on polymeric amides

In 1977 a new concept in chiral GC stationary phases was announced which, like all brilliant inventions, seems obvious once it has been achieved! Frank *et al.* (20) copolymerized dimethylsiloxane with (2-carboxypropyl)methylsiloxane and coupled the resultant polysiloxane's carboxyl group with (*S*)-valine *tert*-butylamide (*Scheme 1*) to obtain a polysiloxane (*Structure XXIII*) with pendant chiral diamide groups. This material displayed such useful resolving powers that it quickly became commercially available under the trade name of Chirasil-Val. It exhibits excellent enantioselectivities and has an upper temperature limit of around 200°C. This performance is well exemplified by the simultaneous separation of all protein amino acids as their *N*-pentafluoropropanoyl isopropyl esters on a 20-metre Chirasil-Val capillary, temperature programmed at 4°C/min (*Figure 6*) (21); as on analogous monomeric diamide phases, the (*S*)-isomers are more retained. Other general classes of compounds

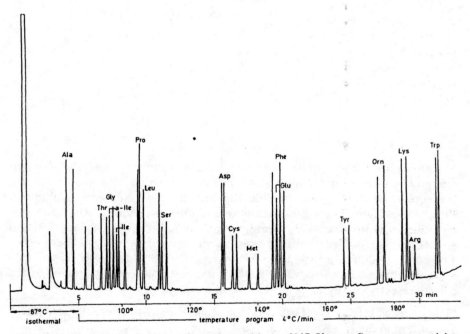

Structure XXIII.

Scheme 1.

Figure 6. Separation of all protein amino acids as N-(O,S)-pentafluoropropanoyl iso-propyl esters on a 20 m Chirasil-Val capillary, temperature-programmed at 4°C/min.

resolved efficiently on this CSP are 2-hydroxycarboxylic acids (22), 3-hydroxy-carboxylic acids (23), 2-amino alcohols (24), and various alcohols and diols (25).

Subsequently, several other chiral polysiloxane CSP for GC were developed. Contrasting procedures are exemplified as follows:

(a) hydrolysis of the cyano groups of the commercially available polysiloxane OV-225, followed by amide coupling with (S)-valine *tert*-butylamide (see *Scheme 2*; ref. 26)

(b) lithium aluminium hydride reduction of the nitrile groups of OV-225 followed by amide coupling of the resultant amino-group to N-benzyl-oxycarbonyl-(S)-valine or -leucine (see *Scheme 3*; ref. 27).

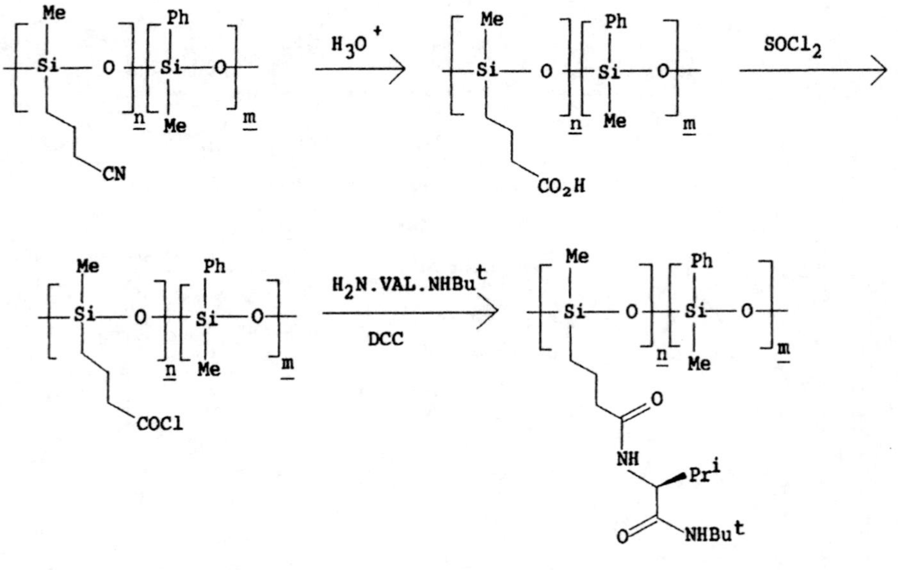

Scheme 2.

Application of the first procedure to another polysiloxane (XE-60) but using (S)-valine as its (R)- or (S)-α-phenylethylamide (28) gave what are claimed to be even more versatile capillary-coating CSPs which, although they have achieved some outstanding separations of a wide range of analytes, including α-hydroxy acids as isopropyl esters and isopropylurethanes (29) and sec-alkylamines as volatile urethanes, seem to have been less widely used. The commercially available form of this type of CSP is XE60-(S)-valine-(S)-α-phenylethylamide (*Structure XXIV*). It has been successfully applied to the enantioselective analysis of the following classes:

(a) isopropyl esters of N-TFA-amino acids and N-*tert*-butylereido derivatives of amino acids (see ref. 30)

(b) methyl esters of N-*tert*-butylureido derivatives of α-alkylated amino acids;

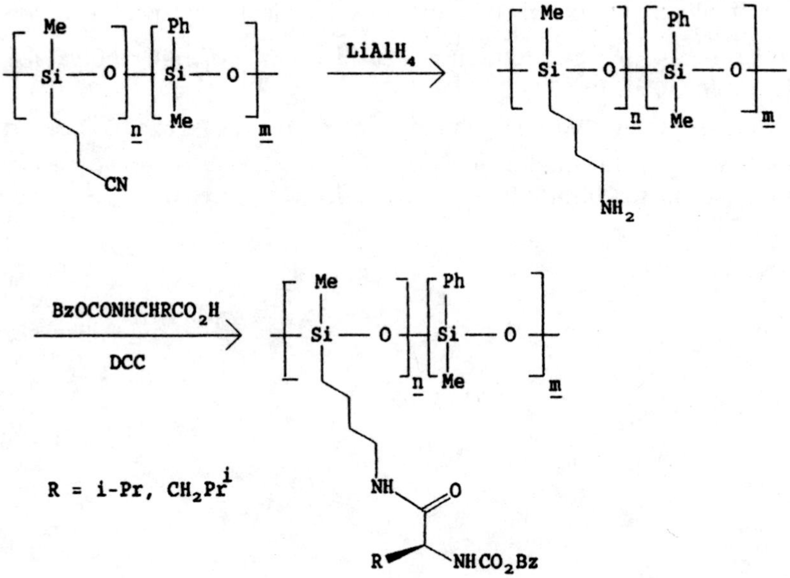

Scheme 3.

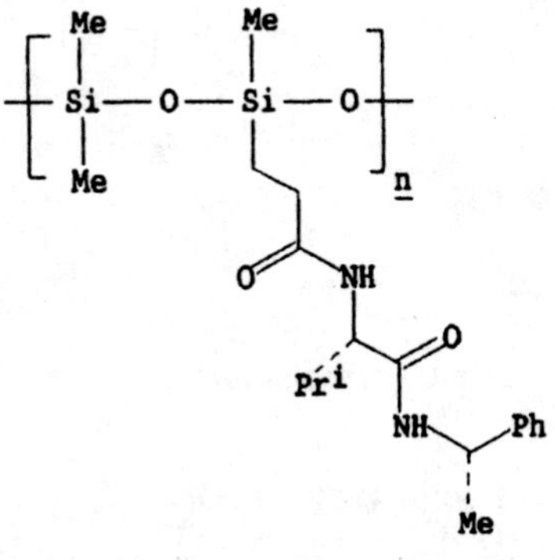

R = i-Pr, CH$_2$Pr^i

Structure XXIV. XE60-L-VAL-(S)-a-PEA

(c) *N,O*-perfluoroacyl derivatives of both commercial β-blockers (31) and aliphatic aminoalcohols (28)

(d) barbiturates (31)

(e) urea derivatives of chiral amines (29)

(f) urethane derivatives of hydroxyacids (29)

(g) TFA-peracylated polyols (32)

(h) isopropylurethanes of chiral alcohols (33)

David R. Taylor

The thermal stability of this type of phase has been improved as follows

(a) by using methyl cyclopolysiloxanes in a chain-opening polymerization (see *Scheme 4*; ref. 34)

(b) by cross-linking them on-column using peroxides or azoalkanes (35)

(c) by a grafting technique in which hydrosilylation of polyhydromethylsiloxane on to chiral allyloxybenzamides was employed (see *Scheme 5*; ref. 35)

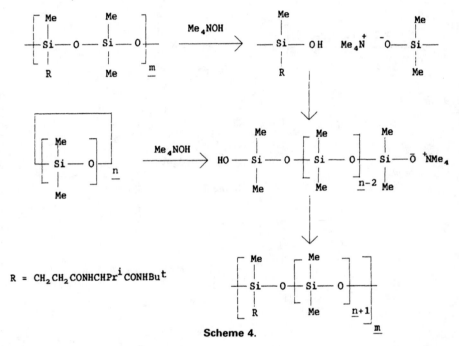

$$R = CH_2CH_2CONHCHPr^iCONHBu^t$$

Scheme 4.

3.2.1 Preparation of chiral polymeric GC phases and the coating of capillaries

Protocol 3 exemplifies the preparation of chiral polymeric GC phases (1) and *Protocol 4* describes the procedure for coating the used silica open tubular columns (1). Although the method is applicable also to borosilicate (glass) columns, such columns are not now commonly employed.

Protocol 3. Preparation of chiral polymeric GC phases

  1. Prepare BOC-(S)-valine-α-phenylethylamide by making a solution of:

    • BOC-(S)-valine   10 mmol

    • DCM   25 ml

2. After cooling to $-10\,°C$ add:
   - Et_3N 10.5 mmol
   - ethyl chloroformate   10.1 mmol

   and after 30 min add dropwise with stirring:
   - (R)- or (S)-α-phenylethylamine   10 mmol

3. After 15 min allow to warm slowly to ambient temperature and continue stirring overnight. Wash the organic layer in sequence with:
   - H_2O
   - $KHSO_4$ aqueous   1% w/v
   - $NaHCO_3$ aqueous   5% w/v
   - H_2O

   and dry over $MgSO_4$.

4. Evaporate to dryness and recrystallize the residue using:
   - diethyl ether-petrol (60–80 °C)

   The product, BOC-(S)-valine-(S)-α-phenylethylamide, has a m.p. of 202 °C.

5. Deprotect the amino group of the BOC-amino acid amide by stirring for 1.5 h the following:
   - BOC-amino acid amide   6.2 mmol
   - glacial acetic acid/HCl gas   35 ml/1.5 mol l^{-1}

6. Evaporate *in vacuo* and shake with:
   - NaOH aq.   15 ml

   to adjust the pH to 10.

7. Extract with ethyl acetate, dry the organic layer over $MgSO_4$, and then filter and evaporate.

8. Recrystallize the residue from diethyl ether-petrol (60–80 °C).[a]

9. To prepare the polysiloxane, reflux for 1 h the following:
   - polysiloxane XE-60   3 g
   - conc. HCl   100 ml
   - dioxane   100 ml

10. Evaporate and dissolve the residue in chloroform. Filter to remove any precipitated or undissolved material. Dry over $MgSO_4$ or Na_2SO_4.

11. For the formation of the COCl group[b], dissolve:
    - the acid from step 10   1.3 g
    - dry toluene   5 ml

Protocol 3. *Continued*
Add slowly:

- oxalyl chloride   4 g

Reflux for 4 h, and evaporate to dryness *in vacuo*.

12. Make a solution of:
    - the acid chloride from step 11   2 mmol
    - DCM   5–10 ml

    Cool to 0 °C, and add sequentially
    - Et$_3$N   0.3 g
    - the (S)-amino acid-(R or S)-α-phenylethylamide   2.2 mmol dissolved in
    - DCM   5 ml

13. Stir at ambient temperature overnight, evaporate, and then stir the residue with dry ether to enable triethylamine·HCl to be filtered off.

14. Wash the ether layer twice with dil. aq. hydrochloric acid and water and then evaporate the dried ether layer.

15. Purify the polysiloxane by chromatography on Sephadex LH20 using *n*-butanol eluant.[c]

[a] The product (S)-valine-(S)-α-phenylethylamide has a m.p. of 52 °C with [α]$_D$ −112.8° (*ca.* 0.8, CHCl$_3$).
[b] Formation of the COCl function is indicated by an IR band at v 1850 cm^{-1}.
[c] The main fraction should have [α]$_D$ −40.7° (*ca.* 0.76, CHCl$_3$) if (S)-valine-(S)-α-phenylethylamide is the intermediate product prepared.

Protocol 4. Coating of fused silica open tubular capillary columns

1. Thoroughly clean fused silica open tubular (FSOT) capillary tubing (i.d. 0.25 mm) obtainable commercially (SGE or J & W Scientific) by repeatedly rinsing with DCM. Fill the capillary with Silanox (0.5%, w/v) suspended in CHCl$_3$ (homogenize by ultrasonication) using the mercury drop technique.

2. Evaporate the CHCl$_3$ by passing oxygen-free cylinder gas (e.g. N$_2$, Ar, or He) for a minimum of 1 h.

3. Install the column (inlet end only) and purge with He for 1 h at 300–310 °C.

4. Coat the FSOT column with stationary phase by the static plug method with a solution of the (S)-amino acid-α-phenylethylamide polysiloxane (0.25%) in DCM as follows:

   (a) fill the FSOT with the solution and close one end with viscous waterglass solution (avoiding air bubbles) and leave overnight

   (b) mount the FSOT in a water bath at 20–25°C and evacuating slowly *via* the open end by attachment to a trapped line leading to a vacuum pump

   (c) repeat the procedure if air bubbles arise forcing a plug of solution from the column

5. Pre-condition the column with a He purge prior to attachment to the detector:

   (a) at 75–100°C for 5 h

   (b) at 150–175°C for 5 h

6. Evaluate for symmetrical peak shape and efficiency with both standard and chiral analytes.[a]

---

[a] Do not use for long periods over 200°C, and monitor t_r and efficiencies intermittently.

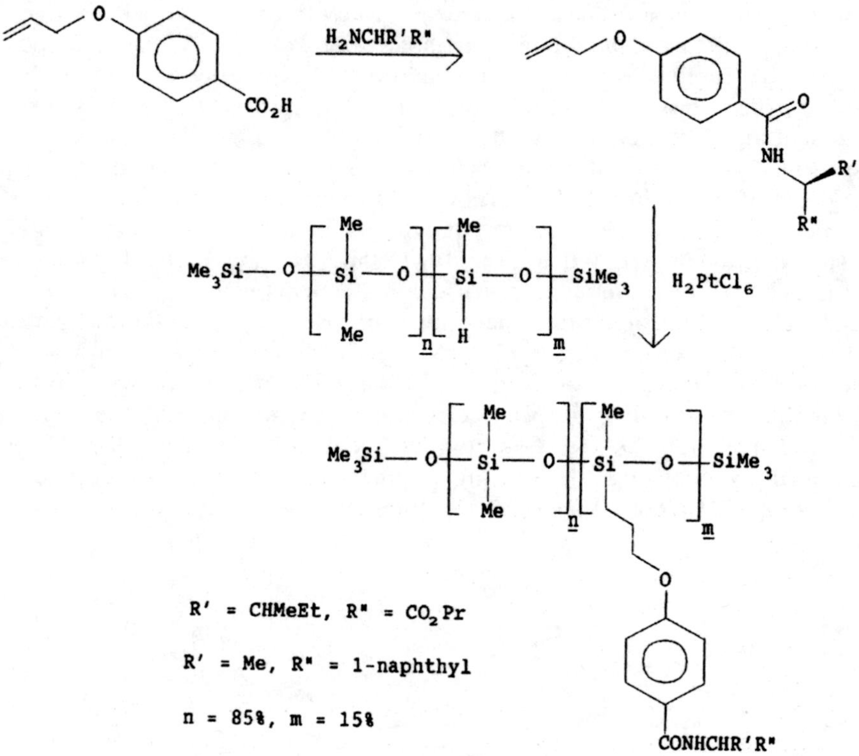

$R' = CHMeEt$, $R'' = CO_2 Pr$

$R' = Me$, $R'' = $ 1-naphthyl

$n = 85\%$, $m = 15\%$

Scheme 5.

259

David R. Taylor

3.3 Chiral GC using metal complexation

Although early versions of complexation chiral stationary phases for gas chromatography had been prepared several years previously, the first successful resolution, that of 3-methylcyclopentene on a squalane solution of the dicarbonyl rhodium (1) β-diketonate of 3-trifluoroacetyl(1*R*)-camphor (*Structure XXV*) was delayed until 1977. A 200 m steel capillary was used in that experiment, which is a significant one in spite of the low α-value reported, because of the minimal functionality of the analyte (37). Subsequently other such CSP were prepared and tested, including the Eu(III)-tris [trifluoroacetyl (1*R*) camphorate] (*Structure XXVI*) (38) and the Ni(II)-bis [trifluoracetylcamphorate] (*Structure XXVII*) (39) and its heptafluorobutanoyl analogue (40–41). Originally, all these phases were dissolved in squalane prior to coating, but subsequently solutions in polysiloxanes were found to offer improved performance with higher operating temperatures.

It rapidly emerged that few alkenes could be resolved on these columns, and attention then focused upon analytes with more chance of resolution via reversible co-ordination, especially oxygen-, nitrogen- and sulphur-containing analytes such as oxiranes, oxetanes, tetrahydrofurans, tetrahydropyrans, thiiranes, thietanes, aziridines, alcohols, and ketones. When deactivated fused silica capillaries replaced steel and nickel capillaries for complexation GC, further classes of compounds, including spiroketals, acetals, esters, and lactones, were shown to be resolvable since higher efficiencies were developed (42). Several review articles by Professor Schurig, the principal architect of this approach to chiral GC, have emerged (43–46).

The role of the heteroatom, frequently oxygen, in these separations is to ensure that the analyte binds reversibly to the metal ion. Generally the metal ion is nickel(II) but other metals such as cobalt(II), manganese(II), rhodium(I), copper(I), and europium(III) have all been used. It is the chiral ligand of the CSP which ensures that this process is enantioselective, provided that the formation constant for each analyte enantiomer's complex is different. Originally the chiral ligand was derived exclusively from (1*R*)-camphor by perfluoroacylation of the C3-anion; subsequently further variation was achieved by changing the starting ketone, for example to (*R*)-pulegone (*Structure XXVIII*) (47) and (1*R*,2*S*)-pinanone (*Structure XXIX*) (21).

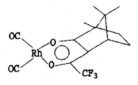

Structure XXV.

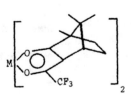

Structure XXVI.

260

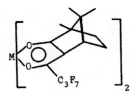

Structure XXVII.   **Structure XXVIII.**   **Structure XXIX.**

The preparation of asymmetric metal camphorate complexes for chiral complexation GC typically proceeds as described in *Protocol 5* (39), and an example of the coating of capillaries with such metal complexes is given in *Protocol 6*.

Protocol 5. Synthesis of metal complexes for complexation GC

1. Mix solutions of the following with stirring at −20°C under argon:
   - (+) − (1*R*)camphor   10.2 g
   - anhydrous ether   100 ml

   with
   - lithium diisopropylamide   67 mmol
   - anhydrous ether   50 ml at −20°C

2. After 30 min, cool to −60°C with vigorous stirring add a chilled solution of:
   - perfluoroacyl chloride   66 mmol

   Adjust the rate of addition to maintain at −60°C.

3. Stir for 1 h, then warm to −20°C before quenching by adding the solution with stirring to:
   - 1 M HCl   150 ml
   - ice   as required

4. Separate, extract the aqueous phase with:
   - ether   4 × 120 ml

   and wash the combined ether layers with:
   - aqueous NaCl   2 × 50 ml

   and then dry over:
   - Na_2SO_4 as required

   and concentrate *in vacuo*.

5. Remove unreacted camphor by sublimation at 40°C using a water pump, and distil the residue at 1 mm/Hg.[a]

Protocol 5. *Continued*

6. Prepare the sodium enolate, by adding under argon with stirring for a total of 3 h a solution of:

   - the perfluoroacylcamphor   7.5 mmol
   - anhydrous benzene[b]   80 ml
     to a suspension of
   - paraffin-free sodium hydride   10 mmol
   - anhydrous benzene   70 ml

7. Concentrate the solution carefully[c] *in vacuo* and dissolve the residue in CHCl$_3$ with gentle warming and then filter.

8. Add an equal volume of dry ether and cool to -5°C to yield the desired gelatinous precipitate of the sodium salt.

9. Isolate by suction and reprecipitate twice from CHCl$_3$/ether and dry at ambient temperature.

10. Mix to dissolve:

    - sodium salt   5 mmol
    - anhydrous ethanol   50 ml

11. Add under argon and reflux overnight:

    - anhydrous powdered nickel(II)chloride   2.6 mmol

12. Remove NaCl by filtering, concentrate *in vacuo,*[c] and sublime the nickel camphorate *in vacuo.*[d]

   [a] Further purification *via* column chromatography may be required (e.g. silica, benzene/*n*-hexane (1/1 v/v)).
   [b] Hazardous chemical.
   [c] Foaming may occur.
   [d] The residue may be extracted with pentane and the extract evaporated to yield more of the nickel(II)camphorate.

Protocol 6. Coating of capillaries with metal complexes (39)

1. Prepare the complexes in squalane on capillaries by:

   (a) mixing with gentle warming the following:

       - complex   15 mg
       - squalane   200 mg
       - ethanol (highest purity)   2 ml
       - CHCl$_3$ (acid-free)   2 ml

   (b) Using the dynamic plug procedure, pass the solution through the capillary (N$_2$ overpressure 0.6 atm) at ambient temperature and continue to pass N$_2$ for 5 h.

2. Connect the capillary to the GC inlet but not the detector and condition by warming to 100°C over 12 h with continuous carrier gas flow.

3. Carefully remove volatiles from the exit end of the column by flaming prior to connecting the detector.[a]

[a] Do not use such a column above 90°C, and note that H_2 must not be used as the carrier gas. Take care to exclude H_2 and O_2 from the coated column at all times. All column effluents should be vented into appropriate fume hoods. Dead volumes at inlet and exit must be avoided since separations are often borderline.

Although these capillaries cannot be obtained commercially, Professor Schurig has offered to provide them on request. Long column lengths have been shown to be unnecessary given appropriate efficiencies, and WCOT of 10–30 m containing these complexes dissolved in SE54 (0.2 μm films) have provided excellent separations in times as short as 2 min. A wide range of analytes have been separated without derivatization on these columns, including oxiranes, aziridines, spiroketals, tetrahydrofurans, ketones, alcohols, and 2-bromocarboxylic acid esters. Diols may be separated as their cyclic boronates or cyclic acetals (48), and halohydrins as their acetates (49).

3.3.1 Synthesis of polysiloxane complexation CSP

Recently, the synthesis of a polysiloxane phase suitable for chiral complexation GC was reported (45). The synthesis involves the hydrosilylation of a 10-methylene-3-perfluoroacyl (1S)- or (1R)-camphor with methyldimethoxysilane, followed after complexation with Ni(II) by hydrolysis of the methoxy groups and copolymerization with hydrolysed dimethoxymethylvinylsilane (*Scheme 6*).

In detail the procedures for the preparation of capillaries coated with such CSP are as described in *Protocol 7* below.

Protocol 7. Preparation of polysiloxane camphorate-coated FSOT capillaries

1. To initiate the preparation of the CSP mix under dry argon at 50°C with stirring overnight:

- methyl dimethoxysilane   5 mmol
- 3-perfluoroacyl-10-methylene (1R) camphor   5 mmol
- hexachloroplatinic acid (freshly prepared)
- in propan-2-ol (1%, w/v)   0.1 ml

and finally add a further amount of:

- methyl dimethoxysilane   1 mmol

and stir for a further 4 h before adding:

- methanol   10 ml

Protocol 7. *Continued*

Reflux for 5 h and then remove methanol *in vacuo* and purify the residue by column chromatography using:

- silica
- eluent-toluene/100–120 petroleum   3/1, v/v.

Collect and evaporate the main fraction, and distil the residue at 0.3 mm/Hg

Collect the fraction boiling at 40–45 °C, $[\alpha]_D$ 11.1° (0.1 dm, neat).

2. Hydrolyse dimethoxymethylvinylsilane (50 ml) with aq. ethanol (25 ml; 20% ethanol) by stirring at 90 °C for 1 hour. Extract the silane with ether and dry azeotropically by distillation with benzene.[a]

3. Prepare the 3-perfluoracyl-10-dimethoxymethylsilyl-nickel(II)camphorate as described for the 10-unsubstituted analogue (*Protocol 5*, steps 6–12), and hydrolyse the crude product with 50 ml aqueous methanol (20%, w/v) with stirring for 10 h at ambient temperature. Extract the product with DCM and evaporate the solvent and dissolve the residue in methanol. Filter and concentrate the solution *in vacuo*.

4. Mix the following:

- hydrolysed diethoxymethylvinylsilane   2.5 g
- trimethylsilanol   0.06 g
- the concentrate from step 3   0.5 g

Evaporate the solvent and polymerize the mixture by adding:

- tetramethylammonium hydroxide (TMAH) in
  methanol (2% w/v)   8 µl

Heat to 100–110 °C and stir for 10 h. Add further TMAH in small portions until the viscosity remains unchanged.

5. Dissolve the polymer in DCM and fractionate three times by addition of MeOH and filtration. Wash the polymer repeatedly with methanol, and end-cap using 1,3-divinyl tetramethyldisilazane in DCM at 20 °C. Evaporate all volatiles and store under argon.

6. Coat diphenyltetramethyldisilazane-deactivated fused silica capillaries by the static method, using

- chiral polysiloxane (0.5–0.8%, w/v)
- DCM/pentane (1/9, v/v)

7. Condition the column in the GC using temperature programming from 60 to 190 °C over 12 h.

---

[a] Hazardous chemical.

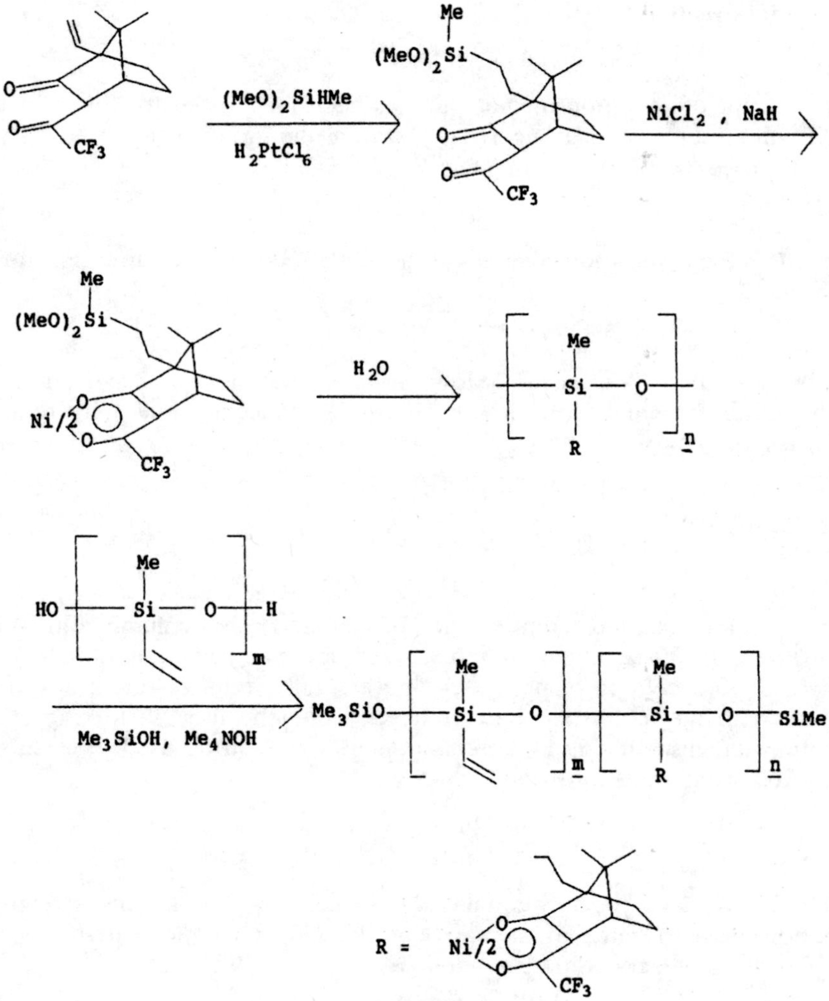

Scheme 6.

3.3.2 Thermodynamic data derived from chiral complexation GC

Using comparative data for columns containing pure siloxane and known molar concentrations (normally in the range 0.05–0.1 molar) of a metal camphorate in the siloxane, useful thermodynamic parameters are obtainable. If we define the difference in free energies between two enantiomers separated on a CSP as $-\Delta_{S,R}\Delta G^{O}$ such that:

$$-\Delta_{S,R}\Delta G^{O} = RT\ln\alpha$$

David R. Taylor

then it follows that:

$$\ln\alpha = -\Delta_{S,R}\Delta H^\circ/RT + -\Delta_{S,R}\Delta S^\circ/R$$

The concept of retention increase R' has been introduced to represent the additional retention resulting from chiral recognition, as opposed to the normal processes of retention. Thus R' is defined as:

$$R' = Ka_A$$

where K is the association constant of the analyte B with the chiral selector A such that

$$K = a_{AB}/(a_A a_B)$$

and the activity of A is a_A, etc. However, since the molar concentrations of analytes are extremely low during GC molar concentrations may replace activities and hence:

$$K = [AB]/[A][B] \text{ and } R' = K[A]$$

The retention increase is given by

$$R' = (t' - t'_o)/t'_o$$

where t' is the adjusted retention time of analyte B on a column containing chiral selector A dissolved in a liquid S, and t'_o is the adjusted retention time of analyte B on a column of pure S. Experimentally, relative retentions of B compared to an achiral reference analyte would be used. The respective retention increases for a pair of enantiomers can then be used to calculate $-\Delta_{S,R}\Delta G^\circ$ using the expression:

$$-\Delta_{S,R}\Delta G^\circ = RT\ln(R'_S/R'_R)$$
$$= RT\ln[(t'_S - t'_o)/t('_R - t'_o)]$$

which, it may be noted, only equals $RT\ln\alpha$ when $t' \gg t'_o$. In terms of relative retentions with reference to an analyte not forming a complex with the metal camphorate, this expression becomes

$$-\Delta_{S,R}\Delta G^O = RT\ln[(r_R - r_o)/(r_S - r_o)]$$

an expression which is independent of [A]. The value of r_o can be determined by extrapolation from the data for each enantiomer on two different columns containing different concentrations of selector A, or since it is the same for both enantiomers, by observing the relative retention of the racemate on a column without chiral selector.

From using such measurements, Schurig's group have established that typical values of $-\Delta_{S,R}\Delta G^\circ$, $-\Delta_{S,R}\Delta H^\circ$, and $-\Delta_{S,R}\Delta S^\circ$ are 60 J/mol, 215 J/mol, and 1.8 entropy units respectively. Furthermore, when chiral geometric isomers having quite different association constants were compared, their thermodynamic parameters of chiral recognition were rather similar. Particularly striking was a tendency for low temperature chiral separations to be enthalpy-controlled and high temperature chiral separations to be entropy-controlled. This latter point has led to the proposal that for some pairs of enantiomers, for which these two factors are opposed, there will exist a temperature where the two terms cancel, the so-called iso-enantioselective temperature. Reversal of elution order would be expected on passing through this temperature, separation factors decreasing as the temperature is approached and increasing subsequently.

Observation of peak coalescence at the iso-enantioselective temperature in complexation GC is fraught with difficulty, since low operating temperature will mean analytes are no longer in the vapour phase, and high temperatures are precluded for such columns. Such a phenomenon only arises where a chemical equilibrium is established, i.e. it would not be expected in inclusion CSP. It was recently observed, together with the anticipated peak reversal, for the enantiomers of (*E*)-2-ethyl-1,6-dioxaspiro[4,4]nonane (*Structure XXX*) on a nickel(II)camphorate CSP, the iso-enantioselective temperature being estimated to be 80°C (*Figure 7*); no peak separation was discernible between 70 and 90°C (50).

Another type of peak coalescence has been reported using complexation CSP; in this case the occurrence of coalescence is revealed by the distinctive peak shape, the main enantiomer signals arising as sharp peaks on either side of a broad plateau (*Figure 8*). Such a peak shape indicates that the enantiomers are undergoing inversion of configuration *during elution* through the GC column. A compound which clearly displays this phenomenon is *N*-chloro-2,2-dimethylaziridine (*Structure XXXI*), which undergoes nitrogen inversion

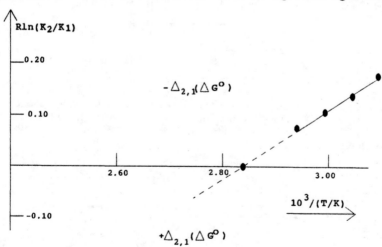

Figure 7. Extrapolation to determine the isoenantioselective temperature for the separation of (**xxx**) on a chiral complexation GC phase (see ref. 40). Enantiomers are: 1, 2*R*, 5*S*; 2, 2*S*, 5*R*.

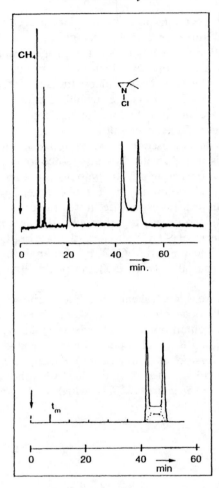

Figure 8. Gas chromatographic coalescence phenomena due to configurational inversion during analysis of (XXXI) at 60°C on a nickel capillary coated with squalane solution of XXVI: M = Ni. (Reproduced with kind permission from reference 51, © 1984 Elsevier Science Pub.)

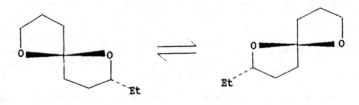

Structure XXX.

8: Chiral separations by gas chromatography

Structure XXXI.

to interconvert the two enantiomers during complexation GC analysis. By means of computerized peak-shape analysis, the first-order rate constant for the inversion process was determined at several temperatures and hence the free energy of activation could be estimated at $\Delta G = 25.1$ kcal mol^{-1} (51).

3.4 Inclusion phases for chiral GC

The cyclodextrins are cyclic oligomers of α-D-glucose, derived by enzymatic hydrolysis of starch by cyclodextrin glycosyl transferases of *Klebsiella pneumonia*, *Bacillus macerans*, etc. The common cyclodextrins contain six (α) seven (β), and eight (γ) D-glucose units. They consist of toroidal (bucket-shaped) molecules as a result of their α-1,4-linkages between adjacent D-glucopyranose units, and are widely used commercially as purification media. For many years their molecular dimensions and characteristics have been well known. Their interiors are hydrophobic, so that they tend to include non-polar molecules of appropriate dimensions: β-cyclodextrin for example will crystallize with cyclohexane included inside the toroidal cavity. Around the narrow rim are the primary CH_2OH groups (one per glucose unit) and around the wider rim are pairs of vicinal secondary hydroxyl groups attached to C-2 and C-3 of the glucose (one pair per glucose unit), each glucose being in a rigid chair conformation. Typical dimensions and physical constants are shown in *Table 4*.

Following pioneering work by the Polish scientist Danuta Sybilska and her colleagues at the Polish Academy of Sciences, who showed how cyclodextrins

Table 4. Dimensions and physical properties of cyclodextrins

Cyclodextrin:	α	β	γ
Number of D-glucose units	6	7	8
Number of chiral CHOH units	12	14	16
Total chiral centres	30	35	40
Molecular mass (daltons)	972.86	1135.01	1297.15
Avg external diameter (pm)	1415	1535	1720
Avg internal diameter (pm)	495	625	800
Cavity volume (nm^3)	0.176	0.346	0.510
Solubility in 100 ml water (g)	14.5	1·85	23.2
Mp/k (may decompose)	551	572	540

David R. Taylor

and their alkyl derivatives could be used as mobile phase additives for chiral HPLC (52), various cyclodextrin-based CSP for HPLC reached commercial availability in the early 1980s. A problem for their use in chiral GC was the lack of a liquid cyclodextrin, since most reported derivatives are crystalline solids below 250°C, above which they begin to decompose. Hence early reports of their use in GC cited packed columns coated with, for example, solutions of α-cyclodextrin in formamide or ethyleneglycol; such columns enabled separations of geometrical isomers of alkenes and arenes, and even the enantiomers of α-pinene, but with very low efficiencies and upper temperature limits as low as 80°C (53). Solid underivatized cyclodextrins were studied as column packings in GSC (54), but better results were obtained in capillary GC when a solid permethylated β-cyclodextrin was used as a solution in OV-1701 (55).

The problem of the high melting point of cyclodextrins was eventually overcome by König *et al.* in Hamburg, who hit upon the idea of using liquid perpentyl-, acetyl-dipentyl-, and butyryl-dipentyl-cyclodextrins (*Structures XXXII–XXXIV*) as chiral GC phases, and have reported numerous successful resolutions on such phases since 1988 (56–61). Column efficiencies are excellent and these phases are now available commercially under the trade name Lipodex (62).

Two interesting variations of this approach have already appeared. Armstrong *et al.* (63) have found interesting reversals of elution order between dipentylcyclodextrins (*Structure XXXV*) and permethyl-(*S*)-hydroxypropyl-cyclodextrins (*Structure XXXVI*). The excellent performance characteristic of these materials has also led to their rapid commercialization, so that in a short space of time a range of different cyclodextrin GC phases (*Table 5*) have become available from a number of suppliers. The most recent development has involved a report that cyclodextrin can be chemically bonded to a silicone polymer (*Scheme 7*), yielding a third type of surface film for such CSP (64–65).

3.4.1 Synthesis of acyl-alkyl- and peralkyl-cyclodextrins

The procedure for preparing the modified cyclodextrins is as described in *Protocol 8*.

n = 6, 7 or 8

R = R' = n-pentyl

R = n-pentyl, R' = acetyl

R = n-pentyl, R' = butanoyl

Structures XXXII–XXXIV. (XXXII) R = R' = *n*-pentyl; (XXXIII) R = *n*-pentyl, R' = acetyl; (XXXIV) R = *n*-pentyl, R' = butanoyl.

OR3
O R2O
O
OR1
n   n = 6, 7 or 8

$R^1 = R^3$ = n-pentyl, R^2 = H

$R^1 = R^2$ = Me, R^3 = (S)-$CH_2CH(OMe)CH_3$

$R^1 = R^3$ = n-pentyl, R^2 = $COCF_3$

Structures XXXV–XXXVII. (XXXV)$R^1 = R^3 = $ n-pentyl, $R^2 = $ H; (XXXVI)$R^1 = R^2 = $ Me, $R^3 = $ (S)-$CH_2CH(OMe)CH_3$; (XXXVII) $R^1 = R^3 = $ n-pentyl, $R^2 = COCF_3$

Table 5. Commercially available cyclodextrin CSP for use in GC

Stationary phase	Ring size	Supplier	Upper limit (°C)
Lipodex A	6	Macherey–Nagel[a]	>200
Lipodex B	6	Macherey–Nagel	>200
Lipodex C	7	Macherey–Nagel	>200
Lipodex D	7	Macherey–Nagel	>200
Chiraldex A-PH[b]	6	Adv Sep Tech[c]	260
Chiraldex B-PH	7	Adv Sep Tech	260
Chiraldex G-PH	8	Adv Sep Tech	260
Chiraldex A-DA[d]	6	Adv Sep Tech	260
Chiraldex B-DA	7	Adv Sep Tech	260
Chiraldex G-DA	8	Adv Sep Tech	260
Chiraldex A-TA[e]	6	Adv Sep Tech	180
Chiraldex B-TA	7	Adv Sep Tech	180
Chiraldex G-TA	8	Adv Sep Tech	180
CP-CD-β-2,3,6-M-19[f]	7	Chrompack[g]	250–275

[a] Macherey–Nagel, Duren, Germany (UK agent: Camlab, Nuffield Rd, Cambridge CB4 1TH).
[b] Permethylated hydroxypropyl derivative.
[c] Advanced Separation Technology Inc, Whippany, NJ 07981, USA (UK agent: Technicol, Brook St, Stockport SK1 3HS).
[d] Dipentyl derivative.
[e] 3-Trifluoroacetyl-2,6-dipentylated derivative.
[f] 2,3,6-Trimethyl derivative dissolved in CPSil-19 siloxane.
[g] Chrompack, Middelberg, Netherlands (UK: 4, Indescon Court, Millharbour, London E14 9TN).

Protocol 8. Preparation of acyl-alkyl- and peralkyl-cyclodextrins

1. To prepare 2,6-di-O-alkylated (CD):

   (a) Treat for 5 days at ambient temperature:

   - α- or β-CD   0.1 mol
   - the bromoalkane   0.3–0.4 mol

Protocol 8. *Continued*

- DMSO 30 ml/g RBr
- pulverized NaOH 0.3–0.4 mol

(b) Add, about half way through the above period, with vigorous stirring the following:

- the bromoalkane 0.3–0.4 mol
- pulverized NaOH 0.3–0.4 mol

(c) Pour the mixture on to water and extract with *tert*-butyl methyl ether (twice). Wash the organic phase with:

- water

and dry the organic phase over:

- anhydrous Na_2SO_4

(d) Evaporate using high vacuum and purify the product (2,6-dialkyl-cyclodextrin) by elution chromatography using:

- silica
- eluant toluene/EtOAc 9:1 v/v

2. To prepare 2,3,6-tri-*O*-alkylated CD, reflux for 4 days:

- the product from step 1 0.1 mol
- the bromoalkane 0.2 mol
- THF 200–500 ml
- sodium hydride suspension 0.1 mol

3. To prepare 2,6-di-*O*-alkyl-3-*O*-acetyl CD:

(a) Reflux for 24 h:

- the product from step 1 0.1 mol
- acetic anhydride 0.7 mol
- Et_3N 0.8 mol
- 4-dimethylaminopyridine 0.0175 mol
- dry DCM 200–500 ml

(b) Add the reflux for a further 72 h:

- Et_3N 0.8 mol
- acetic anhydride 0.7 mol

(c) Evaporate to remove volatiles, add:

- methyl *tert*-butyl ether

and wash successively with:

- H_2O
- Na_2CO_3 dil. aq.

- H_2O
- NaH_2PO_4 dil. aq.
- H_2O

(d) Dry and evaporate under high vacuum at 0.05 Torr.

Purify by chromatography using silica adsorbent.

4. To prepare 2,6-di-O-alkyl-3-O-butanoyl CD:

(a) Reflux for 48 h:

• the product from step 1	0.1 mol
• butanoic anhydride	1 mol
• Et_3N	1.2 mol
• 4-dimethylaminopyridine	0.033 mol
• DCM	

(b) Add further quantities of the following and reflux for a further 8 days:

• butanoic anhydride	1 mol
• Et_3N	1.2 mol;

(c) Remove excess reagent and solvent by passing N_2 through the stirred solution, and dissolve the residue in methyl *tert*-butyl ether.

(d) Wash the solution successively with:

- H_2O
- Na_2CO_3 dil. aq.
- H_2O
- NaH_2PO_4 dil. aq.
- H_2O

Dry the organic layer with:

- anhydrous Na_2SO_4

(d) Evaporate under high vacuum before final purification[a] by elution chromatography using

• silica adsorbent	
• eluent toluene/EtOAc	5:1, v/v

[a] The purity of the derivatised cyclodextrins can be determined by reductive depolymerization followed by GC-MS analysis (see Ref. 66)

3.4.2 Coating the capillaries with derivatized cyclodextrins

Pyrex glass capillaries (e.g. 40 m × 0.2 mm i.d.) may be coated with these materials using 0.2% solution in DCM by the static coating procedure after pretreatment to form a Silanox interlayer [achieved with a 0.3–0.5% suspension of Silanox in CCl_4 using a mercury-drop to advance the suspension,

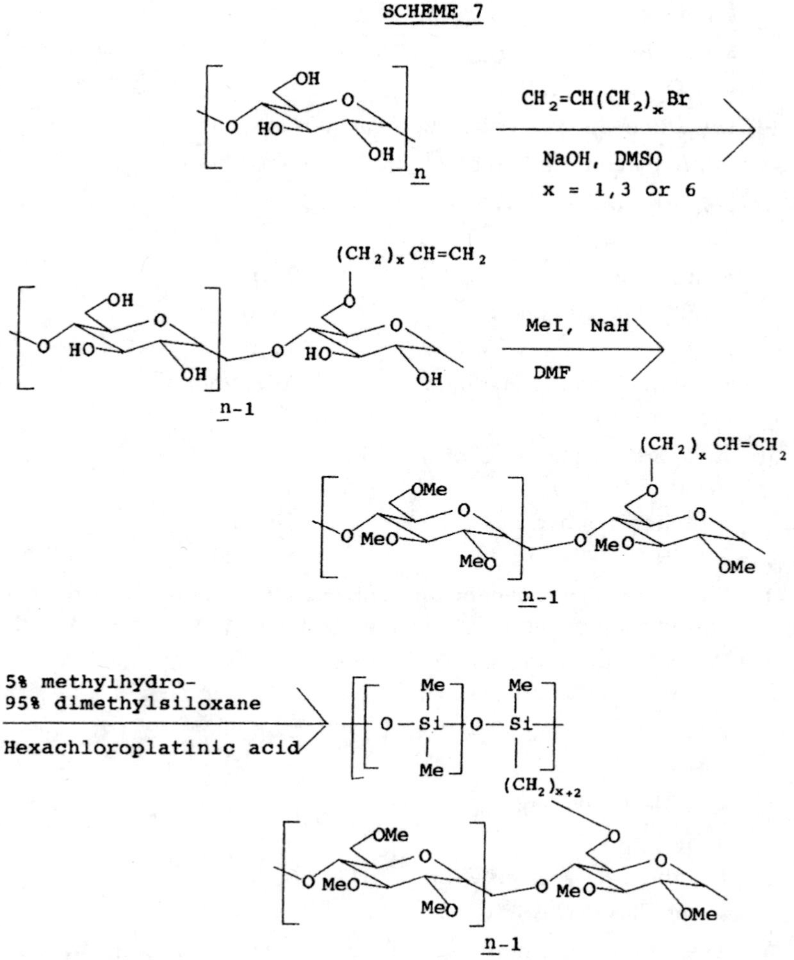

SCHEME 7

Scheme 7.

removing solvent with dry N_2, and heating in the GC for 1 h at 300°C while passing H_2]. They have frequently been used with H_2 carrier gas at temperatures in the range 40–220°C, and may prove to have even higher thermal stability (67).

3.4.3 Permethylated cyclodextrins in OV-1701

An alternative approach to the use of cyclodextrins in GC is that adopted by Schurig's group (55), who used 0.07 mol/l solutions of permethylated α- and β-cyclodextrins in a moderately polar polysiloxane OV-1701 to coat pre-treated Pyrex glass or fused silica capillaries. Very recently the same group

274

showed how a polysiloxane-bonded cyclodextrin could be obtained (65), using the procedure described in *Protocol 9*.

This cyclodextrinated polysiloxane may be coated on to fused silica capillaries, without deactivation of the surface, by the static method using a 0.4% solution in diethyl ether. The resulting capillaries are usable at temperatures lower than proved possible for solutions of permethylated cyclodextrins in OV-1701, which perform poorly below 80°C.

Protocol 9. Preparation of polysiloxane-bonded permethylated cyclodextrin (CD)

1. To prepare 3-*O*-(5-pent-1-enyl)-CD:
   (a) Stir for 1 h at ambient temperature under N_2:
       - anhydrous CD — 3 mmol
       - pulverized NaOH — 15 mmol
       - anhydrous DMSO — 75 ml
   (b) Add and stir under N_2 for 48 h:
       - 5-bromopent-1-ene — 15 mmol
       - DMSO — 10 ml
   (c) Filter and concentrate almost to dryness at 0.01 Torr.
   (d) Dissolve in methanol (10 ml) and precipitate by adding diethyl ether (200 ml).
   (e) Filter and dry the solid *in vacuo* with warming.[a]

2. To permethylate the product from step 1:
   (a) Add cautiously[b] under dry argon a mixture of:
       - pentenyl CD — 1.5 g
       - anhydrous DMF — 100 ml
       to:
       - suspension of paraffin-free NaH — 0.126 mmol
       - *n*-hexane
   (b) Add slowly at ambient temperature, and stir for 30 min:
       - iodomethane — 0.095 mol
   (c) Repeat the two additions using additional:
       - pentenyl CD — 1.5 g
       - iodomethane — 0.095 mol
   (d) Stir for 1 h, decant from unreacted sodium hydride, and pour into water (200 ml).
   (e) Extract successively with:
       - $CHCl_3$ — 3 × 70 ml
       - H_2O — 3 × 15 ml

Protocol 9. *Continued*

and then dry the organic layer with
- anhydrous Na_2SO_4

(f) Concentrate and purify the yellow solid residue[c] repeatedly by elution chromatography using:
- Sephadex LH20
- eluant DCM/MeOH   2:1, v/v

and finally dry *in vacuo* at 40°C over:
- P_2O_5

3. To obtain the final product:
   (a) Reflux under dry N_2
   - dimethylpolysiloxane (5% SiH)   1 mmol
   - di-*O*-methyl-3-*O*-(5-pent-1-enyl)CD   0.72 g (= *ca.* 0.5 mmol)
   - anhydrous toluene   100 ml
   (b) Add repeatedly at intervals of 2.5 h:
   - hexachloroplatinic acid   few mg
   - anhydrous THF   1 ml

   and after 24 h evaporate to dryness.
   (c) Dissolve the residue in
   - anhydrous MeOH   50 ml

   and decant to remove a black lower layer.
   (d) Evaporate and extract the residue with 60–80 petroleum, and finally filter, evaporate, and dry *in vacuo*.

[a] This procedure produces partially pent-1-enylated CD.
[b] H_2 is evolved—care!
[c] Eventually a white dry solid is obtained.

3.4.4 Hydrophilic cyclodextrin-based CSP

In contrast to the previously described hydrophobic peralkyl- and acyl-alkylcyclodextrins, the Armstrong group has developed hydrophilic hydroxy-alkylcyclodextrins for GC (63). Their materials are prepared as described in *Protocol 10* (see also *Scheme 8*).

Protocol 10. Preparation of hydrophilic cyclodextrin (CD) based CSP

1. Mix and stir at 0–5°C:
   - the required CD:
   - NaOH aq.   5% w/w

and then slowly add:

- (S)-propyleneoxide

2. After 6 h, allow to warm to ambient temperature and stir for a further 24 h.

3. Neutralize and then dialyse briefly to remove salts before filtering. Isolate the product by freeze-drying.

4. Permethylate the product from step 3 using:

- iodomethane

- sodium hydride suspension

- DMSO

5. Coat the 20–30 m × 0.25 mm i.d. FSOT capillary by the static method.

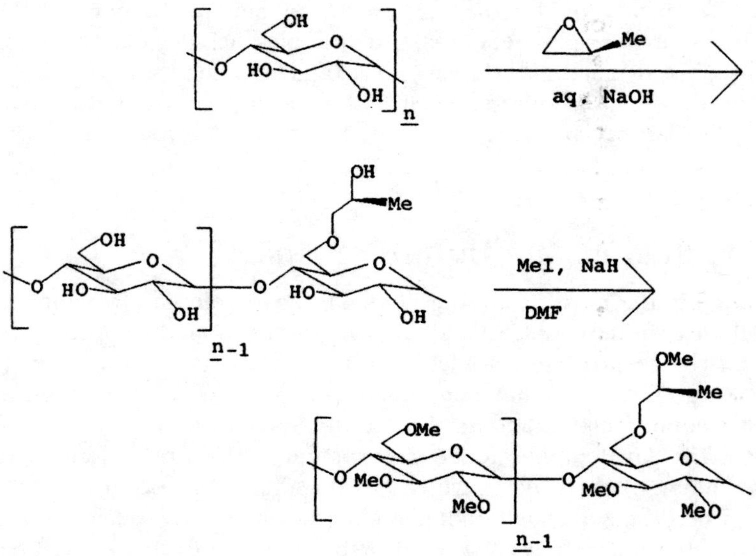

Scheme 8.

Preliminary results with this phase indicate that very polar compounds are advantageously rendered more volatile as acetyl, trifluoracetyl-, and chloroacetyl-derivatives. The type of acylation significantly affects the resolutions which can be achieved (68).

4. Applications of chiral stationary phases in GC

The main problem faced by an analyst who wishes to achieve the chromatographic resolution of a mixture of enantiomers is that of choosing the appropriate column. In chiral GC there is no comparable problem to that found in chiral HPLC, namely choice of mobile phase, but there are a number of other factors specific to GC analysis which require consideration. These include the following:

(a) choice of column temperature regime (isothermal versus programmed)
(b) choice of injection temperature
(c) selection of derivatizing reagent if analytes are not immediately volatile enough for GC analysis

Decisions regarding all of these options must normally wait until after the choice of column, for practical reasons associated with the working temperature range of the column stationary phase, and because any derivatization must not interfere with the key enantioselective interactions between injected analyte and CSP.

The literature provides the best source of guidance regarding the most suitable column for a given analysis, and the tables included have been compiled with this objective in mind. However, trial and error may have to override such considerations if no prior separation analogous to the current objective can be traced. Availability of the desired range of columns may also prove a limiting factor.

5. Conclusions and future prospects

Undoubtedly the most promising phases to have emerged for chiral GC are the cyclodextrin-derived CSP and they will, in the immediate future, offer the best first-choice procedure for chiral GC separations. Of the other phases, those using chiral siloxanes have been available for several years, are still available commercially, and have led to many elegant enantioseparations.

The main development to be anticipated in the future is an improved understanding of the ways in which enantioseparations occur, so that better CSP can be designed. Such understanding of underlying mechanisms should also assist the busy analyst by making the selection of a suitable CSP for a particular separation an easier and more logical process, an area for which expert systems might also be a future development.

References

1. König, W. A. (1987). *The practice of enantiomer separation by capillary gas chromatography*. Huthig, Heidelberg.
2. Schurig, V. (1986). *Kontakte (Darmstadt)*, **1**, 3.
3. Allenmark, S. G. (1991). *Chromatographic enantioseparation: methods and applications* (2nd edn.). Ellis Horwood, Chichester.
4. Bassindale, A. (1984). *The third dimension in organic chemistry*. Wiley, Chichester.

5. Kaye, P. T. and Learmonth, R. A. (1990). *J. Chromatogr.*, **503**, 437.
6. Brückner, H. and Langer, M. (1990). *J. Chromatogr.*, **521**, 109.
7. Deger, W., Gessner, M., Heusinger, G., Singer, G., and Mosandl, A. (1986). *J. Chromatogr.*, **366**, 385.
8. Gil-Av, E., Feibush, B., and Charles-Sigler, R. (1966). *Tetrahedron Lett.*, 1009.
9. Feibush, B. (1971). *J. Chem. Soc. Chem. Commun.*, 544.
10. Charles, R., Beitler, U., Feibush, B., and Gil-Av, E. (1975). *J. Chromatogr.*, **112**, 121.
11. Feibush, B., Gil-Av, E., and Tamari, T. (1972). *J. Chem. Soc. Perkin Trans. II*, 1197.
12. Feibush, B. and Gil-Av, E. (1970). *Tetrahedron*, **26**, 1361.
13. Feibush, B., Balan, A., Altman, B., and Gil-Av, E. (1979). *J. Chem. Soc. Perkin Trans. II*, 1230.
14. Gil-Av, E. and Feibush, B. (1967). *Tetrahedron Lett.*, 3345.
15. Chang, S.-C., Gil-Av, E., and Charles, R. (1984). *J. Chromatogr.*, **289**, 53.
16. Charles, R. and Watabe, K. (1984). *J. Chromatogr.*, **298**, 253.
17. Watabe, K. and Gil-Av, E. (1985). *J. Chromatogr.*, **318**, 235.
18. Ôi, N., Doi, T., Kitahara, H., and Inda, Y. (1982). *J. Chromatogr.*, **239**, 493.
19. Beitler, U. and Feibush, B. (1976). *J. Chromatogr.*, **123**, 149.
20. Frank, H., Nicholson, G. J., and Bayer, E. (1977). *J. Chromatogr. Sci.*, **15**, 174.
21. Nicholson, G. J., Frank, H., and Bayer, E. (1979). *J. High Res. Chromatogr. Chromatogr. Commun.*, **2**, 411.
22. Frank, H., Gerhardt, J., Nicholson, G. J., and Bayer, E. (1983). *J. Chromatogr.*, **270**, 159.
23. Koppenhoeffer, B., Allmendinger, H., Nicholson G. J., and Bayer, E. (1983). *J. Chromatogr.*, **260**, 63.
24. Frank, H., Nicholson G. J., and Bayer, E. (1978). *J. Chromatogr.*, **146**, 197.
25. Koppenhoeffer, B., Allmendinger, H., and Nicholson, G. J. (1985). *Angew. Chem. Int. Ed. Engl.*, **24**, 48.
26. Saeed, T., Sandra, P., and Verzele, M. (1979). *J. Chromatogr.*, **186**, 611.
27. König, W. A. and Benecke, I. (1981). *J. Chromatogr.*, **209**, 91.
28. König, W. A., Benecke, I., and Sievers, S. (1981). *J. Chromatogr.*, **217**, 71.
29. König, W. A., Benecke, I., and Sievers, S. (1982). *J. Chromatogr.*, **238**, 427.
30. König, W. A., Benecke, I., Lucht, N., Schmidt, E., Schulze, J., and Sievers, S. (1983). *J. Chromatogr.*, **279**, 555.
31. König, W. A. and Ernst, K. (1983). *J. Chromatogr.*, **280**, 135.
32. König, W. A. and Benecke, I. (1983). *J. Chromatogr.*, **269**, 19.
33. König, W. A., Francke, W., and Benecke, I. (1982). *J. Chromatogr.*, **239**, 227.
34. Abe, I., Kuramoto, S., and Musha, S. (1983). *J. Chromatogr.*, **258**, 35.
35. Schomburg, G., Benecke, I., and Severin, G. (1985). *J. High Res. Chromatogr. Chromatogr. Commun.*, **8**, 391.
36. Bradshaw, J. S., Aggarwal, S. K., Rouse, C. A., Tarbet, B. J., Markides, K. E., and Lee, M. L. (1987). *J. Chromatogr.*, **405**, 169.
37. Schurig, V. (1977). *Angew. Chem. Int. Ed. Engl.*, **16**, 110.
38. Golding, B. T., Sellars, P. J., and Wong, A. K. (1977). *J. Chem. Soc. Chem., Commun.*, 570.
39. Schurig, V. and Bürkle, W. (1978) *Angew. Chem. Int. Ed. Engl.*, **17**, 132.
40. Schurig, V. and Bürkle, W. (1982). *J. Am. Chem. Soc.*, **104**, 7573.

David R. Taylor

41. Schurig, V. and Weber, R. (1983). *Angew. Chem. Int. Ed. Engl.*, **22**, 772.
42. Schurig, V. and Weber, R. (1984). *J. Chromatogr.*, **289**, 321.
43. Schurig, V. (1988). *J. Chromatogr.*, **441**, 135.
44. Schurig, V. (1983). In *Asymmetric synthesis*, (ed. J. D. Morrison). Academic Press, New York, Vol. 1, p 59.
45. Schurig, V. and Link, R. (1988). In *Chiral separations*, (ed. D. Stevenson and I. Wilson). Plenum Press, New York, p. 91.
46. Schurig, V. (1988). In *Bioflavour '87* (ed. P. Schreier). de Gruyter, Berlin, p. 35
47 Weber, R. and Schurig, V. (1981). *Naturwissenschaften*, **68**, 330.
48. Schurig, V. and Wistuba, W. (1984). *Tetrahedron Lett.*, **25**, 5633.
49. Joshi, N. N. and Srebnik, M. (1989). *J. Chromatogr.*, **462**, 458.
50. Schurig, V., Össig, J., and Link, R. (1989). *Angew. Chem. Int. Ed. Engl.*, **28**, 194.
51. Burkle, W., Karfunkel, H., and Schü rig, V. (1984). *J. Chromatogr.*, **288**, 1.
52. Debowski, D., Jybilska, D., and Jurczak, J. (1982). *J. Chromatogr.*, **237**, 303.
53. Koscielski, T., Sybilska, D., and Jurczak, J. (1983). *J. Chromatogr.*, **280**, 131.
54. Mraz, J., Feltl, L., and Smolkova-Keulemansova, E. (1984). *J. Chromatogr.*, **286**, 17.
55. Schurig, V. and Nowotny, H.-P. (1988). *J. Chromatogr.*, **441**, 155.
56. König, W. A., Lutz, S., Mischnick-Lubbecke, P., and Brassat, B. (1988). *J. Chromatogr.*, **447**, 193.
57. König, W. A., Lutz, S., Wenz, G., and von der Bey, E. (1988). *J. High Res. Chromatogr. Chromatogr. Commun.*, **11**, 506.
58. König, W. A., Lutz, W., Colberg, C., Schmidt, N., Wenz, G., von der Bey, E., Mosandl, A., Gunther, C., and Kustermann, A. (1988). *J. High Res. Chromatogr. Chromatogr. Commun.*, **11**, 621.
59. König, W. A., Lutz, S., Hagen, M., Krebber, R., Wenz, G., Baldenius, K., Ehlers, J., and im Dieck, H. (1989). *J. High Res. Chromatogr.*, **12**, 35.
60. König, W. A., Krebber R., and Mischnick, P. (1989). *J. High Res. Chromatogr.*, **12**, 732.
61. König, W. A., Krebber, R., and Wenz, G. (1989). *J. High Res. Chromatogr.*, **12**, 790.
62. König, W. A., Krebber, R., and Wenz, G. (1989). *J. High Res. Chromatogr.*, **12**, 641.
63. Armstrong, D. W., Li, W., and Pitha, J. (1990). *Anal. Chem.*, **62**, 214.
64. Schurig, V. and Nowotny, H.-P. (1990). *Angew. Chem. Int. Ed. Engl.*, **29**, 939.
65. Mischnick-Lübbecke, P. and Krebber, R. (1989). *Carbohydrate Res.*, **187**, 197.
66. Fischer, P., Aichholz, R., Bölz, U., Juza, M., and Krimmer, S. (1990). *Angew. Chem. Int. Ed. Engl.*, **29**, 427.
67. König, W. A., Lutz, S., Mischnick-Lübbecke, P., Brassat, B., von der Bey, E., and Wenz, G. (1988). *Starch/Stärcke*, **40**, 472.
68. Armstrong, D. W. and Jin, H. L. (1990). *J. Chromatogr.*, **502**, 154.
69. Weinstein, S., Feibush, B., and Gil-Av, E. (1976). *J. Chromatogr.*, **126**, 97.
70. Ôi, N., Kitahara, H., Inda, Y., and Doi, T. (1981). *J. Chromatogr.*, **213**, 137.
71. Parr, P. and Howard, P. Y. (1972). *J. Chromatogr.*, **66**, 141.
72. Corbin, J. A., Rhoad, J. E., and Rogers, L. B. (1971). *Anal. Chem.*, **43**, 327.
73. Andrawes, F., Brazell, R., Parr, W., and Zlatkis, A. (1975). *J. Chromatogr.*, **112**, 197.
74. Parr, W. and Howard, P. Y. (1973). *Anal. Chem.*, **45**, 711.

75. König, W. A. and Nicholson, G. J. (1972). *Anal. Chem.*, **47**, 951.
76. König, W. A. Stolting, K., and Kruse, K. (1977). *Chromatographia*, **10**, 444.
77. Parr, W., Yang, C., Bayer, E., and Gil-Av, E. (1970). *J. Chromatogr. Sc.*, **8**, 591.
78. Stolting, K. and Konig, W. A. (1976). *Chromatographia*, **9**, 331.
79. Abe, I., Kohno, T., and Musha, S. (1978). *Chromatographia*, **11**, 393.
80. Lochmüller, C. H. and Stouter, R. W. (1974). *J. Chromatogr.*, **88**, 41.
81. Horiba, M., Kitahara, H., Yamamoto, S., and Ôi, N. (1980). *Agric. Biol. Chem.*, **44**, 2987.
82. Saeed, T., Sandra, P., and Verzele, M. (1980). *J. High Res. Chromatogr. Chromatogr. Commun.*, **3**, 35.

9

Environmental analysis using gas chromatography

GERRY A. BEST and J. PAUL DAWSON

1. Introduction

In the past decade, there has been a marked increase in public awareness of pollution of the environment. This has put pressure on regulating authorities and research organizations to produce information about the levels of contamination and their significance. As a result, a greater amount of analysis is being carried out on all types of environmental samples. Probably the most rapid development in the analytical measurements of pollutants has taken place in the quantification of organic constituents. This is largely because of the great advances that have taken place in the technique of gas chromatography.

For many years, the concentration of organic matter was assessed using only non-specific tests such as chemical oxygen demand (COD) or biochemical oxygen demand (BOD). These techniques are, however, no longer sufficient to investigate modern day pollution problems because of the vast range of artificial contaminants that are entering the environment. Some of these contaminants are extremely toxic at low concentrations e.g. dioxins, whilst others are now known to be persistent and bio-accumulable, and will be found in environmental samples many decades after their use has ceased, e.g. DDT and PCBs.

1.1 Pathways to the environment

It has been estimated that new compounds are being invented and brought into use at a rate of over 1000 per year. Many of these will find their way into the environment. *Figure 1* illustrates the various pathways for entry of material into the receiving environment and also shows the many different types of samples that may need to be taken to investigate the fate of a pollutant.

This chapter shows how these samples can be examined for the major organic contaminants and how their concentration is determined using gas chromatography.

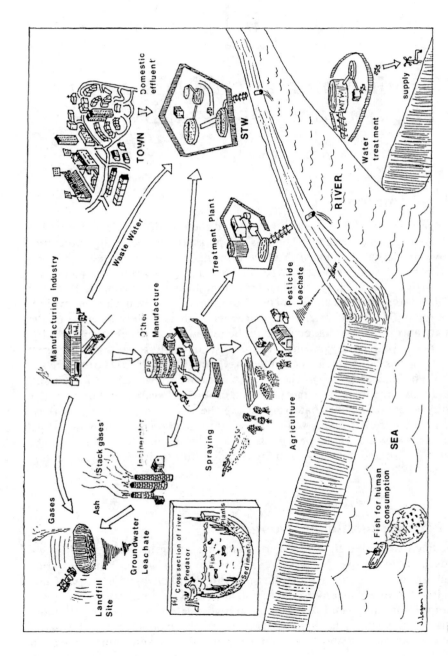

Figure 1. Illustration of the variety of pathways by which material can enter the receiving environment. Also indicated are the many different samples that may be required to be taken in order to investigate a pollutant.

1.2 Instrumentation

The most rapid development in GC in recent years has been the design of columns. Although wide-bore packed columns still have a useful role in GC work, the greater sensitivity and separating power of narrow-bore and wide-bore capillary columns have made these the favoured option for most chromatographers. The selection and design of columns are discussed in detail in Chapters 3 and 4.

For the most part, to carry out the techniques described in this chapter it is assumed that the reader has a 'standard' GC fitted with either a flame ionization detector (FID), electron capture detector (ECD), and/or nitrogen/phosphorus detector (NPD). However, the introduction in recent years of GCs coupled to 'bench-top' mass spectrometers or ion-trap detectors at affordable prices has meant that these combined facilities are now becoming commonplace in laboratories involved in environmental analysis. These detectors are very sensitive and, more importantly, highly specific. They determine the mass of fragment ions and molecular ions produced from the substances eluting from the GC column and in many instances can produce unique mass fragmentation patterns.

Interpretation of such a pattern can yield virtually unambiguous identification of the majority of organic compounds commonly found in environmental samples. The identification is assisted by the mass spectral data base data of about 59 000 organic compounds (NIST library) which can be called up to compare with the actual mass spectra obtained from the sample chromatogram.

2. The need for GC analysis of environmental samples

2.1 Pollution problems

Referring to *Figure 1*, there are many places in the pathways where an organic compound can enter the environment and reach concentrations that are unacceptable with resultant adverse effects, e.g. ill health or death of organisms in that part of the environment. There are many examples in literature giving details of how an incident was investigated and how the pollutant was brought under control. Some recent papers have described the analysis of pesticides in sea water (1), shrimps (2), and the leachate from timber yards (3), the concentration of chlorinated solvents in groundwater (4), and the drainage of herbicides into an agricultural catchment (5). It is not the intention in this book to give extensive lists of references. Those cited just illustrate the wide variety of sample types that are being examined with GC.

2.2 Statutory regulations

With the public demand for a cleaner environment and an improvement in the knowledge of the effect of various pollutants, there have been increasing controls applied to discharges by the regulatory authorities such as the National Rivers Authority (NRA), Her Majesty's Inspectorate of Pollution (HMIP), and the Scottish River Purification Boards (RPBs). The conditions imposed on a discharge into the atmosphere or receiving water by the regulating authority will sometimes specify the upper limit for an organic constituent. The quality of the discharge will then have to be checked at regular intervals by sampling and analysis to ensure that the discharge conditions are being complied with.

Much of the UK's environmental legislation now emanates from the European Community (EC) in the form of Directives with which each member country must comply. Some of the Directives contain specific limits for organic constituents in discharges and also recommend analysis by GC. The relevant EC Directives relating to the aquatic environment are shown in *Table 1*.

More recently, at a meeting of the states bordering the North Sea, agreement was reached at reducing the load of a range of pollutants entering this body of water. At the third North Sea Conference, the UK Government published a list of what it considered were the most dangerous of these pollutants. This list became known as the Red List of substances and the UK Government committed itself to reducing the estimated loading being discharged to one-half of the 1985 baseline by 1995. Since the list has been published, more substances have been added to it and, for some substances, the target is now for a 75% reduction in loading. The present Red List is shown in *Table 2*, whilst *Table 3* lists the so-called first priority substances which will be incorporated into the Red List at a later date.

From the above, it is clear that the requirements for the GC analysis of environmental samples has grown rapidly. Many laboratories are now hard pressed to become familiar with new techniques and to provide the data requested on time and with confidence in the quality of the results.

3. Analytical quality control of GC data

Before any analysis is carried out on environmental samples, it is very important to incorporate checks on extraction efficiency, losses on clean-up, purity of blanks, and consistency of separation on the column, through the use of analytical quality control (AQC) (1).

With each batch of samples that have been collected from the environment, the following quality control should be included.

(a) Internal standards at known concentrations should be added to the samples, such as *trans*-heptachlorepoxide (*trans*-HE), decachlorobiphenyl (DCBP), and ε-hexachlorocyclohexane (εHCH). Each of these standards are eluted from the clean-up columns in the different fractions to check on losses. These internal standards rarely occur in natural

Table 1. Directives specifying limits on organic compounds to be measured by GC

No.	Title	Organic components	Limits (μg/litre)
75/440/EEC	Surface water directive	Cyfluthrin	0.001
		Hydrocarbons	50–1000[a]
		Permerthrin	0.01
		Pesticides	1–5.0[a]
		Phenols	1–100[a]
		Polycyclic aromatic hydrocarbons	0.2–1.0[a]
80/778/EEC	Quality of water intended for human consumption	Pesticides individually	0.1
		Pesticides total	0.5
		Benzo-3,4-pyrene	0.01
		Tetrachloromethane	3.0
		Trichloroethane	30.0
		Tetrachloroethane	10.0
		Trihalomethanes	100.0
		Phenols	0.5
		Polyaromatic hydrocarbons	0.2
76/659/EEC	Quality standards for freshwater required to support fish	Cyfluthrin	0.001[b]
		Flucofuron	1.0[b]
		PCSDs and PAD	0.05[b]
		Permethrin	0.01[b]
		Sulcofuron	25.0[b]
76/464/EEC	Dangerous Substances Directive: list 1 substances	Carbon tetrachloride	12.0[c]
		Chloroform	12.0[c]
		DDT	0.025
		p-DDT	0.01
		total 'drins	0.03
		Endrin	0.005
		Aldrin	0.01
		Dieldrin	0.01
		Isodrin	0.005
		Hexachlorobenzene (HCB)	0.03
		Hexachlorobutadiene (HCBD)	0.01
		Hexachlorocyclohexane	
		HCH or Lindane	0.11
		Pentachlorophenol	2.0
		1,2-Dichloroethane (DCE)	10.0[a]
		Perchloroethylene (PER)	10.0[a]
		Trichlorobenzene (TCB)	0.1[a]
		Trichloroethylene (TRI)	10.0[d]

[a] Limits depend upon amount of treatment required.
[b] To come into force in 1992.
[c] Applies to fresh and saline water.
[d] Proposed water quality standards—not yet finalized.

Table 2. Red list of substances which are regarded as hazardous and whose entry into the environment must be reduced

Mercury	Simazine
Cadmium	Atrazine
Copper	Tributyl tin compounds
Zinc	Triphenyl tin compounds
Lead	Azinphos ethyl
Arsenic	Azinphos methyl
Chromium	Fenitrothion
Nickel	Fenthion
'Drins (Aldrin, Dieldrin, Endrin)	Malathion
γHCH (Lindane)	Parathion
DDT	Parathion methyl
Pentachlorophenol (PCP)	Dichlorvos
Hexachlorobenzene (HC)	Trichloroethylene
Hexachlorobutadiene (HCBD)	Tetrachloroethylene
Carbon tetrachloride	Trichlorobenzene
Chloroform	1,2-Dichloroethane
Trifluralin	Trichloroethane
Endosulphan	Dioxins

Table 3. Candidate compounds for adding to Red List

Demeton O	Cyanuric chloride
Dimethoate	1,1-Dichloroethylene
Mevinphos	1,3-Dichloropropan-2-ol
2,4-D	2-Chloroethanol
Linuron	Ethyl benzene
Pyrazon	Biphenyl
2-Amino-4-chlorophenol	1,4-Dichlorobenzene
Anthracene	1,3-Dichloropropene
Chloracetic acid	Hexachlorethane
4-Chloro-2-nitrotoluene	1,1,1-Trichloroethane

2,4-D = 2,4-dichlorophenoxyacetic acid

samples so their measured concentrations will reflect that which was added to the samples.

(b) One of the environmental samples should be analysed in duplicate to check on the consistency of the procedure.

(c) If the environmental samples are of water or effluent, the blank should be deionized water. If, however, these are of tissue or sediment, then reference blanks can be obtained from a supplier of reference materials.

Each of these AQC samples is treated identically to the environmental samples. In addition, the GC instrument is calibrated at the beginning and end of each run of samples with at least four different concentrations of the target compounds made up as standard solutions in solvent together with a known concentration of the internal standard. The calibration curves are drawn using either peak height, or peak area if an integrator is available, and the concentration of samples calculated from these curves. For repeated analysis of a similar type of sample, it is necessary to check on the calibration by analysing only two concentrations of standard solutions.

Once the analysis is compiled, the following AQC criteria should be met otherwise the analysis should be repeated.

(a) When the concentration of the internal standard (e.g. *trans*-HE), which has been through the full analytical procedure, is compared with the concentration of the internal standard in the calibrated solutions, the recovery should be between 50–130%.

(b) The percentage recovery of each of the target compounds added to the control sample and the spiked environmental sample should be calculated after a correction factor (the response factor—see below) is obtained for the internal standard. The recoveries should be between 80–120%.

(c) The results for the analyses of the duplicated environmental samples should be within 25% of each other.

(d) The blank solutions should give no detectable levels of the target compounds.

4. Isolation of target compounds from the sample matrix

Before any GC separation and quantification can be carried out on an environmental sample, the target compounds must first be isolated from the sample matrix and interfering compounds removed by a clean-up procedure. For some matrices, the procedure is quite involved and time consuming, and losses of the target compounds will occur. The extent of losses must be assessed so that the concentration determined by the GC analysis is adjusted for the lower yield.

Referring to *Figure 1*, the types of environmental samples that may be collected are as follows:

- *water*—river water, sea water or potable supply
- *effluent*—industrial or sewage
- *sewage sludge*
- *sediment*—fresh water or marine
- *biological tissue*—from a variety of organisms such as fish, invertebrates, birds

- *gases*—from stack emission, landfill of industrial and domestic wastes, or workplace environments
- *oil*—from pollution incident or from suspected source

Apart from gases and oil, the most likely pollutants that will need to be determined are pesticides, solvents, polyaromatic hydrocarbons (PAHs), and polychlorinated biphenyls (PCBs). The pesticides may be either insecticides or herbicides and are usually organochlorine (OC), organophosphorus (OP), or organonitrogen (ON) compounds. There are an enormous number of these compounds and a comprehensive guide is available from the Royal Society of Chemistry (6). It is not the intention in this chapter to describe the techniques for extracting and quantifying each one from a variety of matrices. Only the broad principles will be described for the most commonly occurring substances, first organochlorine compounds, second organophosphorus, and third organonitrogen compounds. Methodologies for the determination of many synthetic organic contaminants in environmental samples have been clearly set out elsewhere (7, 8).

These substances will generally be present at low concentrations in environmental samples, typically in the pg to μg range (10^{-12}–10^{-6}/litre), and rarely at the mg level (10^{-3}/litre). Gas chromatography is an extremely sensitive technique, especially when an ECD is used as the detector.

Because of the sensitivity, the technique is very prone to contamination and high background interference.

4.1 Contamination

Contamination of samples, sampling equipment, and instruments is costly in analytical time and is very frustrating for the analyst. Some of the most common types of interference in the methods are summarized below.

4.1.1 Materials

Any material which contains a semi-volatile organic compound which could come in contact with the sample will cause interference. Such materials include plastic tubing other than PTFE, plastic bottle tops, liners in bottle caps, GC vial septa, rubber components in gas regulators, impure carrier gas, and impurities dissolved in solvents from the laboratory atmosphere among others.

4.1.2 Carry over

Interfering compounds may occur if a sample containing a low concentration of target compounds is analysed immediately after analysis of a sample containing a high concentration of the same compounds.

4.1.3 Common interferants

Two phthalate compounds are nearly always present in every chromatogram at the mg/l level. These compounds are di-*n*-butylphthalate and di(2-ethyl hexyl)phthalate.

4.1.4 Glassware cleaning

A suitable cleaning procedure for all glassware is as follows.

(a) Wash thoroughly with hot water and a proprietary detergent such as Decon.

(b) Rinse three times with tap water followed by deionized water and dry in an oven at 105°C for one hour.

(c) Cool to room temperature.

(d) Rinse sample bottles and equipment with acetone followed by hexane, before use.

4.2 Procedure for the extraction of organochlorine compounds and PCBs from water

The water sample is put into a prepared 1.2 litre glass bottle fitted with either a glass stopper or a plastic cap lined with a PTFE insert. Further information on sampling techniques and the design of sampling programmes are found in a number of other sources (9, 10) and it is not intended in this chapter to deal with the topic. For the extraction of the target compounds from the sample, the procedure outlined in *Protocol 1* should be followed.

Protocol 1. Solvent extraction of water sample for organochlorine pesticides

Materials and equipment

- HPLC grade or glass distilled hexane (GDG)
- anhydrous sodium sulphate (prepared by heating at 500°C for 4 h)
- 2 litre separating funnel
- 100 ml measuring cylinder
- 2 litre measuring cylinder
- 2 × 250 ml round-bottomed flask
- filter funnel with Whatman No. 1 filter paper
- rotary evaporator
- a nitrogen supply*

This supply* should end in a fine jet of glass tubing with a controllable tap connected to the supply line fitted with pellets of molecular sieve (type 13x) and 15–40 mesh silica gel. Alternatively, an air pump (e.g. aquarium pump) can be used providing the two filters are included in the air line.

As an alternative to the latter two items the following can be used.

- Kuderna–Danish evaporator (7)
- micro Snyder column (7)

Protocol 1. *Continued*

Method

1. Shake the sample bottle to ensure that all particulate matter is suspended and transfer the contents to a pre-washed 2 litre separating funnel.

2. Add two internal standards to the sample in the funnel. Suitable ones are 40 μl of 1000 μg/litre octachloronaphthalene and 40 μl of 1000 μg/litre ε-hexachlorocyclohexane (εHCH).

3. Add 50 ± 0.5 ml HPLC-grade hexane to the sample bottle, stopper, and shake for about 3 min and transfer the contents to the separating funnel.

4. Shake the separating funnel vigorously, venting the pressure at regular intervals into a fume cupboard.

5. Allow the layers to separate and run the lower aqueous layer back into the sample bottle. Collect the hexane layer in a 250 ml round-bottomed flask.

6. If an emulsion has formed, this is best broken up by centrifugation. Place the hexane extract and emulsion into pre-cleaned centrifuge tubes, spin at 400 *g* for about 5 min, and pipette off the aqueous layer. Put the hexane into the round-bottomed flask and rinse out the centrifuge tubes with hexane, putting the rinsings into the flask also.

7. Extract the aqueous sample in the sample bottle with a further 50 ± 0.5 ml hexane, allow to separate. Pour the aqueous layer into a measuring cylinder and note the volume extracted.

8. To the combined hexane extracts, add approximately 3 g anhydrous sodium sulphate and shake for about 30 sec. Allow the sodium sulphate to settle and then filter the dried hexane through a No. 1 filter paper into another pre-cleaned 250 ml round-bottomed flask. Rinse the sodium sulphate with 10 ± 0.5 ml hexane and transfer this residual extract to the other hexane extract via the filter funnel. Add 1 ml of iso-octane to the hexane as a 'keeper' to prevent evaporation losses.

9. Using a rotary evaporator under vacuum, reduce the volume of the hexane extract to about 10 ml. Avoid evaporating to less than 10 ml because loss of organochlorine compounds will take place. Transfer this volume, with rinsing, to a 10 ml graduated test-tube and reduce it further to less than 1 ml by evaporating it slowly (i.e. small disturbance on surface only) in a fume cupboard using the nitrogen or compressed air line.

Figures 2 and 3, respectively, illustrate the Kuderna–Danish evaporator and micro Snyder column facilities.

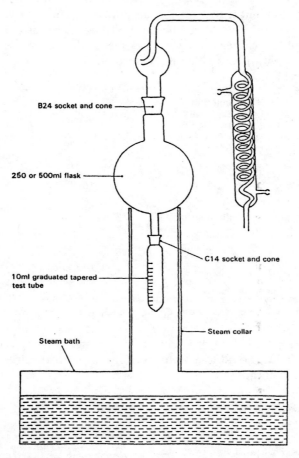

B24 socket and cone

250 or 500ml flask

C14 socket and cone

10ml graduated tapered
test tube

Steam collar

Steam bath

Figure 2. Kuderna–Danish evaporator system for concentration of pesticide solutions.

4.3 Procedure for the extraction of organochlorine compounds from samples of effluent

This procedure is essentially the same as for clean water samples and thus *Protocol 1* is followed. An emulsion is more likely to form with these types of sample. Its formation can be discouraged by adding 1 g Analar grade sodium chloride in step 2 along with the internal standards.

Effluent samples, however, may contain sulphur compounds which interfere with the chromatography by producing extra peaks. These compounds are more likely to be present in dirty samples if they have aged in any way, such as when originating from a stagnant area and can be removed using either mercury or tetrabutylammonium sulphate (TBAS). The efficiency of the removal of sulphur compounds by these methods has recently been investigated (11) and it was found that both procedures work satisfactorily. The TBAS method is preferred because of the hazards associated with mercury and the difficulties with the disposal of the metal. The method for removal of sulphur is described in *Protocol 2*.

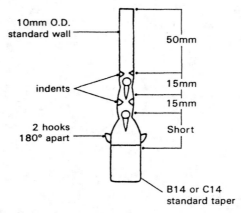

Figure 3. Micro Snyder column. (Reproduced from J. A. Burke, P. A. Mills, and D. C. Bostwick (1966). *J. Assoc. Offic. Anal. Chem.*, **49**, 999 (with kind permission).)

Protocol 2. Removal of sulphur compounds from dirty effluent samples

Materials and equipment

- propan-2-ol
- TBAS solution (prepared by dissolving 3.5 g tetrabutylammonium sulphate in 100 ml deionized water, extracting any contaminants with hexane, discarding the hexane, and saturating the solution with sodium sulphite)
- sodium sulphite
- hexane (GDG)

- deionized water (hexane-extracted)
- anhydrous sodium sulphate (heated at 500°C for 4 h)
- 2 × 25 ml stoppered test tube
- Pasteur pipettes
- filter funnel fitted with a Whatman No. 1 filter paper

Method

1. To the concentrated extract contained in a 50 ml test-tube obtained from step 9 in *Protocol 1*, add 100 μl of propan-2-ol, 2 ml TBAS solution, and 200 mg sodium sulphite.

2. Shake the mixture for 3 min by hand and allow to separate.

3. Remove the aqueous layer with a Pasteur pipette and put into a 25 ml stoppered test-tube. Add to this layer 5 ml hexane, shake for 1 min, and allow to separate. Take off the hexane layer with a Pasteur pipette and add to the hexane in the other tube.

294

4. Rinse the hexane extracts by adding 5 ml deionized water and shaking for 3 min. Allow to separate, and remove and discard the aqueous layer.

5. Dry the hexane by adding approximately 0.5 g anhydrous sodium sulphate to the hexane and shake for about 30 sec.

6. Filter the hexane through a No.1 filter paper in a filter funnel and collect in a clean 25 ml test-tube. Rinse the sodium sulphate with 5 ml hexane and transfer this via the filter funnel to the other hexane. The extract is now ready for the next stage.

4.4 Extraction of organochlorine compounds from sediment samples

Care must be taken when sampling sedimentary material because of its variation in character in different parts of the river or lake. Over 90% of the organochlorine compounds are associated with the finest particle sized material so a sample should be taken of this fraction. In a river, the sediment should be sampled from a depositing zone rather than an eroding zone. Similarly, in a lake, the edge material should be avoided as it is subject to wave action. The sample should be scooped up in a stainless steel or glass container and then sieved through a 63 µm mesh metal sieve. The fine fraction is collected and retained.

If possible, samples should be extracted without being dried, because the drying process can lead to the loss of pesticides and other organic materials. If the sediment cannot be extracted at once, it should be stored at low temperature in a deep freeze. Sediments may contain sulphides and carbonates which must be removed prior to the extraction of the organochlorine compounds from the matrix.

There are two procedures for extracting organochlorine compounds from sediments, either by agitating the sediment and hexane with ultrasound or else repeated extractions using the soxhlet method. Both methods are satisfactory and the preferred one depends upon the availability of the equipment.

A check on extraction efficiency is made by adding internal standards which are extracted and quantified. However, this does not necessarily reflect the true extraction efficiency of the method because the organochlorine compounds in the sample may be more tightly bound to the sediment particles than the added standards. The amount of organochlorine compounds in the sediment needs to be related to the amount of dry matter and to the organic content of the sample. These should be determined separately by determining the moisture content of the sediment by drying a weighed amount in an oven at 105°C, while the total organic content is determined by igniting in a muffle furnace or chemical oxidation with chromic acid. *Protocol 3* describes the procedure used.

Protocol 3. Extraction of organochlorine compounds from sediments with ultrasonics

Materials and equipment

- glacial acetic acid
- deionized water, hexane-extracted
- propan-2-ol
- 100 ml conical flask
- ultrasonic bath
- mechanical shaker
- Pasteur pipettes
- centrifuge

Method

1. Centrifuge the sample immediately or once it has been thawed out. Pipette off the excess water with a Pasteur pipette.
2. Weigh out accurately about 5 g of wet sediment into a 100 ml pre-washed conical flask. Remove sulphides and carbonates by adding 5 ± 0.5 ml of propan-2-ol. Shake well and vent off gases.
3. Using a microlitre syringe add the internal standards such as 40 μl of 1000 μg/litre ε-HCH.
4. Clamp the unstoppered flask into an ultrasonic bath and agitate with ultrasound for about 30 min.
5. Add 15 ml hexane, stopper the flask, and shake in a mechanical mixer for 5 min.
6. Using a Pasteur pipette, transfer the hexane extract and the emulsion that will have formed, into a 100 ml centrifuge tube.
7. Balance the tube with another containing water and centrifuge at 15000 r.p.m. for 5 min.
8. Remove the hexane layer with a Pasteur pipette and dispense into a 60 ml stoppered tube.
9. Repeat the hexane extraction from step 5 twice more and add the extracts to the 60 ml tube. Discard the sediment.
10. Remove sulphur compounds in the extract by carrying out the procedure in *Protocol 2*.

Protocol 4 describes extraction of OCs using soxhlet extraction.

Protocol 4. Extraction of organochlorine compounds from sediments with soxhlet extraction

Materials and equipment

- pre-washed pestle and mortar
- soxhlet extraction equipment
- stoppered glass test-tube (glass stopper or PTFE-lined)
- rotary evaporator
- nitrogen or compressed air blower
- extraction thimbles 20 mm × 120 mm
- methyltertiarybutylether (MTBE)

Method

1. Remove any supernatant water from the sample with a Pasteur pipette and weigh up to 10 g to an accuracy of 0.01 g of the wet sediment in a watch glass, and transfer this to the mortar.

2. Grind up the sediment with enough anhydrous sodium sulphate until a dry, free flowing powder is produced. Add the internal standards to the mixture.

3. Transfer the powder to the soxhlet thimble and place it in the extraction apparatus charged with 120 ± 10 ml ether/hexane mixture (20:80, v/v).

4. Allow the sample to be extracted for at least 10 cycles; about 3 h will be required.

5. Transfer the hexane to the rotary evaporator and concentrate it to no less than 10 ml.

6. Remove sulphur compounds using the procedure outlined in *Protocol 2*.

4.5 Extraction of organochlorine compounds from tissue samples

This procedure is the same as those described for sediment samples, i.e. ultrasonic or soxhlet extraction, except for the preparation of the material. The amount of material extracted depends upon the lipid content because the more fatty material in the tissue the more pre-injection clean-up is required. Thus, for a fish liver sample, about 2 g of material is extracted. For muscle tissue, which contains less fatty material, the sample weight can be increased to about 5 g.

The lipid content is determined on a separate amount of sample by extracting the fats with petroleum ether and evaporating the ether to constant weight over a steam bath. The tissue sample for extraction is weighed wet after it has been broken up by homogenizing, using a suitable homogenizer such as Ultra turex. An alternative method for preparing the biological tissue, is to freeze-dry the material and then grind it to a fine powder.

5. Clean-up procedures

In all of the extraction techniques described so far, it is quite likely that the extractant solvent will contain interfering substances which will produce additional peaks in the chromatogram. This occurrence is particularly so in the extracts from samples of effluent, sediment, and tissue which contain fats, oils, and other naturally occurring substances. The long-established procedure for the removal of these substances has been to pass the extract through an alumina column and then to separate the target compounds into different batches by passing the clean-up extract through a silica column. Recently, the clean-up and separation has been streamlined by the use of commercially available solid-phase extraction cartridges or discs. These three techniques will be described.

5.1 Clean-up and separation of extracts using alumina and silica columns

This technique removes interfering compounds by passing the extract through basic and acidic alumina and then separating the organochlorine compounds on a silica gel column.

5.1.1 Preparation of solid adsorbents

The preparation of solid adsorbents—basic alumina, acidic alumina, and silica gel is as follows.

i. Basic alumina

The starting material to be used is Merck No. 1097 or equivalent. About 100 g alumina is placed in a silica dish, heated in a muffle furnace for 4 h at 800 °C, cooled to about 200 °C and then to room temperature in a desiccator. An aliquot is weighed into a stoppered glass container (such as a conical flask with a PTFE-lined screw top) and deionized water equivalent to 4% of the alumina weight is added. The sample is agitated to mix thoroughly and stored well-sealed.

ii. Acidic alumina

A portion of the alumina is washed with 1 M HCl by making a slurry in beaker filtered through a sinter funnel, then dried in a silica dish at 150 °C for 4 h and cooled in a desiccator. Water (4%, w/w) is added to a weighed portion in a stoppered flask, the sample mixed well, and stored sealed.

These alumina preparations will slowly deactivate on exposure to air and so should be discarded after two weeks.

iii. Silica gel

About 100 g silica gel (Merck Kirsel Gel 60 or Merck 7754) is heated in a silica dish in a muffle furnace for 2 h at 500 °C, cooled and then placed in a desiccator. A portion is weighed into a stoppered glass container and water equivalent to 3% of the silica weight added. The sample is shaken well to mix thoroughly and stored sealed. The silica gel deactivates more rapidly than the alumina and should preferably be prepared daily.

The procedure for clean-up and separation is given in *Protocol 5*.

Protocol 5. Clean-up and separation of organochlorine
compounds using alumina and silica columns

Materials and equipment

- HPLC grade or glass-distilled hexane
- chromatographic columns—6 mm internal diameter and about 100 cm long. These columns can either be tapered at one end and plugged with a prewashed glass wool plug

or fitted with a PTFE-lined stop-cock (see *Figure 4*)
- Pasteur pipette
- 2 × 25 ml glass stoppered tubes

Method

1. Prewash the columns with acetone followed by glass distilled or HPLC grade hexane and allow to dry.

2. Prepare the acid/base alumina column by first adding 2 ± 0.1 g acidic alumina then 1 g basic alumina to the column and tap the column to settle.

3. Prepare the silica column by adding 2.5 ± 0.1 g deactivated silica gel to the column and tap to settle.

4. Wet the alumina column by passing through 10 ± 1 ml hexane and run off excess until the hexane meniscus is level with the column material.

5. Transfer the hexane/iso-octane extract solution on to the column with a Pasteur pipette. Allow it to be adsorbed on to the column. Wash the sample test-tube with 1 ml hexane and add this to the column.

6. The organochlorine compounds are now eluted from the column in three different fractions by adding hexane. Collect the first fraction of 4 ml eluate in a glass test-tube, the second fraction of the next 6 ml in another tube and label it eluate 3, and the third fraction of the next 20 ml and label it eluate 4.

7. Blow down the first fraction to 1 ml and then further split by transferring it to the dry silica gel column and allowing it to be adsorbed by the column material and rejecting any excess. Elute the organochlorine compounds in two fractions by passing through hexane and collecting the first 6 ml and then the next 7 ml. Label these eluates 1 and 2.

8. Evaporate each of the eluates on a blower slowly down to about 0.5 ml. Store in small glass-stoppered bottles with the volume made up to 1 ml by rinsing out the evaporating tube. These samples are now ready for injection into the gas chromatograph.

If present in the original sample, the organochlorine compounds contained in the eluates are listed in *Table 4.*

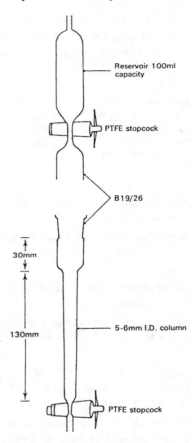

Figure 4. Schematic diagram of an adsorption chromatography column.

5.2 Modified method for clean-up and separation using alumina/silica nitrate and silica gel

A modified form cf this clean-up and separation technique involves splitting the eluates into two fractions instead of four (12). The concentrated extract is cleaned-up on an alumina/silver nitrate column followed by silica gel. The preparation of the sorbents is described below.

5.2.1 Preparation of solid adsorbents

i. Alumina

About 100 g alumina (Woden 200 neutral or similar) is heated in a silica dish at 500°C for 4 h and then cooled. To a weighed portion in a stoppered glass container, deionized water equivalent to 7% (w/w) of the alumina weight is added and the sample agitated to mix thoroughly. The alumina is kept in a

Table 4. List of components in eluates from *Protocol 5*

		Eluate number	
1	**2**	**3**	**4**
Hexachlorobenzene (HCB)	pp-DDT	pp-TDE	Heptachloroepoxide
Hexachlorobutadiene (HCBD)	op-DDT	HCH	Dieldrin
Heptachlor	Endosulphan B	Endosulphan A	Chloropyrifos
Aldrin			Endrin
pp-DDE			
Methoxichlor			
PCBs			

sealed container. This adsorbent is stable for only about a week once re-exposed to atmosphere.

ii. Alumina/silver nitrate
A batch of material for adding to the column is prepared by dissolving 0.75 g AgNO₃ in 0.75 ± 0.1 ml water and then adding 4 ± 0.2 ml acetone. To this solution in an unstoppered conical flask 10 ± 0.2 g of dried alumina is added and the sample shaken to mix. The acetone is allowed to evaporate and the preparation then stored in the dark until ready for use. The adsorbent should be freshly prepared daily.

iii. Silica gel
The silica gel is prepared as described in Section 5.1.1 iii.

5.2.2 Procedure for the clean-up and separation of organochlorine compounds
Protocol 6 describes the procedure for the clean-up and separation of organochlorine compounds using alumina/silica nitrate and silica gel columns.

Protocol 6. Clean-up and separation of organochlorine compounds using alumina/silver nitrate and silica gel columns

Materials and equipment

- activated alumina
- alumina (silver nitrate)
- silica gel
- hexane (GDG)
- diethyl ether/hexane mixture (20:80, v/v)
- glass chromatographic columns
- Pasteur pipette
- glass-stoppered tubes
- 100 ml round-bottomed flask
- rotary evaporator

Protocol 6. *Continued*

Method

1. Plug the base of a chromatographic column with hexane-washed glass wool or cotton wool and add 15 ml hexane. Pour in 1 ± 0.2 g alumina/ silver nitrate and allow to settle, and follow this by 2 ± 0.2 g alumina (7% water) and, again, allow to settle. Charge with a little anhydrous sodium sulphate.

2. Run off the excess hexane until the liquid level is at the top of the column. Add the concentrated hexane extract, with rinsing, to the top of the column.

3. Pass 30 ± 1 ml hexane through the column and collect the eluate in a 100 ml round-bottomed flask.

4. Concentrate the extract in a rotary evaporator down to 10 ml. Transfer this to a test-tube and blow slowly down to 1 ml.

5. Prepare a silica gel column by adding 2 ± 0.1 g silica gel to a plugged chromatographic column and top up with a little anhydrous sodium sulphate.

6. Add the concentrated hexane eluate from the alumina/silver nitrate column, with rinsing, to the silica gel column and allow it to be adsorbed.

7. Add 10 ± 0.2 ml hexane to the top of the column and collect the first 7 ml eluate in a test-tube (eluate 1).

8. Add 12 ± 1 ml diethyl ether/hexane mixture (20:80, v/v) and collect all the eluate in a second test-tube (eluate 2).

9. Concentrate each eluate down to 1 ml with a blower and keep in a sealed vial until injection into the gas chromatograph.

Table 5 lists the possible compounds in eluates 1 and 2.

Table 5. List of components in eluates 1 and 2—*Protocol 6*

Eluate number	
1	**2**
Aldrin	Dieldrin
pp-DDE	Endrin
op-DDT	Chlorpyrifos
PCBs	HCH
Endosulphan B	Heptachlor epoxide
HCB	op-DDT
HCBD	pp-TDE
Heptachlor	Endosulphan A

5.3 Clean-up and separation of extracts using solid-phase extraction (SPE) cartridges

There are a variety of cartridges available (e.g. Bond Elut from Analytichem International). Each is filled with a chemically modified silica adsorbent and the appropriate one selected according to the nature of the material to be extracted. For separation and clean-up of organochlorine compounds, the most appropriate one is aminopropyl (catalogue number 611303.).

The clean-up procedure is speeded up by the use of a syringe which is fitted to the disposable cartridge as shown in *Figure 5*.

Protocol 7 describes the procedure using SPE.

Syringe   Adapter   Cartridge

Figure 5. Solid-phase extraction cartridge and syringe. (Reproduced with kind permission from Analytichem International.)

Protocol 7. Clean-up and separation of extracts using silica

Materials and equipment

- Bond Elut aminopropyl cartridge
- 10 ml glass test-tubes
- silica gel column
- nitrogen or air blower
- methanol—HPLC grade
- HPLC grade or glass-distilled hexane
- MTBE/hexane (20:80, v/v)

Method

1. Take a Bond Elut aminopropyl cartridge, put the black adaptor piece on top, and fit a 10 ml glass syringe onto the adaptor.
2. Put 5 ml methanol into the syringe and pass it through the tube. Do not allow the tube to go dry.
3. Wash the excess methanol from the tube 5 ml hexane; again ensure that the tube does not go dry.
4. Detach the syringe, add hexane/iso-octane extract to the top of the tube, and allow it to pass through. Collect the clean extract in a glass tube.
5. Pass a further 3 ml hexane through the tube with the syringe and collect the eluate in the same test-tube. This should remove any strongly coloured contaminants from the extract. If the eluate is still coloured, pass it through another Bond Elut cartridge.

Protocol 7. *Continued*

6. Dispose of used tubes as these should not be re-used.

7. Evaporate the extract to 1 ml by blowing down.

8. The extract should now be cleaned of interfering compounds but before GC analysis it needs to be subdivided on a silica gel column. Prepare a silica gel column as in *Protocol 5* by adding 3 g 3% (w/w) deactivated silica gel to the column.

9. Put the cleaned-up eluate on top of the column and allow it to be adsorbed.

10. Elute the column with 7 ml hexane and collect the eluate in a glass test-tube. Label the tube eluate 1.

11. Put another test-tube under the column and add 20 ml of ether/ hexane (20:80, v/v) to the column. Collect the eluate and label the tube eluate 2.

Table 6 lists the components that may be contained in eluates 1 and 2 , obtained according to *Protocol 7*.

5.4 Extraction of semi-volatile organic compounds from water samples using extraction discs

This method has recently been developed and validated by the Environmental Health Laboratories in Indiana, USA, for the US Environmental Protection Agency (13). It is suitable only for clean water samples. The organic compounds are removed from the water as it is filtered through the extracting disc and then the compounds are eluted with a suitable solvent.

Protocol 8 describes the extraction of organochlorine compounds from water using extraction discs.

Table 6. List of possible components in eluates 1 and 2—*Protocol 7*

Eluate number	
1	**2**
HCB	HCH
HCBD	Endosulphan A
Aldrin	Dieldrin
Heptachlor	Endrin
pp-DDE	Heptachlor epoxide
op-DDE	op-DDT
Endosulphan B	pp-TDE
PCBs	Chlorpyrifos

Protocol 8. Extraction of organochlorine compounds from water using extraction discs

Materials and equipment

- Empore 3M C_{18} extraction disc (Analytichem International)
- extraction funnel with fritted glass support
- 1 litre vacuum conical flask
- 50 ml glass test-tubes
- vacuum pump
- filter funnel with Whatman No.1 filter paper
- nitrogen or air blower
- methylene chloride/ethyl acetate mixture (50:50, v/v; GDG)
- deionized water, hexane-extracted
- ethyl acetate (GDG)
- anhydrous sodium sulphate

Method

1. To 1 litre sample in a pre-cleaned glass bottle add 50% HCl until pH is less than 2.

2. To the acidified sample add internal standards.

3. Place the extraction disc into the filtration apparatus (see *Figure 6*) and rinse the disc with 10 ml methylene chloride/ethyl acetate mixture. Allow to dry for 1 min.

4. Rinse the disc with 10 ml methanol, apply the vacuum, and follow this with 10 ml hexane-extracted deionized water. Remove the vacuum before the disc goes dry and discard the filtrate.

5. Pass 1 litre of the sample through the disc with vacuum applied so that the flow rate is approximately 50 ml/min. Discard the filtrate.

6. Place a 50 ml collection tube into the vacuum flask so that it is beneath its filter funnel (see also *Figure 6*).

7. Add 5 ml ethyl acetate to the sample bottle and rinse it out. Pour the solvent into the filter funnel. Rinse the inside walls of the filter funnel with a little ethyl acetate.

8. Apply vacuum until the eluate starts to flow through the disc. Remove vacuum and let the solvent penetrate the disc for about 1 min.

9. Reapply a little vacuum so that the eluate enters the collection tube at a rate of 1–2 drops per min.

10. Repeat steps 7–9 with a further 5 ml ethyl acetate.

11. Remove the collection tube and dry the eluate by passing it through a filter paper containing about 3 g anhydrous sodium sulphate.

12. Blow the solvent down to about 0.5 ml and transfer, with rinsing with methylene chloride/ethyl acetate mixture, to a sample vial, and make up volume to 1 ml. The extract is now ready for quantification by GC.

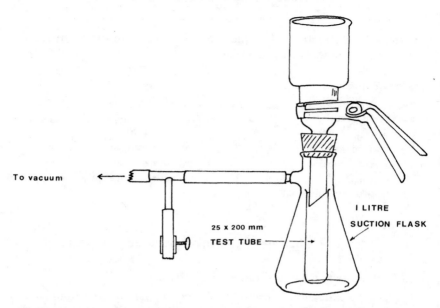

To vacuum

25 x 200 mm

TEST TUBE

I LITRE

SUCTION FLASK

Figure 6. Filtration apparatus fitted with an extraction disc showing collection vessel. (Reproduced with kind permission from Analytichem International.)

6. Analysis of pentachlorophenol

Pentachlorophenol is a common organochlorine compound that is found in environmental samples. It is used as a preservative for wood and glue as well as an insecticide, herbicide, and as a defoliant. It has been found in effluent from the paper industry, tanneries, and textile plants.

Before it can be extracted from water samples by a solvent, a derivative is formed by reacting it with acetic anhydride (see *Protocol 9*). This reagent hydrolyses if left on the open bench so the bottle should be kept stored in a dark cupboard.

Protocol 9. Extraction from water samples and the derivatization of pentachlorophenol (PCP)

Materials and equipment

- Analar disodium tetraborate (Borax)
- acetic anhydride (Fluka Chemie)
- HPLC grade or glass-distilled iso-octane
- 50 ml screw-capped tube
- Pasteur pipettes

Method

1. Put 40 ml sample (or 20 ml for sewage effluent or trade effluents) in a 50 ml stoppered tube.

---

2. Add 0.2 g borax and 0.2 ml analytical grade acetic anhydride.

3. Shake the tube to mix the reagents.

4. Add 2 ml of a 10 μg/litre iso-octane solution of α-Hexachlorocyclo-hexane (αHCH) as an internal standard.

5. Shake the tube, with venting, to extract the PCP acetate into the hexane layer. If an emulsion forms in the hexane layer, transfer it to a centrifuge tube, and centrifuge at 1000 r.p.m. for 2 min or use an ultrasonic bath.

6. Transfer about 1 ml of the hexane layer to a sample vial ready for GC analysis.

7. Determination of non-persistent pesticides present in water samples

As a result of the accumulating evidence that some organochlorine pesticides are persistent in the environment and bio-accumulate in the tissue of some target organisms, pesticide manufacturers formulated new products which were more 'environment friendly'. These pesticides are largely based on nitrogen and phosphorus as illustrated in *Structures I* to *IV* below, for simazine, azinphosethyl, fenitrothion, and dichlorvos.

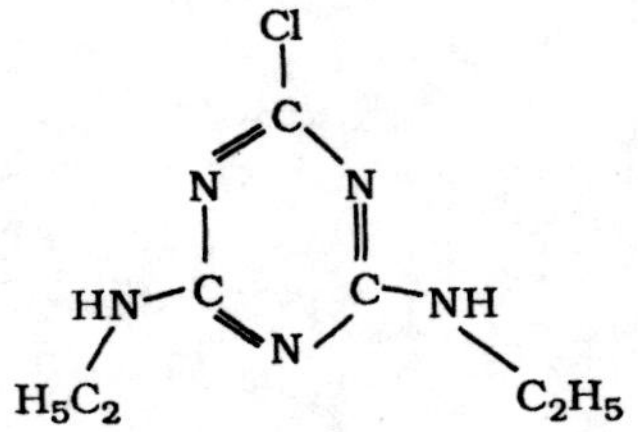

Structure I.

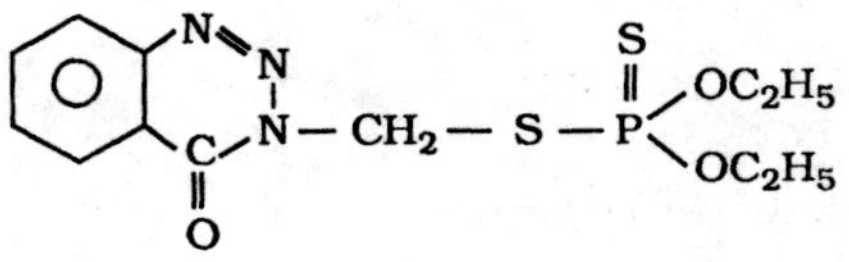

Structure II.

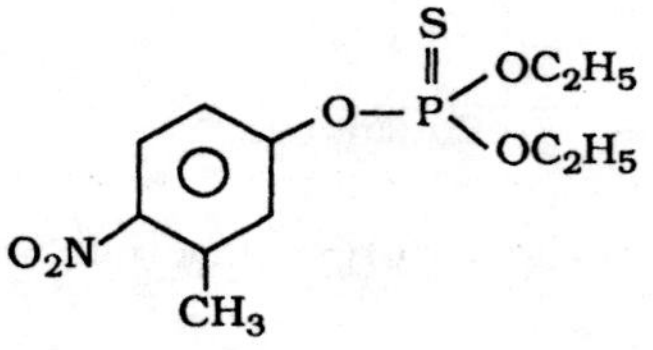

Structure III.

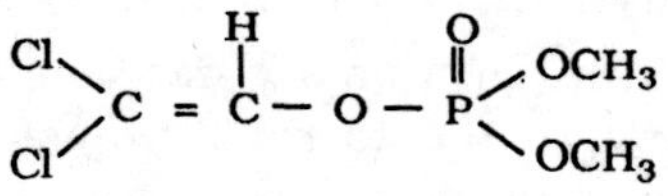

Structure IV.

Another group of pesticide materials is based on insecticidal properties of the naturally occurring pyrethroids which are extracted from the pyrethreum plant (*Chrysanthemum cinerariaefolium*). An example of a synthetic analogue pyrethroid is permethrin (*Structure V*).

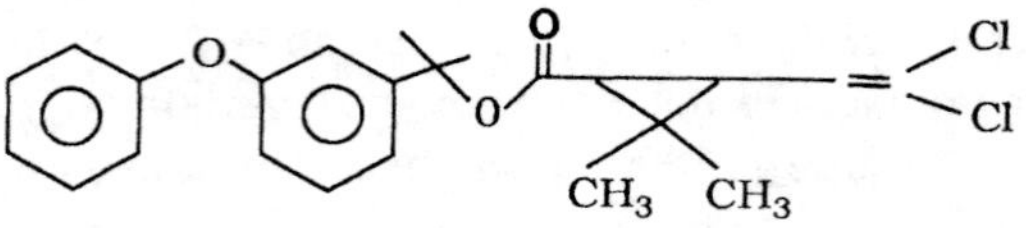

Structure V.

A third group of pesticides are herbicides based on 2,4-dichlorophenoxy-acetic acid (2,4-D). They have been widely used as selective weedkillers for many years and act by stimulating rapid growth in the weeds, causing them to go through their life cycle in a shorter time. The main compounds in this group are illustrated in *Structures VI* to *X*, 2,4,D, dicambra, MCPA, silvex, and 2,4,5T.

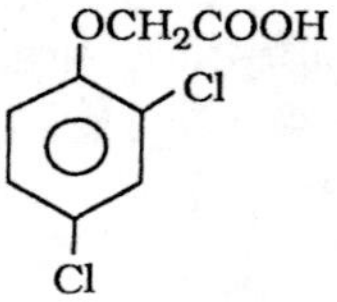

Structure VI.

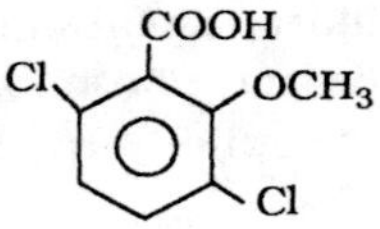

Structure VII.

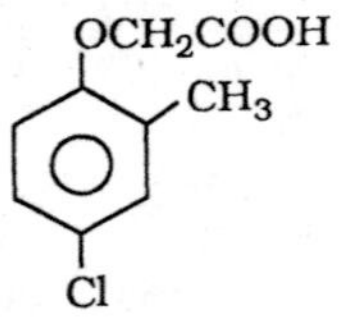

Structure VIII.

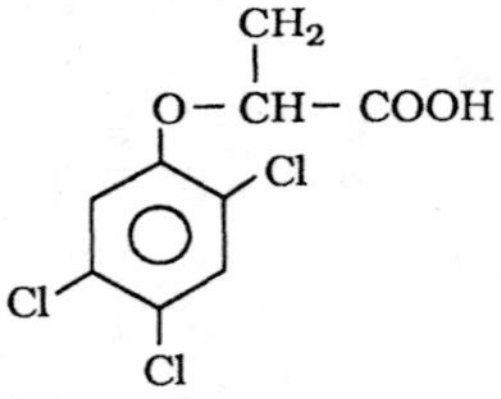

Structure IX.

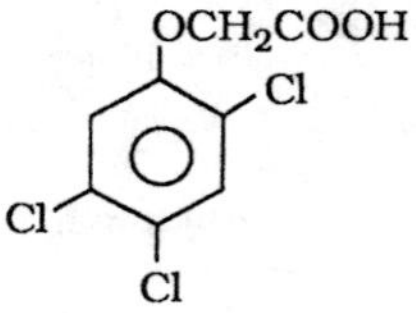

Structure X.

These compounds are normally present as their sodium salts in weedkilling formulations.

7.1 Extraction and determination of organophosphorus and organonitrogen compounds in water

These substances are extracted from a water sample with dichloromethane; the extract is dried, evaporated to small volume, and analysed by GC. Dichloromethane is dangerous if adsorbed through the skin or inhaled, and thus operators should wear gloves and eye protection, and work in a fume cupboard. *Protocol 10* describes the procedure for extraction and clean-up prior to GC analysis.

Protocol 10. Extraction and clean-up of OP and ON pesticides

Materials and equipment

- analytical grade sodium chloride
- dichloromethane—glass-distilled grade
- anhydrous sodium sulphate
- 2 × 1 litre glass separating funnels
- 1 litre glass measuring cylinder
- stoppered glass tubes
- filter funnel fitted with Whatman No. 1 filter paper

- all glass rotary evaporator
- nitrogen or air blower

The last two items can be replaced by:

- Kuderna–Danish evaporator concentrator
- micro Snyder column

or

- Turbo vap concentrator

Method

1. Transfer the water sample[a] to a 1 litre separating funnel set up in a fume cupboard.

2. Add 30 ± 2 g sodium chloride to the sample and shake the funnel until it dissolves.

3. Wash the sample bottle with 50 ± 1 ml dichloromethane and transfer this to the separating funnel. Shake the funnel gently at first with frequent venting and then vigorously for 2 min. Allow the layers to separate.

4. Transfer the lower organic layer to a second funnel and extract the sample with a further 50 ml dichloromethane. Transfer this extract to the second funnel.

5. If the combined extract is an emulsion, shake vigorously to break it up, and allow layers to separate.

6. Filter the solvent through a No. 1 filter paper containing 3 g anhydrous sodium sulphate into a rotary evaporator flask.

7. Reduce the solvent to about 10 ml in the rotary evaporator with vacuum applied.

8. Transfer the solvent extract to a graduated test-tube and reduce the volume carefully to 1 ml by blowing a stream of clean air or nitrogen (see *Protocol 1*) over it. The extract is now ready for analysis by GC.

[a] Water samples are collected in prewashed glass bottles with glass stoppers or with PTFE-lined caps. The contents should be extracted as soon as possible, if not they should be stored in a refrigerator.

Table 7 lists the pesticides that can be determined by this method.

Table 7. List of organophosphorus and organonitrogen pesticides determined according to *Protocol 10*

Dichlorvos	Fenitrothion
Trifluralin	Malathion
Simazine	Fenthion
Atrazine	Chlorfenvinphos
Propazine	Chlordazon
Azinphos ethyl	

7.2 Extraction and clean-up of permethrin from water samples

Permethrin is increasingly being used to tackle a variety of problems requiring an effective insecticide that will not cause adverse effects in the receiving environment. It is used, for example, to disinfect mains pipes of crustaceans, as a moth-proofing agent in carpet manufacture, and in garden sprays to combat a variety of pests.

Clean water samples can be extracted with hexane, the hexane dried and then concentrated before quantification by GC. The method given here (*Protocol 11*) for dirtier samples uses solid-phase extraction cartridges, as described in *Protocol 6*.

Protocol 11. Extraction and clean-up of samples for the analysis of permethrin

Materials and equipment

- hexane (GDG)
- anhydrous sodium sulphate
- methanol (GDG)
- ether (GDG)
- ether/hexane mixture (5:95, v/v)
- Bond Elut C$_{18}$ octyl (No 606303)
- Bond Elut silica (No. 601303)

- 1 litre separating funnel
- 1 litre measuring cylinder
- glass filter funnel with Whatman No. 1 filter paper
- 10 ml graduated test-tube
- 5 ml test-tube
- blower

Method

1. Transfer the contents of the sample bottle (500 ml) to a 1 litre separating funnel. Acidify to pH 2. Rinse out the bottle with 50 ml hexane. Repeat with a further 50 ml hexane.

2. Stopper the funnel, shake vigorously for 3 min, and allow to separate. If an emulsion forms, break this up by centrifugation as described in *Protocol 3*.

3. Run off the lower aqueous layer into a measuring cylinder and note the volume.

310

4. Pass the hexane layer, through a No. 1 filter paper containing about 2 g anhydrous sodium sulphate, into a graduated test tube.

5. Blow dry the hexane extract down to about 1 ml.

6. Prepare a C_{18} octyl cartridge by washing with 1 ml methanol, followed by 20 ml hexane, and discard the wash solvents.

7. Add the 1 ml of the hexane extract to the top of the tube and allow it to be adsorbed.

8. Place a 5 ml test-tube under the Bond Elut tube. Put 5 ml hexane into syringe and force through the tube. Collect only the first 2 ml in the test-tube.

9. Evaporate the 2 ml eluate down to 1 ml.

10. Prepare a silica Bond Elut tube by passing through 3 ml ether followed by 20 ml hexane. Discard the wash solvents.

11. Add the 1 ml of eluate to the top of the silica tube and allow it to be adsorbed.

12. Wash the tube with 10 ml hexane and discard the eluate. Do not allow tube to go dry.

13. Elute with ether/hexane (5:95, v/v) mixture by adding about 6 ml to the syringe. Force the solvent through into a test tube. Discard the first 2.5 ml and collect the next 3 ml. This latter volume contains any permethrin that may be present.

14. Concentrate to 1 ml slowly using a blower. The extract is now ready for GC quantification.

7.3 Extraction and determination of phenoxyacetic acid type of herbicides

The herbicide is converted to its acidic form, extracted with ether, and evaporated to dryness. The dried material is converted to a butyl ether and the derivative is extracted with hexane (14). The procedure is described in *Protocol 12*.

Protocol 12. Determination of phenoxyacetic acid herbicides in water samples

Materials and equipment

- glass-distilled diethyl ether
- 5% potassium hydroxide solution
- hexane (GDG)
- anhydrous sodium sulphate
- A.R. grade sulphuric acid (conc.)

- 5 M sulphuric acid
- dichloromethane (GDG)
- butan-1-ol
- acetone (GDG)
- 2 × 1 litre separating funnels

Protocol 11. *Continued*

- 100 ml separating funnel
- filter funnel fitted with Whatman No. 1 filter paper
- 2 × round-bottomed flasks (250 ml and 50 ml)
- glass-stoppered test-tubes
- Pasteur pipettes
- rotary evaporator
- nitrogen or air blower drier

Method

1. To the water sample, which has been taken and stored in a 1.2 litre glass-stoppered bottle, add 5 ml 5 M sulphuric acid so that the pH is less than 2.

2. Add 150 ± 5 ml diethyl ether to the sample bottle and shake vigorously for 2 min, venting often into a fume cupboard. Transfer the contents to a 1 litre separating funnel and allow the phases to separate.

3. Rinse the sample bottle with a further 5 ml diethyl ether and add this to the separating funnel.

4. Run off the aqueous layer into another 1 litre separating funnel and then dry the diethyl ethyl layer by passing it through a filter funnel containing about 5 g anhydrous sodium sulphate in a filter paper. Collect the diethyl ether in a round-bottomed flask.

5. Extract the aqueous layer with a further 50 ± 5 ml diethyl ether. Measure the volume of the aqueous layer in a 1 litre cylinder, then discard it. Pass the diethyl ether layer through the anhydrous sodium sulphate and collect the filtrate in the round-bottomed flask.

6. Connect the flask to a rotary evaporator and evaporate down to about 5 ml. Transfer the concentrated extract with rinsing to a test-tube and blow to dryness with the blower drier.

7. Add 5 ml 5% potassium hydroxide solution to the tube, swirl to mix, and then place the tube in a beaker of hot water in a water bath for about 1 h.

8. Transfer the solution to a 100 ml separating funnel and rinse the tube out with a little distilled water. Add the rinsings to the funnel.

9. Clean the aqueous solution by extracting twice with 10 ml hexane and discard the hexane layer.

10. Add 2 ± 0.2 ml concentrated sulphuric acid and mix to form the acidic form of the herbicide. Add 20 ± 2 ml dichloromethane to the funnel and shake for 2 min. Allow the layers to separate and pass the dichloromethane extract through a fresh batch of anhydrous sodium sulphate in the filter funnel. Collect the dried organic layer in a round-bottomed flask.

11. Evaporate the solvent down to 5 ml, transfer to a test-tube, and blow it to dryness.

> **12.** Add 0.5–1.0 ml butan-1-ol followed by 2 drops concentrated sulphuric acid. Place the tube in a beaker on a water bath for 1 h, then cool.
>
> **13.** Add a little distilled water and 1 ± 0.1 ml hexane. Stopper the tube and shake for 30 sec. Allow the layers to separate and pipette off the hexane layer into a stoppered tube for quantification by GC.

8. GC Separation and quantification of target compounds

8.1 Choice of column

As mentioned in earlier chapters, there are now many choices of columns and stationary phases for the chromatographer to use. The selections summarized in the following sections are not a definitive list, but experience has shown that these columns provide good separation of the target compounds. With the introduction of capillary columns, the choice of stationary phase is not so critical as with packed columns and most components can be satisfactorily separated on either a non-polar (OVI) or a polar (Carbowax 20 M) column.

8.1.1 Narrow-bore capillary columns and their utilization in the determination of target compounds

Examples of narrow-bore capillary columns which are widely used include 30 m × 0.25 mm DB5 (J & W Scientific) and SP608 (Supelco Ltd) with d_f of 0.15 or 0.25 μm. Several illustrations of the uses are summarized below.

For organic chlorine compounds typical GC conditions are:

- detection mode   ECD
- carrier/make-up gas,   helium, 1 ml/min;
  flow rate   nitrogen, 60 ml/min
- temperature programme   60°C hold for 1 min
  up to 140°C at 20°C/min, hold for 1 min
  up to 190°C at 3°C/min, hold for 1 min
  up to 280°C at 15°C/min, hold for 1 min
- injection volume   1 μl
- injection mode   cold on-column.

For OP and ON pesticides the appropriate GC conditions are:
- detection mode   NPD flow rates as above
- temperature programme   35°C, hold for 1 min
  up to 160°C at 10°C/min, hold for 2 min
  up to 200°C at 2°C/min, hold for 1 min
  up to 300°C at 10°C/min, hold for 5 min
- injection volume   3 μl
- injection mode   splitless

8.1.2 Wide-bore capillary columns and their utilization for the determination of organochlorine pesticides

Although wide-bore capillary columns are not as widely employed as the narrow or medium bore variety, these columns can be used, in general, for pesticides and priority pollutants where fast analysis is essential as an alternative to packed columns, and examples are 30 m × 0.53 mm DB5; DB608 (J & W Scientific) with d_f of 1.0 μm.

Typical GC conditions are as follows:

- carrier gas; flow rate — helium; 6 ml/min
- make-up gas; flow rate — nitrogen; 60 ml/min
- temperature programme — 140°C, hold for 4 min
  up to 225°C at 5°C/min
  up to 250°C at 2°C/min, hold for 10 min
- injection volume — 4 μl

8.1.3 Application of packed columns to target compound analysis

Packed columns can be used selectively for a variety of target compounds without specialized knowledge because their preparation is less technically difficult than for capillary columns. However, if high performance and efficient separation is a priority as in complex mixtures such columns would not be recommended. Also the range of components analysable on one column would be limited.

For phenoxyacetic acid herbicides appropriate GC column and conditions are:

- glass column 1.5 m × 3 mm packed with 80–100 mesh Chromosorb WHP containing 4% DC200 stationary phase
- carrier gas; flow rate — nitrogen; 60 ml/min
- injector temperature — 250°C
- oven temperature — 200°C
- detector temperature — 210°C
- injection volume — 5 μl

For organochlorine compounds the following GC column and conditions are appropriate:

- glass column 1.5 m × 2 mm packed with 8–100 mesh Chromosorb WHP containing 5% OV-1-1
- carrier gas; flow rate — nitrogen; 60 ml/min
- injector temperature — 240°C
- oven temperature — 200°C
- detector temperature — 320°C
- injection volume — 5 μl

8.2 Calculation of concentration in samples using the internal standard method

In this method, the peak heights or areas of the target compounds are normalized to that of the internal standard for both the calibration solutions and the sample solutions. The concentration of the target compounds is calculated using the normalization factor, K, where

$$K = \frac{C_{IS} \cdot H_S}{C_S \cdot H_{IS}}$$

where C_S is the concentration of standard, C_{IS} the concentration of internal standard, H_S the peak height of standard, and H_{IS} is the peak height of internal standard.

To calculate the concentration of the target compound (C_t) in the sample a rearranged form of the above formula is used.

$$C_t = \frac{C_{IS} \cdot H_t}{K \cdot H_{IS}}$$

where H_{it} is peak height of target compound and C_t its concentration. H_t, H_S, and H_{IS} can be replaced by A_t, A_S, and A_{IS}, the areas of the relevant peaks as appropriate.

The following data provide a worked example of the quantification of GC data using the internal standard normalization method. For a concentration of standard solution of permethrin of 0.2 mg/litre and a concentration of internal standard of decachlorobiphenyl of 0.02 mg/litre, the following peak heights were measured:

Permethrin standard	22 units
DCBP standard	33 units

$$K = \frac{0.02 \times 22}{0.2 \times 33} = 0.067$$

For the sample solution of same initial volume containing 0.2 mg/litre decachlorobiphenyl (DCBP) and the same amount of extract injected into the GC:

Peak height for permethrin sample	18 units
Peak height for 0.02 mg/litre DCBP in sample	36 units

$$\text{Concentration of permethrin in sample} = \frac{0.02 \times 18}{0.067 \times 36} = 0.15 \text{ mg/litre}$$

8.3 Typical GC chromatograms for target compound analysis

To illustrate the type of work carried out on environmental samples, a number of illustrative figures have been included to indicate the type of chromatographs obtained. *Figures 7* to *10* highlight the organochlorine compounds and unknown compounds extracted from a Paisley sewage treatment works effluent. Traces correspond to eluates 1 to 2 in the separation scheme outlined in *Protocol 7* and summarized in *Table 6*. *Figures 11* and *12* illustrate chromatograms of a standard solution of pentachlorophenol (with αHCH as internal standard) and an extract from a Glasgow sewage work final effluent. *Figures 13* and *14* show chromatograms of a mixed standard solution containing permethrin and an extract of sewage works effluent, both separated on a narrow-bore capillary column. *Figure 15* highlights a chromatogram of a standard solution of OP and ON herbicides and pesticides separated on a narrow-bore capillary column.

9. Sampling and analysis of gases and vapours

Environmental samples of gases and vapours arise from three main sources:

(a) atmosphere at the workplace or in homes (e.g. gas leaks)

(b) stack emissions from incinerators, power stations, factory effluent

(c) gases from landfill sites, spillages, deliberate emissions (regular monitoring of the air was carried out at the time of the Gulf war when there were fears of chemical warfare)

9.1 Sampling

Each of these sources requires special sampling equipment before any analysis is carried out. For some atmospheres, the pollutant may be present at very low concentrations and a pre-concentration technique is used.

9.1.1 Atmospheres

If the target pollutant is likely to be present in the p.p.m. range, the usual sampling device is a gas bulb typically of 250 ml capacity. This apparatus is fitted with two valves, one at each end (A and B) and has a sampling septum in the centre (C), as shown in *Figure 16*. In order to obtain a representative sample of the atmosphere, the gas bulb, with its valves open, is connected to a small vacuum pump. Sufficient of the atmosphere is passed through the bulb to give at least a 5 vol. exchange. The period of time required to achieve this condition is calculated from the extraction rate of the vacuum pump. The valves are closed and the bulb taken to the laboratory for analysis. A sample for GC analysis is obtained by pushing a gas syringe through the septum and extracting 1–2 ml. For lower concentrations of pollutants the volatile component has to be concentrated by being adsorbed on to a suitable adsorbent material.

There are three main types of adsorption tube as follows:

(a) activated carbon from which the adsorbed material is removed by solvent extraction

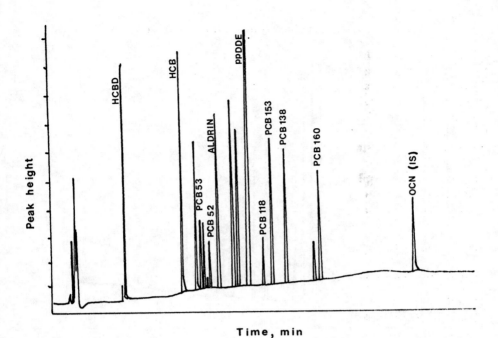

Figure 7. Eluate 1 of mixed standard solution from *Protocol 7* procedure showing HCBD, HCB, aldrin, pp-DDE, and PCBs.

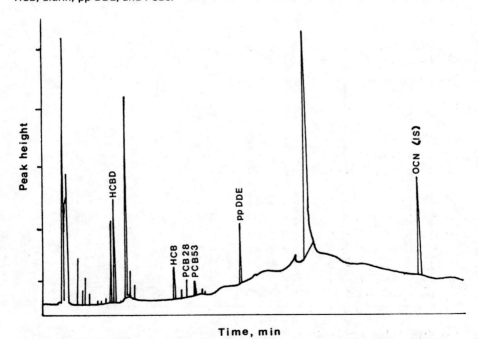

Figure 8. Eluate 1 of extract from a Paisley sewage treatment works final effluent from *Protocol 7* procedure showing presence of HCBD, HCB, pp-DDE, and some PCBs.

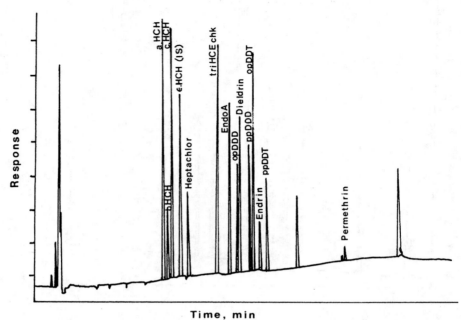

Figure 9. Eluate 2 of mixed standard solution from *Protocol 7* procedure showing HCH, heptachlor epoxide, endosulphan A, op-DDD, op-DDT, endrin, pp-DDT, and permethrin.

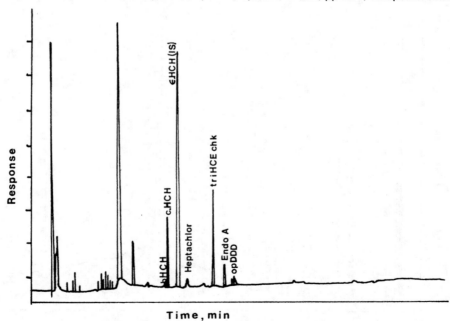

Figure 10. Eluate 2 of extract from Paisley sewage treatment works final effluent from *Protocol 7* procedure showing presence of HCH, heptachlor epoxide, endosulphan A, and op-DDD.

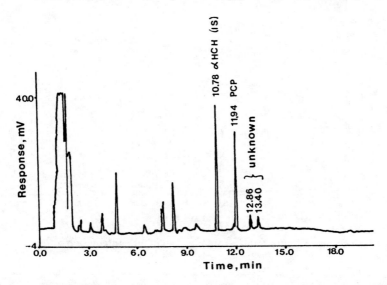

Figure 11. Chromatogram of a standard solution of PCP with αHCH as internal standard.

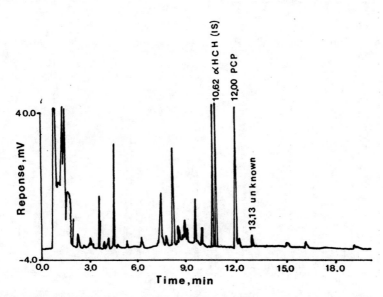

Figure 12. Chromatogram of an extract from a Glasgow sewage works final effluent showing presence of PCP with αHCH as internal standard.

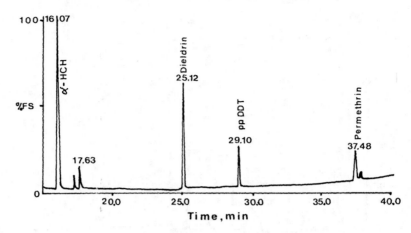

Figure 13. Chromatogram of a mixed standard solution containing permethrin.

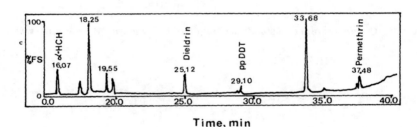

Figure 14. Chromatogram of an extract from a sewage works indicating the presence of permethrin.

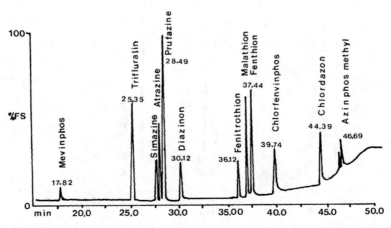

Figure 15. Chromatogram of a standard solution of organophosphorus and organo-nitrogen herbicides and pesticides separated on a narrow-bore capillary column.

320

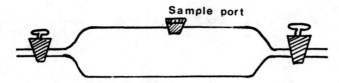

Sample port

Figure 16. Gas bulb for sampling atmosphere.

Figure 17. Adsorption tube for sampling atmospheres.

(b) Tenax tubes which are thermally desorbed and so have the advantage of no solvent peak (15)

(c) graphite tubes, also thermally desorbed.

The atmosphere may also be sampled using a cryogenic trap (16). These accessories are very effective at trapping the most volatile components, but they are complicated to operate and are a specialized technique.

Depending on the concentration of the target compound, the adsorption tube may be used to collect the pollutant either by passive diffusion or by pumping the atmosphere through the tube. The higher concentrations are measured by diffusion. If a workplace atmosphere is being monitored, the worker will have an adsorption tube (see *Figure 17*) attached to a lapel during the working day.

For lower concentrations, the atmosphere is sampled by drawing it through the tube fitted with a pump. The amount of air sampled can be calculated by measuring the pumping rate. If an individual worker's exposure is being assessed, portable pumps are available that are connected to a lapel-worn adsorption tube.

Sampling of stack emissions is a specialized technique and has been well-described in official methods (16, 17). It is important to sample the fumes in the chimney at the same rate at which they are being ejected and there are various techniques to ensure this. The sample is cooled by passing it through condensers and, if it is a gas, it is passed into a gas bulb whilst vapour is adsorbed on to Tenax-GC or other adsorbent material.

Landfill gases, particularly methane, C_2–C_4 hydrocarbons, hydrogen, and oxygen, are sampled via a sampling probe pushed into the deposited refuse (*Figure 18*). The gases at the required depth are evacuated via a vacuum pump and a sample put into a gas bulb for GC analysis.

For vapours present in the landfill emissions in the range of 0.1–400 mg/m^3, the preferred method (18) is adsorption on to Tenax-GC. Usually 25 ml

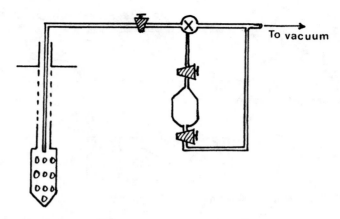

Figure 18. Sampling of gases in a landfill.

sample is sufficient and this is adsorbed on the first part of the adsorption material. To ensure that no material is lost in transit to the laboratory, the Tenax-GC should be securely sealed with swagelok caps. The tubes should be cooled before removing the caps prior to GC analysis after heat adsorption.

Tenax-GC tubes are unsuitable for alcohols, acid, and amines in the C_1–C_4 range. The preferred method for these compounds is cryogenic cooling of about 30 litres gas to obtain about 1 ml condensate. The level of these substances is in the range of 0.1–2000 mg/m^3.

Quality control of the GC data is very important and, for atmospheric sampling, an internal standard should be used to check on recoveries and the performance of the sampling equipment. A suitable compound for atmospheric work is anisole. A known concentration dissolved in methanol should be injected into the adsorption tube before sampling is started. The method is standardized by sampling a prepared standard atmosphere.

9.2 Desorption of components

The trapped atmospheric pollutants are released from the adsorption tube by heating in a stream of nitrogen attached to the GC (*Figure 19*) or, if activated carbon is used, is eluted with a suitable solvent. For most conditions, Tenax-GC is the preferred adsorbent and is suitable for a wide range of volatile compounds. The amount of heat that is applied to the tube depends upon the volatility of the adsorbed compounds.

A more satisfactory chromatogram can be obtained by holding the desorbed compounds in a cold trap prior to GC analysis as shown in *Figure 20*. The cold trap is cooled either with ice, solid carbon dioxide, or liquid nitrogen, depending upon the volatility of the compound. Once in the cold trap, the compounds are then released into the carrier gas stream by rapidly heating the trap up to 300 °C.

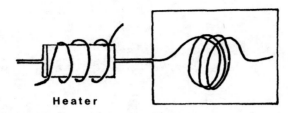

Figure 19. Thermal desorption of target compounds from an adsorption tube.

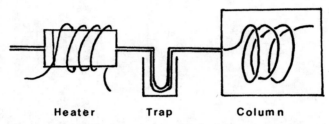

Figure 20. Thermal desorption and cold trapping of target compounds from an adsorption tube.

Two sample chromatograms are given showing the results of these techniques. In *Figure 21* a 2 litre sample of ambient atmosphere taken from a flat above a shoe repair shop was adsorbed on to a 0.13 g Tenax-GC tube and desorbed via a cryogenic trap. The GC column and conditions are:

- column   15 m × 0.32 mm i.d. silica capillary with OV 1701; d_f 1 μm
- temperature programme   20°C for 5 min then up to 32°C followed by a rise to 200°C at 4°C/min
- detector   MS with EI source
- split injection   10:1 split

In *Figure 22* the chromatogram is of 25 ml landfill gas adsorbed on to 0.13 g Tenax-GC. The conditions are the same as for the chromatogram in *Figure 21*.

10. Determination of types of oil in pollution samples

10.1 Oil fingerprinting

Oil is a common pollutant in waterways and the sea and can originate from many sources, e.g. tanker spillages, storage tanks overflowing or puncturing, breaks in pipelines. The various constituents in the oil can be separated by GC and an examination of the overall appearance of the chromatographic

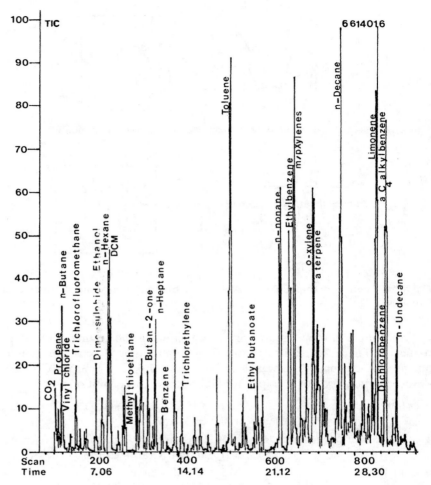

Figure 21. Sample of ambient atmosphere of a flat above a shoe repair shop, the occupant of which complained of solvent odours. Sample: 2.05 litres of air on 0.13 g Tenax TA with 40 ng internal standard (in this sample the internal standard is overshadowed by the relatively high level of the pollutants). Injection: thermal desorption (200°C via a cryogenic trap using a split ratio of 10: 1). Column: 25 m × 0.32 mm FSOT OV1701, d_f 1.0 μm. Temperature programme: initial temperature 20°C for 5 min, 32°C to 200°C at 4°C/min. Detector: mass spectrometer with EI and fast scanning modes. Bracketed numbers indicate concentrations in mg/m^3.

trace (the oil 'fingerprint') gives an indication of the type of oil involved in the pollution. The GC trace of the pollution sample can be matched against GC traces of possible sources of the oil in the vicinity of the incident. The trace from the source with the greatest similarity in appearance is likely to be the cause of the pollution. The oil may be exposed to the atmosphere for some

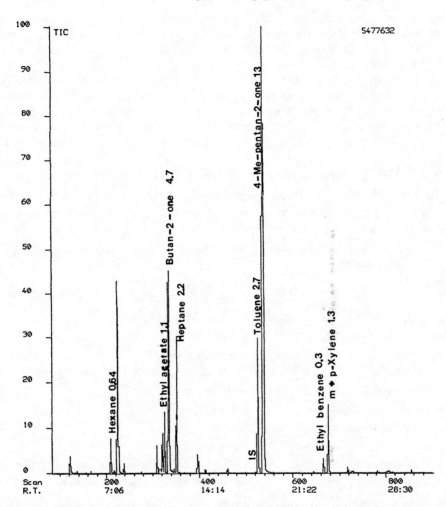

Figure 22. Organic vapour analysis of a landfill gas emission (at source). Sample: 25 ml on 0.13 g Tenax TA with 200 ng internal standard. Injection, column, and detector details identical to those indicated in *Figure 22*.

time before it is discovered and many of the more volatile components will then have evaporated into the atmosphere. For these weathered samples, the GC trace compared with possible sources will be markedly different because of the absence of the light low molecular weight fractions from the chromatogram. However, an examination of the ratio of some of the normal hydrocarbon peaks to those of adjacent isoprenoid peaks, for chromatograms run on capillary columns, can often be used to track down a source of pollution.

An oil fingerprint is best obtained using a packed column coupled to a flame ionization detector (FID), whilst a capillary column linked to FID will provide the details in the trace to discriminate between peaks for the normal hydrocarbons and those for the isoprenoid. Details of GC conditions are given below.

In general, the oil is injected directly into the GC from the sample. For some samples however, the oil may need to be removed from solid material, e.g. birds' feathers or stones, or other samples such as surface water, may contain only a trace. In these instances, the oil is removed by either petroleum ether or hexane and the solvent solution of the oil injected into the GC.

10.2 Oil fingerprinting—GC operating conditions

10.2.1 Columns and conditions for lubricating and fuel oils

For packed columns used in analysis of lubricating and fuel oils, the column and GC conditions are as follows:

- stainless steel column packed with 5% OV101 phase on Chromosorb W HP 80–100 mesh
- carrier gas flow rate   nitrogen; 25 ml/min
- fuel gas flow rate   hydrogen; 25 ml/min air; 250 ml/min
- temperature programme   75°C up to 280°C at 10°C/min
- injection volume   0.3 μl

A typical chromatogram is illustrated in *Figure 23*.

For analysis using capillary GC, the column and GC conditions are:

- 30 m × 0.25 mm DB5 with d_f of 0.25 μm
- carrier gas; flow rate   helium; 1 ml/min
- split injection ratio   60:1
- make up gas; flow rate   nitrogen; 25 ml/min
- fuel gas; flow rate   hydrogen; 25 ml/min, with air, 250 ml/min
- temperature programme   100°C up to 220°C at 4°C/min
- injection volume   0.3 μl

A typical chromatogram is shown in *Figure 24*.

Capillary columns are suitable for analysis of lubricating and fuel oils, and column and GC conditions are as follows:

- short capillary column, 15 m × 0.25 mm DB5 with d_f of 0.25 μm
- carrier gas; flow rate   helium, 1 ml/min
- split injection rates   60:1
- make up gas; flow rate   nitrogen; 25 ml/min
- fuel gas; flow rate   hydrogen; 25 ml/min with air; 250 ml/min
- temperature programme   100°C up to 290°C at 4°C/min
- injection volume   0.3 μl

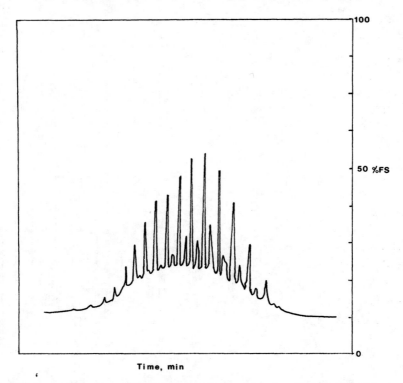

Figure 23. Typical chromatographic profile obtained for standard diesel oil under low resolution GC conditions.

References

1. Hinckley, D. A. and Bidleman, T. F. (1989). *Environ. Sci. Technol.*, **23**(8), 995.
2. Murray, H. E. and Beck, J. N. (1990). *Bull. Environ. Contam. Toxicol.*, **44**(5), 789.
3. McNeill, A. (1990). *J. Inst. Water Environ. Manag.*, **4**(4), 330.
4. Rivett, M. O., Lerner, D. N., and Lloyd, J. W. (1990). *J. Inst. Water and Environ. Manag.*, **4**(3), 242.
5. Williams, R. J., Bird, S. C., and Clare, R. W. (1991). *J. Inst. Water and Environ. Manag.*, **5**(1), 80.
6. RSC (1987). *The agrochemicals handbook*, (2nd edn). Information Services, Nottingham.
7. HMSO. *Methods for the examination of water and associated materials*. HMSO, London.
8. US-EPA (1988). *Methods for the determination of organic compounds in drinking water and waste water*, PB 89–220–461, Series 500 and 600. Methodologies. Environmental Monitoring and Support Laboratory, Cincinnati, OH.

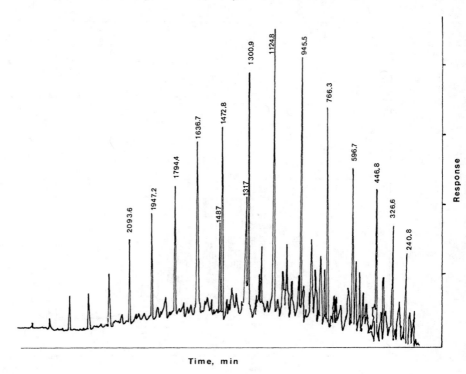

Figure 24. Typical chromatographic profile obtained for standard diesel oil under high resolution GC conditions. 1317, Pristane; 1487, phytane.

9. Hunt, D. R. E. and Wilson, A. L. (1986). *The chemical analysis of water, general principles and techniques*. RSC, London.
10. Ellis, J. C. (1989). *Handbook on the design and interpretation of monitoring programmes*. WRC Report NS 269, Medmenham.
11. Brannon, J. M. and Karn, R. (1990). *Bull Environ. Contam. Toxicol.*, **44**(4), 542.
12. HMSO (1986). Methods for the examination of water and associated materials. *Chlorobenzenes in water, organochlorine pesticides and PCBs in turbid water, halogenated solvents and related compounds in sewage sludge and water*. HMSO, London, pp. 1–44.
13. Environmental Health Laboratories. (1989). *Analysis of semi-volatile organic chemicals in water by capillary gas chromatography/mass spectrometery*. EPA Method 525, 3M Empore Disk Extraction, Version 1.0,. Environmental Health Laboratories,
14. HMSO (1985). *Methods for the examination of water and associated materials. Chlorophenoxy acidic herbicides, trichlorobenzoic acid, chlorophenols, triazines and glyphosphate in water*. HMSO, London, pp. 1–50.
15. US-EPA (1988). *EPA compendium of methods for the determination of toxic*

organic compounds in ambient air. PB 87-168688. Environmental Monitoring and Support Laboratory, Cincinnati, OH.
16. Hawksley, P. G. W., Badzioch, S., and Bracket, J. H. (1977). *Measurement of solids in flue gases* (2nd edn). British Coal Utilisation Research Association (BCURA), The Institute of Fuel, London.
17. US-EPA (1971). *Standards of performance for new stationary sources. Federal Register*, **36**(247), US-EPA, Springfield, VA.
18. Brookes, B. I. and Young, P. J. (1983). *Talanta*, **30**, 665.

$$\boxed{10}$$

The role of gas chromatography in petroleum exploration

GARETH E. HARRIMAN

1. Introduction

It is now widely accepted that petroleum has a biogenic origin, having derived from plant or animal remains, or a mixture of both. After death, most organisms and plant debris decay and return (through oxidative processes) to the atmosphere as carbon dioxide. A small amount (<1%) of this organic matter, however, is deposited in aquatic environments that are not oxidizing but reducing, with the net result that some of the organic matter is preserved within the sedimentary record. As further sedimentation takes place, and the burial depth of the organic-rich sediments increases, the organic matter undergoes biological and chemical changes (diagenesis), which can lead to the formation of petroleum source rocks. These source rocks, under appropriate conditions of temperature and pressure, may generate and expel liquid hydrocarbons which can subsequently migrate to suitable traps where they pool as petroleum accumulations.

GC is a phenomenally powerful separation tool that is capable of separating the hundreds of components that go to make up petroleum, and for many years geochemists have applied GC techniques to the characterization of petroleums and source rock extracts. Using flame ionization detectors (FIDs) and by varying column lengths, phases, and temperature programmes it was found that oils frequently produced different gas chromatograms, each yielding its own set of geochemical information. Most of the chromatograms generated in this way are dominated by normal alkanes and it is the distribution of these alkanes that allows the geochemist to distinguish gross differences between oils and source rocks, and to surmise the nature of the source materials responsible for an oil.

Although geochemists have, in the past, concentrated on the information afforded by these alkane distributions, attention in recent years has shifted towards the more complex compounds that are to be found in oils and source rocks at much lower concentrations than the predominant alkanes. Steranes and triterpanes are two such classes of compounds, commonly referred to as *chemical fossils* or *biomarkers*, and because these compounds provide valuable information regarding product–precursor relationships they are widely used to assess the geohistory of an oil and to provide an insight into sourcing, maturity, and migration of petroleum. Because these compounds are present in petroleum at very low concentrations (in the low p.p.m. range), their detection and subsequent quantification only becomes practical when the FID of the gas chromatograph is replaced with a mass spectrometer. This combination of high-resolution capillary gas chromatography and high-resolution mass spectrometry, together with powerful laboratory computers, has provided a major boost to the geochemists' understanding of petroleum formation and generation.

This chapter reports on analyses of crude oils utilizing gas chromatographs fitted with fused silica columns. Detection and identification is achieved either by flame ionization or computerized mass spectrometry, where the mass spectrometer acts, in effect, as a highly structurally specific GC detector.

2. Composition of crude oils and source rock extracts

Petroleum and the soluble fractions from source rocks comprise a wide range of hydrocarbons and non-hydrocarbons. The hydrocarbons are usually the most abundant components (except in thermally immature source rock extracts and heavily degraded oils), with the saturated hydrocarbons normally dominant. Straight-chain alkanes with lesser amounts of the iso- and cyclo-alkanes make up most of the saturate fraction of an oil, with benzene, toluene, naphthalenes, and phenanthrenes dominating the aromatic hydrocarbon fraction. The resins and asphaltenes which make up the non-hydrocarbon fraction of an oil or source extract comprise high-molecular-weight polycyclic compounds frequently containing nitrogen, sulphur, and oxygen. The distribution of these components within an oil or source rock extract will vary according to the source and maturity of the sample but can also be affected by secondary alteration processes such as biodegradation.

Saturated hydrocarbons are the most important of the four main constituents of a crude oil and, in order to characterize them by gas chromatography, it is necessary to isolate them from the aromatic and non-hydrocarbon fractions. This is achieved by first separating the asphaltenes from the crude oil by precipitating them with a light alkane (often pentane or heptane). The remaining pentane-soluble fraction is then subjected to liquid chromatography using activated silica and alumina as the stationary phase. The saturated hydrocarbons are eluted from the column with a non-polar solvent such as hexane. Aromatic hydrocarbons are eluted with a more polar solvent such as toluene or dichloromethane, and the resin fraction is eluted with methanol.

An analytical scheme for the evaluation of oils, source rock extracts, and their isolated fractions is outlined in *Figure 1*.

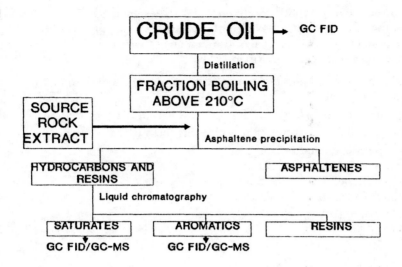

Figure 1. A typical fractionation scheme for the separation of oils and source rock extracts.

3. Whole oil GC

Careful sample collection and handling is essential to obtain the maximum information from whole oil gas chromatography. Drill stem test (DST) samples, collected when proving the productive possibilities of a well, are the best samples for GC analysis, as these have been collected at the temperature and pressure of the producing horizon. Unfortunately, sampling oil in this way is expensive, requiring the use of sophisticated pressure vessels, and because of this many samples are collected at the well head under atmospheric pressure. This inevitably leads to some loss of the lighter alkane fraction, but losses can be minimized if the sample is transferred to a gas-tight bottle and stored at 4°C prior to GC analysis. Samples stored in this manner will retain much of the volatile hydrocarbon fraction and the relative abundance and composition of the gasoline components, as determined by GC, can provide information which will help the geochemist to assess the maturity, source, or even the degree of alteration of a crude oil.

Protocol 1 gives a procedure for obtaining information on the gasoline components of an oil and generating a chromatogram on the whole oil fraction, all in a single analysis. Better separations for the gasoline components may be achieved by using columns with different (and maybe thicker) phase coatings and cryogenic temperature programming, but this is usually at the expense of good column resolution for the longer ($>C_{15+}$) chain hydrocarbons and often two analyses of the same oil, using different columns, are required. The method described here is designed to yield the maximum information on the composition of a crude oil in a single GC analysis.

Whole crude oils are analysed by means of capillary GC according to the operating procedures described below.

- gas chromatograph: any commercial chromatograph designed for use with capillary columns
- column: 60 m × 0.32 mm i.d. fused silica capillary column[a]
- phase coating: DB-5 0.25 μm film thickness (d_f)
- temperature conditions
  initial: 25°C, held for 3 min
  programme: 7°C,/min to 310°C
  final: 310°C, held for 15 min
- injector: 310°C
- detector: 310°C
- carrier gas: helium
- head pressure: 50 KPa
- sample vol.: 0.1 μm
- injection mode: splitless
- detector: FID

Protocol 1. Gas chromatographic analysis of whole crude oils

1. With GC conditions as described above, open the GC oven door and immerse the first few inches of capillary column (injector end) into a dewar containing liquid nitrogen.[a]

2. After allowing the liquid nitrogen to freeze the GC column for 2–3 min inject the sample using a graduated syringe (1 μm capacity) on to the column through the heated injector port.

3. Allow 2 min for complete transference of sample on to the column before removing liquid nitrogen.

4. Close oven door and start oven programming sequence.

[a] Caution, as flexible fused silica columns will break if handled roughly.

A whole oil gas chromatogram corresponding to a 41° API gravity[b] oil from Siberia is shown in *Figure 2*. The chromatogram reveals a preponderance of

[a] Fused silica columns in which the liquid phase is actually bonded to the fused silica are most suitable for this analysis as the column can be washed with solvents to remove build-up of polar compounds which can cause poor column resolution.

[b] API gravity = American Petroleum Institute; it is a measure of density for petroleum, calculated from equation: API = 141.5°/specific gravity at 16°C − 131.5°. Heavy oils are <25° API, medium oils 25 to 35° API, light oils >45° API.

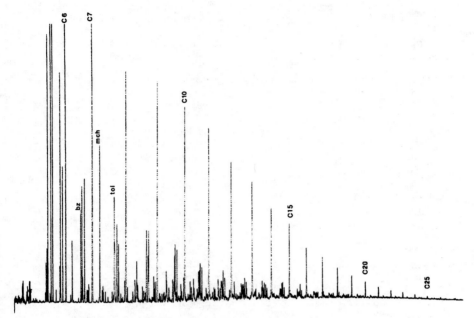

Figure 2. Whole oil chromatogram of a 41° API oil from Siberia.

the low-molecular-weight gasoline (C_4–C_7) and kerosene (C_7–C_{11}) hydro-carbons with the distribution of the $>C_{12+}$ hydrocarbons falling off sharply. This hydrocarbon 'fingerprint' is consistent with that of a light (high gravity), unaltered crude oil, and under the GC conditions outlined for *Protocol 1* baseline separation of the individual peaks can be achieved for nearly all of the early eluting hydrocarbons. These components can be identified by refer-ence using a combination of Kovats' indices and internal standards and the distribution of the gasoline hydrocarbons (as listed in *Table 1*) reported either as absolute abundances or, as is more usual, percentage composition.

Although the initial composition of a crude oil is controlled by the nature of the organic matter in the source rock, the temperature at which the hydro-carbons are generated has a significant effect on the composition of the lighter hydrocarbons. This thermal maturation of the organic matter occurs not only in the source rocks but also in the reservoir crudes and ultimately leads to the formation of lighter (higher gravity) crude oils with a subsequent reduction in the concentration of the branched, cyclo, and aromatic hydrocarbons.

Thus the concentration ratios of certain low-molecular-weight hydro-carbons can be used as maturity parameters for unaltered crude oils (1). For example, the ratios *n*-hexane/methylcyclopentane and *n*-heptane/ methyl cyclohexane will reveal high values for mature oils but lower values for less mature oils.

Table 1. Gasoline range hydrocarbons common to most crude oils

Peak no.	Component
1	Isobutane
2	*n*-Butane
3	Isopentane
4	*n*-Pentane
5	2,2-Dimethylbutane
6	Cyclopentane
7	2,3-Dimethylbutane
8	2-Methylpentane
9	3-Methylpentane
10	*n*-Hexane
11	Methylcyclopentane
12	2,2-Dimethylpentane
13	2,4-Dimethylpentane
14	Benzene
15	Cyclohexane
16	3,3-Dimethylpentane
17	1,1-Dimethylcyclopentane
18	2-Methylhexane
19	3-Methylhexane
20	1,c,3-Dimethylcyclopentane
21	1,*t*,3-Dimethylcyclopentane
22	1,*t*,2-Dimethylcyclopentane
23	*n*-Heptane
24	Methylcyclohexane
25	Toluene

Significant ratios	Significance
Heptane value $= 100 \times \dfrac{23}{15,\ 17-24}$	Maturity parameter
Isoheptane value $= \dfrac{18 + 19}{20 + 21 + 22}$	Maturity parameter
Peak 10/11	Maturity parameter
Peak 23/24	Source/maturity parameter
Peak 19/14	Water washing
Peak 24/25	Water washing
Peak 9/10	Biodegradation

Thermal maturity, however, is not the only factor that governs the quality of crude oil and where oil accumulations come into contact with flowing waters their compositions can be altered by aerobic bacteria which degrades petroleum (2), ultimately leaving a residue which is of little economic value. The extent to which an oil is biodegraded depends, to some degree, on the depth and temperature of the oil pool and relates to the survival temperature of the micro-organisms. In the early stages of biodegradation, the normal alkanes are consumed, with the result that the relative concentration of the branched and cyclic chain hydrocarbons increases. This process is always accompanied by water washing, although water washing can proceed without biodegradation taking place. When this happens, the oil compositional changes that take place are most obvious within the gasoline hydrocarbon fraction where the higher solubility of some of the gasoline components means that water washing can be assessed by the preferential removal of benzene, toluene, and the alkanes prior to the naphthenes. Thus, monitoring the degree of water washing can be achieved with the use of gasoline ratios which are sensitive to water washing and biodegradation. For example, the 3-methylpentane/benzene and methylcyclohexane/toluene ratios are observed to increase with increasing water washing. Similarly, the ratios of methyl-butane/n-pentane and 3-methylpentane/n-hexane can also reflect changes in oil composition due to the effects of water washing/biodegradation. Some of the more commonly used gasoline ratios are listed in *Table 1*.

The composition of an oil, as determined by GC can also yield information relating to the source of the oil. *Figure 3* presents another example, this time of a 39° API oil from Pakistan. Unlike the Siberian oil, the crude oil from Pakistan is a solid at room temperature, although it flows freely at only slightly elevated temperatures. The cause of the relatively high pour point of this oil can be seen from the gas chromatogram which reveals a much higher proportion of the longer ($>C_{15}$) chain hydrocarbons than for the oil depicted in *Figure 2*. These alkanes, particularly those in the nC_{20}–nC_{35} molecular range which show a marked predominance of odd carbon number alkanes to even carbon alkanes, derive from higher plant waxes and thus indicate that a significant amount of higher plant material contributed to the source of this oil.

4. GC analysis of fractions isolated from crude oils and source rock extracts

While whole oil GC is an extremely useful technique for comparing crude oils and assessing their maturity and degree of alteration, the resulting chromato-grams are often complex and difficult to interpret. Moreover, since most crude oils are dominated by the light ($<C_{12}$) fraction, much valuable informa-tion relating to the heavier ($>C_{15}$) fraction is lost. To overcome this problem, the geochemist performs GC on the saturated and aromatic hydrocarbon fractions isolated via the analytical scheme outlined in *Figure 1*. Because these fractions derive from the oil fraction distilling above 210°C, they are frequently termed '$>C_{15}$ fractions'. *Figure 4* shows a typical gas chromato-gram of a $>C_{15}$ saturates fraction using the GC conditions given below.

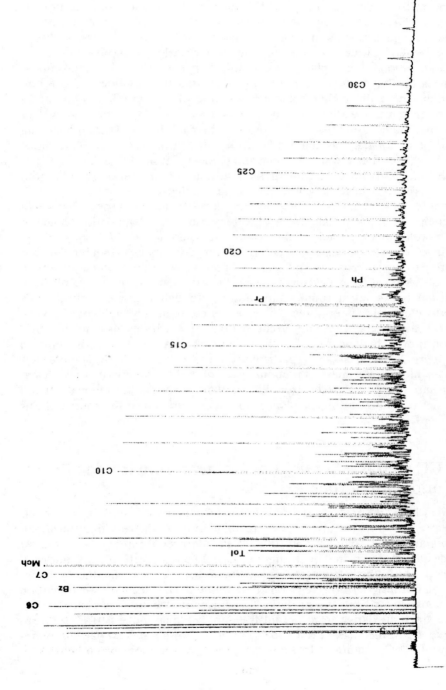

Figure 3. Whole oil chromatogram of a 39° API oil from Pakistan. Pr, pristane; Ph, phytane.

- gas chromatograph: any commercial chromatograph designed for use with capillary columns
- column: 30 m × 0.32 mm i.d. FSOT column
- phase coating and thickness: DB-1 0.25 μm d_f
- oven temperature programme: 60°C to 310°C at 7°C/min
- oven temperature final: 310°C, held for 10 min
- injector: 310°C
- detector: 310°C
- carrier gas: helium
- head pressure: 50 kPa
- injection mode: splitless
- detector: FID

Because the hydrocarbon distributions detected by GC reflect gross differences between the types of source materials responsible for a sample, the $>C_{15}$ saturate chromatograms act as a 'fingerprint' for a particular oil or source rock extract. For example, as demonstrated by the crude oil from Pakistan, oils derived from terrestrial source environments are characterized by high-molecular-weight (C_{25}–C_{35}) alkanes, often (especially if immature) exhibiting an odd carbon preference (3). In contrast, oils derived from marine source rocks are generally dominated by alkanes in the C_{15} to C_{20} molecular range (algal-derived) with lesser amounts of the heavier alkanes. *Figures 4* and *5* illustrate the differences between the saturated hydrocarbons isolated from two such oils. *Figure 4* corresponds to an oil generated mostly from marine organic matter that was deposited in marine setting, while *Figure 5* shows the distribution pattern of alkanes isolated from source rocks deposited in a fluvio-deltaic source environment to which only terrestrially derived organic matter was contributed. The much higher abundance of higher molecular weight alkanes ($>C_{20}$) for the *Figure 5* oil reflect the 'waxy' nature of the higher-plant-derived alkanes. Also the isoprenoids pristane and phytane, which occur as distinctive doublets eluting after the C_{17} and C_{18} alkanes, respectively, can—based on the premise that phytane (Ph) predominates in highly reducing (anoxic) environments while in less reducing circumstances pristane (Pr) prevails—be used as an indicator of depositional environment (4, 5). Over the years, many attempts have been made to apply Pr/Ph ratios to distinguish marine from non-marine oils. Unfortunately, because the ratio is influenced by changes due to the increased levels of stress experienced by the oils as they mature and can also be affected by other processes, caution must be exercised when applying Pr/Ph ratios to source identification.

The overall distribution of the *n*-alkanes in a saturate chromatogram (particularly for source rock extracts) can also provide information relating to the relative maturity of the sample. In shallow immature sediments, alkanes with

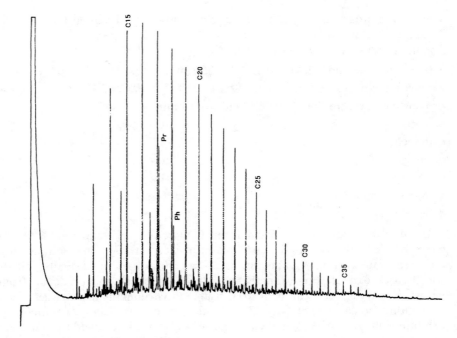

Figure 4. Saturate GC of marine oil.

odd carbon numbers are observed to predominate over the even carbon
number alkanes. As the source rocks mature, this preference is reduced until
eventually the even and odd alkanes are present in equal amounts. This
carbon preference may be monitored using any of the carbon preference
indices (CPI) publshed in the literature, but the one presented below is one
of the most commonly used (6), with values greater than 1 corresponding
to immature sediment extracts and values of 1 indicating an extract to be
mature.

$$\tfrac{1}{2} \cdot \frac{(C_{25} + C_{27} + C_{29} + C_{31} + C_{33})}{(C_{24} + C_{26} + C_{28} + C_{30} + C_{32})} + \frac{(C_{25} + C_{27} + C_{29} + C_{31} + C_{33})}{(C_{26} + C_{28} + C_{30} + C_{32} + C_{34})}$$

Unfortunately the geochemical information provided by the *n*-alkane distribution is often misleading or can be reduced to insignificance by the effects of
biodegradation and contaminants introduced during drilling.

Figures 6a and *b* illustrate the effects on the *n*-alkane distributions caused
by biodegradation. Oil A, collected from an Eocene reservoir in the North
Sea, has been subjected to both water washing and biodegradation, with the
result that all of the alkanes and most of the isoalkanes have been removed.
What remains are the cyclic alkanes, which are more resistant to bacterial
attack, and the naphthenes (responsible for the characteristic hump of so

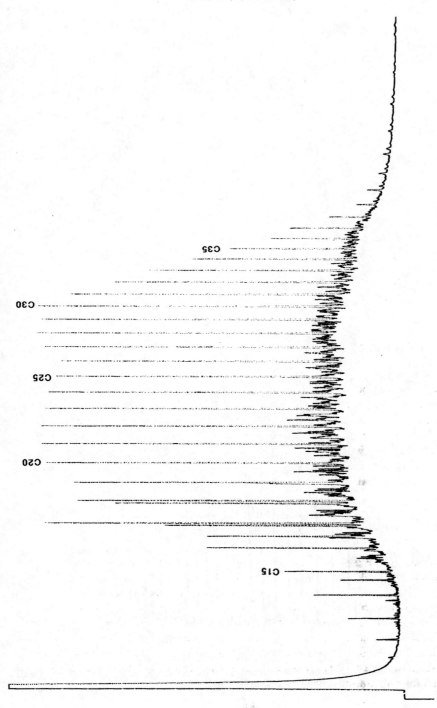

Figure 5. Saturate GC of non-marine oil.

Gareth E. Harriman

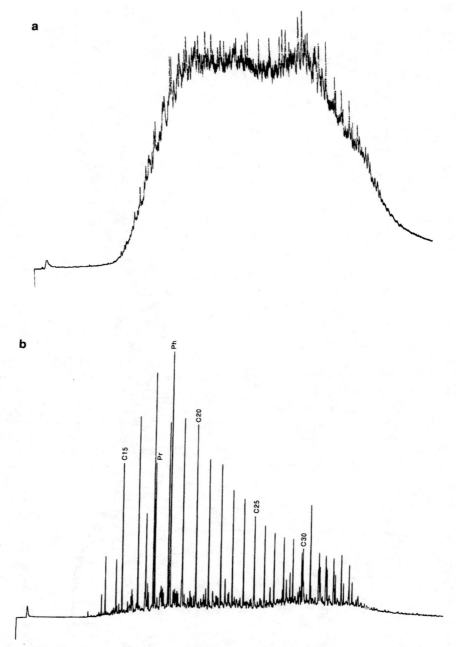

Figure 6. GC of oil A (a), biodegraded oil collected from the UK sector of the North Sea, and oil B (b), mildly biodegraded oil from the Middle East.

many crude oils) which are indigestible by the bacteria. Oil B, an oil from the Middle East, has only suffered mild biodegradation, the net result of which has been a reduction in the abundance of the longer chain alkanes and a relative increase in the concentration of the steroidal and pentacyclic terpenoid hydrocarbons. Normally these classes of biomarkers are not observed in the GCs of mature and unaltered crude oils due to the fact that they are present at much lower concentrations and, as shown in *Figure 6b*, their structural complexity and elution pattern is such that their identification by routine GC techniques, as described above, is virtually impossible. However, because these classes of biomarker are less susceptible to secondary alteration processes such as biodegradation, they can provide far more geochemical information than the n-alkanes and it is, therefore, often desirable to monitor their distributions within the oil or source rock extract.

Using the latest technology, the geochemical analyst can, by combining the separation powers of high resolution fused silica columns with a mass spectrometer as the detector, obtain structural information not only on the components that are responsible for the chromatographic peaks on a GC but also from those compounds that appear to be hidden in the baseline.

5. GC–MS analysis of fractions isolated from crude oils and source rock extracts

By replacing the FID as the detector of a GC with a MS, compounds, that are separated by the GC column elute into the ion source of the MS where they are bombarded by a high-energy electron beam that ionizes the compound of interest, breaking it up into a series of fragment ions. These ions are then accelerated through a mass analyser where they are separated according to their mass/charge ratio (m/z) before being detected and processed by the MS data system. Complete mass spectra can be collected in this way, allowing for the identification of individual compounds by comparing the recorded spectra against library spectra or authentic standards. Alternatively, ions that are diagnostic of certain hydrocarbon classes can be selectively monitored. For example, terpanes fragment with two major ions. The first, and most diagnostic, is the fragmentation resulting from the break up of the A and B rings (*Structure I*) producing an ion at m/z 191. A second fragmentation involving rings D and E results in the formation of a second fragmentation ion whose mass depends on the substituent at C_{21} (R = H or C_nH_{n+1}) (R).

Similarly, the steranes (*Structure II*) fragment with a characteristic ion at m/z 217 and by monitoring the relative intensities of both the m/z 217 and 191 ions during the course of a GC–MS analysis, the distribution of steranes and triterpanes within a sample may be assessed. This technique, known as selective ion recording (SIR) is more sensitive, by several orders of magnitude, than full scanning experiments, and can be applied to many different classes of biomarkers by simply choosing the appropriate diagnostic ions. *Table 2*

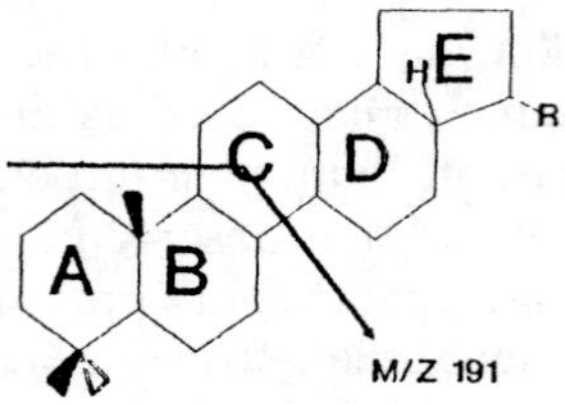

Structure I.

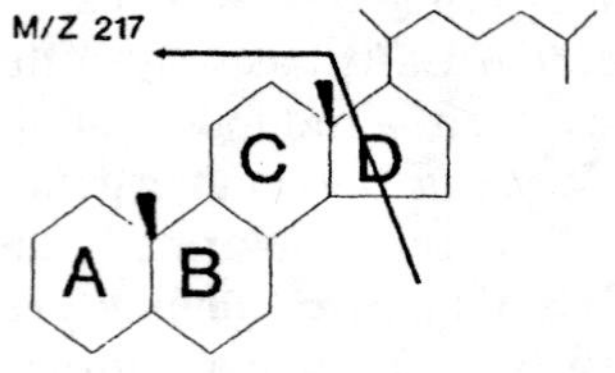

Structure II.

lists the masses monitored during a typical biomarker analysis of a saturate hydrocarbon fraction, with the second column showing the corresponding compound classes.

A procedure for low (unit mass) resolution SIR GC–MS analysis of saturate hydrocarbon fractions is given in *Protocol 2*. The procedure describes analyses carried out using a VG TS250 mass spectrometer coupled to an HP5890 series GC, but it is suited to accommodate any of the modern fast-scanning bench-top GC–MS systems which are sensitive for low mass analysis.

Protocol 2. Low resolution SIR GC–MS analysis of saturate hydrocarbons isolated from crude oils or source rock extracts

Conditions

- gas chromatograph: HP5890
- column: 30 m × 0.32 mm i.d. FSOT capillary column[a]
- phase coating: DB-5; 0.25 μm d_f
- oven temperature programme: 60°C to 310°C at 4°C/min
- final: 310°C, held for 10 min
- injector: 310°C

- carrier gas: He
- column head pressure: 50 kPa
- GC interface (a) temperature: 250°C
- mass spectrometer: VG TS250
- source temperature: 210°C
- electron energy: 70 eV
- filament current: 500 μA

Method

1. Tune the mass spectrometer to give a combination of good sensitivity and Gaussian peak shape, using a reference compound such as perfluorotributylamine. This calibrant should be introduced to the ion source via the direct inlet and tuning carried out according to the instrument manufacturer's users' manual.

2. For magnetic sector instruments such as the TS250, set the instrument resolution to 500. It may be necessary to readjust the tuning at this stage.

344

3. Calibrate the instrument over the desired mass range (this will vary depending on the masses monitored). Follow the instructions in the users' manual for calibration.

4. Both saturate and aromatic hydrocarbon fractions can be analysed by this method and should be diluted (1 mg to 1 ml) in dichloromethane prior to analysis.

5. Injection.[b] To reduce sample losses during injection use a splitless injection technique. Load a 1–3 μl aliquot of the sample on to a solids injector and allow the solvent to evaporate. Then start the GC–MS acquisition and, using the solids syringe, introduce the sample to the hot injector. During this procedure the purge line to the GC injector is closed for a predetermined period (usually 30 sec) before being re-opened for the remainder of the analysis. The use of a solids syringe prevents excessive amounts of solvent entering into the ion source of the MS. For best results one should try to avoid 'fast' injections by leaving the syringe in the injector for 15–20 sec, to allow all of the sample to be vaporized and swept on to the GC column. If an auto-sampler is fitted to the GC, samples should be suitably diluted before loading on to the carousel.

6. Data acquisition may, depending on the system in use, be initiated from the GC or the MS data system but should always (for reproducibility of data) correspond to time of injection.

[a] Fused silica capillary columns may be pushed right through the GC interface so that they protrude into the ion source. This reduces the risk of decomposition of the compounds in the interface and eliminates any loss of chromatographic resolution but it must be remembered to withdraw the column before removing the ion source for cleaning.
[b] GC injectors need regular cleaning. A dirty injector can lead to a loss of sensitivity.

The sterane and terpane distributions of a typical North Sea crude oil are shown in *Figures 7* and *8*. Mass fragmentograms like these are invaluable in oil–oil and oil–source correlation studies but can also provide the geochemist with information relating to source input, maturation, and migration (7). For example, it is well-documented that the steranes present in crude oil originate from algae, plankton, or higher plants with the relative distributions of the C_{27}, C_{28}, and C_{29} sterane members reflecting the contributions from each of the different biological sources. Because marine organisms generally contain steranes in which the C_{27} components dominate, and organic matter derived from higher plants is characterized by C_{29} steranes, a ternary plot of the sterane distributions (8) should prove useful in separating marine oils from non-marine oils. Unfortunately, this is an over-simplification as the assessment of the sterane distributions is often hampered by the co-elution of diasteranes of higher carbon number with regular steranes of lower carbon number. Moreover, both brown and green algae—common to the marine

Table 2. Diagnostic ions for biomarker analysis using SIR

	Ion (m/z)	Compound class
Saturate fraction	85, 127	*n*-Alkanes
	109, 123	Diterpanes, Bicyclic
	179, 193	alkanes
	191	Terpanes (tricyclic) (tetracyclic) (pentacyclic)
	177	Demethylated hopanes
	205	Methyl hopanes
	217, 218	Steranes
	231	4-Methyl steranes
	259	rearranged steranes
Aromatic fraction	178	Phenanthrene
	192	Methyl phenanthrenes
	206	Dimethyl phenanthrenes
	184	Dibenzothiophene
	198	Methyl dibenzothiophene
	212	Dimethyl dibenzothiophene
	231	Triaromatic steranes

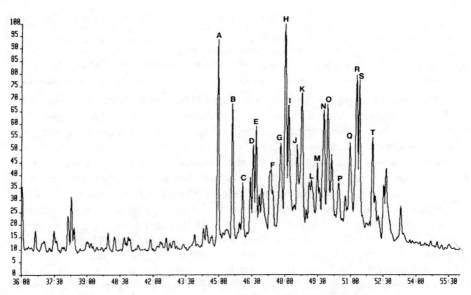

Figure 7. Distribution of steranes in a typical North Sea oil. See *Table 3* for identification of compounds A–T.

346

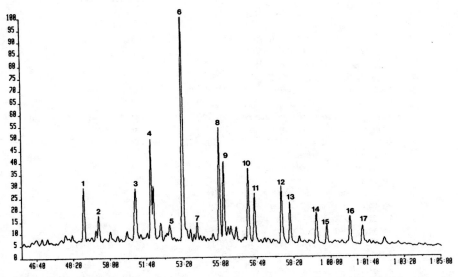

Figure 8. Distribution of terpanes in a typical North Sea oil. See *Table 4* for identification of compounds 1–17.

environment but relatively uncommon in non-marine sediments—are characterized by C_{29} steranes (9) and these can make a substantial contribution to the biomass. This is illustrated in *Figure 9*, which shows the dominance of C_{29} steranes in the m/z 217 mass chromatogram of an oil which is derived from algal marine source rocks which were laid down before the emergence of land plants.

The stereochemistry of the steranes is also important as it can provide the geochemist with information relating to the maturity of an oil or source rock extract. The most commonly used steranes for maturity assessment are the C_{29} steranes because these are relatively unaffected by co-elution from other sterane isomers. In thermally immature samples, the biological sterane isomers having $\alpha\alpha\alpha20R$ stereochemistry predominate, but with increasing maturity these isomers are transformed to steranes with $\alpha\alpha\alpha20S$ stereochemistry. As thermal maturity increases further the $20S$ epimers convert to steranes with $\alpha\beta\beta$ stereochemistry, until eventually thermodynamic equilibrium is reached.

Changes in the stereochemistry of the hopanes can also be used to monitor the differences in crude oil maturity. For example, in source rock extracts of low maturity the 17β hopanes (moretanes) are amongst the more abundant hopanes, as are the higher (C_{31}–C_{35}) hopanes exhibiting a favoured $22R$ stereochemical configuration. With increasing maturity, the 17β hopanes decrease in concentration, the 17α hopanes become a lot more abundant in concentration, and the $22R$ epimer of the C_{31}–C_{35} higher hopanes decreases

347

Gareth E. Harriman

Table 3. Compound identification for steranes (*Figure 7*)

A   13β,17α-Diacholestane (20*S*)
B   13β,17α-Diacholestane (20*R*)
C   13α,17β-Diacholestane (20*S*)
D   13α,17β-Diacholestane (20*R*)
E   24-Methyl-13β,17α-diacholestane (20*S*)
F   24-Methyl-13β,17α-diacholestane (20*R*)
G   24-Methyl-13α,17β-diacholestane (20*S*) + 14α,17α-cholestane (20*S*)
H   24-Ethyl-13β,17α-diacholestane (20*S*) + 14β,17β-cholestane (20*R*)
I   14β,17β-Cholestane (20*S*) + 24-methyl-13α,17β-diacholestane (20*R*)
J   14α,17α-Cholestane (20*R*)
K   24-Ethyl-13β,17α-diacholestane (20*R*)
L   24-Ethyl-13α,17β-diacholestane (20*R*)
M   24-Methyl-14α,17α-cholestane (20*S*)
N   24-Methyl-14β,17β-cholestane (20*R*) +
    24-ethyl-13α,17β-diacholestane (20*R*)
O   24-Methyl-14β,17β-cholestane (20*S*)
P   24-Methyl-14α,17α-cholestane (20*R*)
Q   24-Ethyl-14α,17α-cholestane (20*S*)
R   24-Ethyl-14β,17β-cholestane (20*R*)
S   24-Ethyl-14β,17β-cholestane (20*S*)
T   24-Ethyl-14α,17α-cholestane (20*R*)

in concentration as the 22*S* epimer increases, until the distribution of the isomers achieve an equilibrium status similar (around 60%) to that of the steranes. Other terpane ratios may also be used for maturation purposes. The T_s/T_m ratio (defined in Table 4), in particular, is an excellent parameter for monitoring subtle changes in crude oil maturity but the ratio is also source dependent and its usefulness is, therefore, restricted to oils derived from the same source rocks. *Table 5* summarizes some of the biomarker ratios commonly used for the assessment of crude oil and source rock extract maturity.

As with the steranes, certain hopanes are useful as source indicators. Peak 3 in *Figure 8*, for example, has been identified as 28,30-bisnorhopane, a C_{28} terpane which—although its concentration varies with maturity—is a ubiquitous marker for North Sea oils that have been generated from Upper Jurassic (Kimmeridge Clay Formation) source rocks. Similarly, gammacerane is a terpane that is commonly associated with oils that derive from hypersaline, lacustrine, or carbonate-evaporite source environments, while perhydro-carotene is thought to be a non-marine marker characterizing non-marine source beds such as those found in the Mid-Devonian of the Inner Moray Firth and in the Beatrice crude oil (10).

Sometimes mass fragmentograms that are acquired via the procedures laid out in *Protocol 2* are misleading or are of little use because interference from other, stronger, compounds within the sample can mask the compounds of interest. Alternatively, if the concentration of the biomarkers is particularly

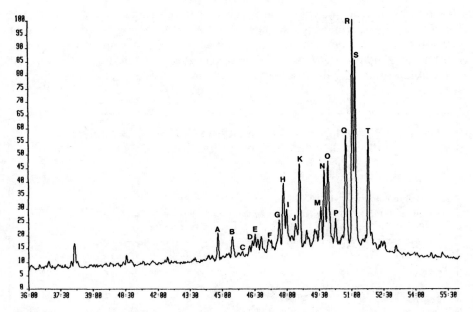

Figure 9. Sterane distribution of an oil derived from algal marine source rocks. See *Table 3* for identification of compounds A–T

Table 4. Compound identification for terpanes (*Figure 8*)

1 18α(*H*)-22,29,30-Trisnorhopane (T_s)
2 17α(*H*)-22,29,30-Trisnorhopane (T_m)
3 17α(*H*)-28,30-Bisnorhopane
4 17α(*H*)-Norhopane
5 17β-Normoretane
6 17α-Hopane
7 17β-Moretane
8 C_{31} Homohopane (22*S*)
9 C_{31} Homohopane (22*R*)
10 C_{32} Bishomohopane (22*S*)
11 C_{32} Bishomohopane (22*R*)
12 C_{33} Trishomohopane (22*S*)
13 C_{33} Trishomohopane (22*R*)
14 C_{34} Tetrakishomohopane (22*S*)
15 C_{34} Tetrakishomohopane (22*R*)
16 C_{35} Pentakishomohopane (22*S*)
17 C_{35} Pentakishomohopane (22*R*)

---

low, the chemical background signals become a significant problem leading to spurious or co-eluting peaks or rising baseline (column bleed). These problems hinder both the identification and quantification of the compounds of interest and can only be overcome by pre-concentrating the biomarker fraction by molecular sieving, or urea clathration, or by analysing the sample on a GC–MS system using a higher (about 2500) resolving power. This latter option allows for the evaluation of whole crude oils or source rock extracts with little or no pre-treatment of the sample. *Protocol 3* describes the analysis of whole crude oils, again using a TS250 mass spectrometer, but with the resolving power set to 2500.

Protocol 3. Medium (2500) resolution SIR GC–MS analysis of whole crude oils or source rock extracts

Conditions

- gas chromatograph: HP5890
- column: 30 m × 0.32 mm i.d. FSOT capillary column
- phase coating: DB-5; 0.25 μm d_f
- oven temperature
- programme: 60 °C to 310 °C at 4 °C/min
- final: 310 °C for 10 min
- injector: 310 °C
- carrier gas: He
- source temperature: 210 °C
- electron energy: 70 eV
- filament current: 1000 μA

Method

Use the same procedure as described in *Protocol 2* except as follows:

1. Set instrument resolution to 2500.
2. Enter into acquisition system, accurate masses for fragment ions to be monitored (see *Table 4*).
3. Insert additional lock mass to compensate for any drift in the system that may occur during analysis (can result from unstable magnetic field).
4. Dilute oil or source rock extract to be analyses 1:10 with dichloromethane before injecting between 0.1–0.2 μl of sample at a split ratio of 10:1.

The advantages of using higher resolution mass spectrometry than low resolution (as per *Protocol 2*) for the GC–MS analysis of untreated oils is based on the mass spectrometer's ability to separate ions of different mass. This resolution is defined (11) as the mass difference (Δm) between two overlapping mass spectral peaks m_1 and m_2. These peaks are considered to be resolved when 100 h/H is equal to or less than 10, where H is taken as the height of the peaks and h is the depth of the valley between the two peaks. Thus the resolution when $100h/H = 10$ is the value of m_1/Δm. Separation of the diagnostic fragment ions 231.1174 (triaromatic steranes) from 231.2113 (methyl steranes) therefore requires a resolution of 231.1174/0.0939, i.e. 2460.

In a typical biomarker analysis, the ions m/z 191, 217, and 218 are moni-

Table 5. Biomarker maturation parameters

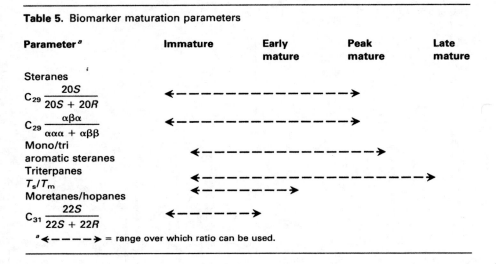

Parameter[a]	Immature	Early mature	Peak mature	Late mature
Steranes				
$C_{29} \dfrac{20S}{20S + 20R}$	←– – – – – – – – – – – – – – –→			
$C_{29} \dfrac{\alpha\beta\alpha}{\alpha\alpha\alpha + \alpha\beta\beta}$	←– – – – – – – – – – – – – – –→			
Mono/tri aromatic steranes		←– – – – – – – – – – – – –→		
Triterpanes				
T_s/T_m		←– – – – – – – – – – – – – – – –→		
Moretanes/hopanes		←– – – – – –→		
$C_{31} \dfrac{22S}{22S + 22R}$	←– – – – –→			

[a] ←– – – –→ = range over which ratio can be used.

tored and, when the analysis is performed on an isolated saturate hydro-carbon fraction or a sample where the alkanes have been removed by molecular sieving, the resulting mass chromatograms are interference-free and representative of the ions monitored. The chemical work procedures prior to GC–MS adversely can, however, affect the biomarker distributions, often giving misleading results. Moreover, it is often desirable to monitor the aromatic hydrocarbons at the same time as collecting data on the saturate markers and, because of the chemical interference problems outlined above, this is not possible with conventional low-resolution GC–MS. High-resolution GC–MS, however, overcomes this problem, allowing for the collection of all relevant mass chromatograms[a] in a single analysis. The accurate mass fragment ions listed in *Table 6* represent some of the more common ions monitored during the GC–MS analysis of whole oil or source rock extracts.

Figure 10c shows how, by using high resolution GC–MS, the monoaromatic steranes (*m/z* 253.1956)—which are normally masked by the alkanes (*m/z* 253.2895) at low resolution—can be monitored providing valuable information for source identification (12). *Figures 10* and *11* are examples of GC–HRMS analysis of an oil from North Africa. The mass fragmentograms for the triterpanes (*Figure 10a*), steranes (*Figure 10b*), mono- and triaromatic steranes (*Figures 10c* and *d*), and phenanthrenes (*Figure 11*) are derived from one analysis, thus yielding the maximum information regarding the source and maturity of the sample analysed.

[a] Note that the compound classes not only include the important saturated hydrocarbon classes such as the steranes and triterpanes but also the aromatic steranes and phenanthrene series of compounds that provide additional information on source and maturity.

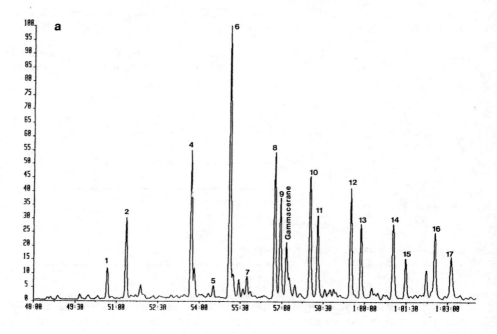

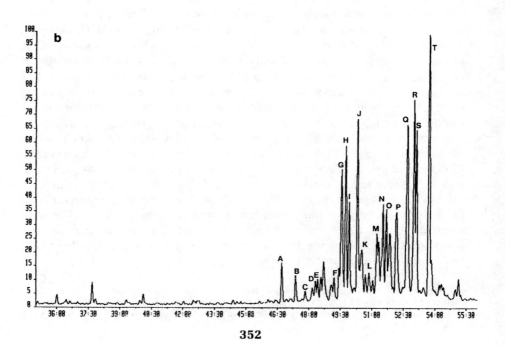

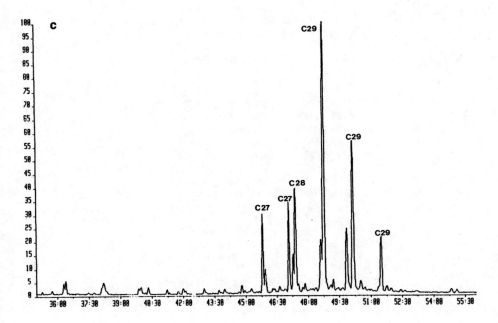

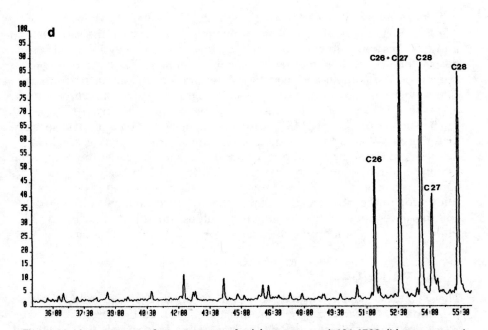

Figure 10. Accurate mass fragmentogram for (a) terpanes, *m/z* 191.1798, (b) steranes, *m/z* 217.1956; (c) monoaromatic sterane, *m/z* 253.1956; (d) triaromatic steranes, *m/z* 231.1200.

Table 6. Accurate masses of ions used for medium resolution SIR GC–MS

Ion (m/z)	Compound class
85.1017	Alkanes
123.1174	Diterpanes, bicyclic alkanes
177.1642	Demethylated hopanes
178.0782	Phenanthrene
184.0344	Dibenzothiophene
191.1798	Terpanes
192.0928	Methyl phenanthrenes
198.0501	Methyl dibenzothiophenes
205.1956	Methyl hopanes
206.1088	Dimethyl phenanthrenes
212.0457	Dimethyl dibenzothiophenes
217.1956	14-α steranes
218.2034	14-β steranes
231.1174	Triaromatic steranes
231.2113	Methyl steranes
253.1956	Monoaromatic steranes
259.2427	Rearranged steranes

The advantages of GC–HRMS as described above are obvious, especially when large numbers of samples are to be analysed for both the saturate and aromatic markers. However the main disadvantage of analysing whole crude oils in this way is that the heavy asphaltic components of the oils tend to remain on the column and lead to a deterioration of column performance. This effect can be prevented, to some degree, by use of a pre-column, although it is just as easy to deasphaltene the oil prior to analysis.

The success of GC–HRMS is also dependent on the type of mass spectrometer used. Whatever the choice of instrument, the advantages gained by increasing the resolution are offset by a loss in sensitivity. On the TS250, the loss in sensitivity going from 500 resolution to *ca.* 2500 is greater than 90% and for some oils the sensitivity proved inadequate for GC–HRMS. Moreover calibration of the mass spectrometer at the higher resolution is often unstable and this necessitates either running the analysis with a lock mass inserted for calibration correction during the analysis or recalibrating after each analysis.

6. Conclusions

This chapter has demonstrated the usefulness of capillary GC and GC–MS as a method for characterizing petroleums and organic extracts from rocks. The type of column used depends on the complexity of the sample to be analysed and the nature of the components to be resolved. Crude oils, like those

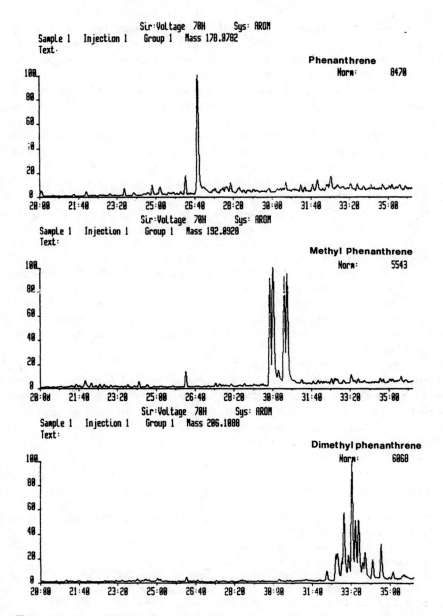

Figure 11. Aromatic hydrocarbons in crude oil as detected by accurate mass SIR.

depicted in *Figures 2* and *3*, contain many hundreds of components and require analysis on long columns with high efficiency. For less complex mixtures, shorter columns, which offer the advantages of shorter analysis time and less column bleed, can be used. Such columns, when coupled to a selective detector like an MS, become highly specific analytical tools allowing for the separation and identification of compounds which provide the explorationist with a better understanding of the processes behind petroleum formation, generation, and migration.

The importance of capillary GC and GC–MS in petroleum exploration has been, and will continue to be, in the identification and characterization of hydrocarbon source rocks and crude oils, and in the determination of oil–oil and oil–source correlations. More recently, gas chromatography has found a role in production geochemistry in characterizing reservoir fluids to assess reservoir continuity, evaluate co-mingled oils, and identify non-productive zones. Furthermore, improvements in column and instrument technology have opened up new areas of research for the geochemist, allowing for the investigation of high molecular markers ($>C_{40}$) using aluminium-clad columns which are much more stable at high temperature than conventional columns. In addition to this capability, the recent development of MS/MS techniques (13) has enabled the geochemist to resolve complex mixtures of biomarkers such that the relative concentrations of various isomers with the same carbon number can be measured. Moreover the technique allows for the identification of compounds (C_{30} steranes, for example) that are otherwise obscured by more abundant co-eluting compounds.

Geochemists are now focusing their interest on the sulphur- and nitrogen-containing compounds that are found in fossil fuels and it is in these areas of geochemical research that gas chromatography coupled to sulphur-selective detectors and mass spectrometers will continue to play a significant role in the future.

References

1. Welte, D. H., Kratochvil, H., Rullkoter, J., Ladwein, H., and Schaefer, R. G. (1982). *Chem. Geol.*, **35**, 33.
2. Bailey, N. J. L., Jobson, A. M., and Rogers M. A. (1973). *Chem. Geol.*, **11**, 203.
3. Blumer, M. (1975). *Angew Chem.*, **14**, 507.
4. Brooks, J. D., Gould, K., and Smith, J. (1969). *Nature*, **222**, 257.
5. Powell, T. and McKirdy, D. M. (1973). *Nature*, **243**, 37.
6. Bray, E. E. and Evans, E. D. (1961). *Geochim. Cosmochim. Acta*, **22**, 2.
7. Seifert, W. K. and Moldowan, J. W. (1978). *Geochim. Cosmochim. Acta*, **42**, 77.
8. Huang, W-Y. and Meinschien, W. G. (1979). *Geochim. Cosmochim. Acta.*, **43**, 739.
9. Goodwin, T. W. (1973) In *Lipids and biomarkers of eukaryotic microorganisms* (ed. J. A. Erwin). Academic Press, New York, pp 1–40.

10. Bailey, N. J. L., Burwood, R., and Harriman, G. E. (1990). *Org. Geochem.*, **16,** 1157.
11. Rose, M. E. and Johnstone, R. A. W. (1982). *Mass spectrometry for chemists and biochemists*. Cambridge University Press, Cambridge.
12. Moldowan, M. J., Seifert, W. K., and Gallegos, E. J. (1985). *Am. Assoc. Petrol. Geol.*, **69,** 1255.
13. Harriman, G. E., Owen, R., Parr, V. C., Weir, O., and Wood, D. (1989). In *Novel techniques in fossil fuel mass spectrometry, ASTM STP 1019* (ed. T. R. Ashe and K. V. Wood). American Society for Testing and Materials, Philadelphia, p. 59.

Combined gas chromatography–mass spectrometry

RICHARD P. EVERSHED

1. General considerations

Gas chromatography–mass spectrometry (GC–MS) remains the most effective technique for the separation, detection, and characterization of the components of complex organic mixtures (1). GC and MS are regarded as the 'natural combination', compared to other combined chromatographic–mass spectrometric techniques, as both display optimum performance with volatile or semi-volatile materials when sample sizes per component are in the nanogram range (2). Considered in its simplest form, a mass spectrometer (MS) is a device for producing and mass measuring ions. In the case of organic molecules, the mass and relative abundance of molecular ($M^{+\cdot}$) or pseudomolecular ions (e.g. $[M + H]^+$) and fragment ions are a direct reflection of molecular structure. Hence, mass spectrometry is unique in its ability often to yield complete structure assignments without needing to use other physicochemical techniques. Although isomeric structures may be distinguished on the basis of their mass spectra alone, such distinctions are often best made by considering elution orders in combined chromatographic/mass spectrometric analyses.

The complexity of mass spectrometric instrumentation has meant that analyses have traditionally been conducted by highly trained operators. However, the increasing availability of less sophisticated computer-controlled bench-top GC–MS instruments enables the relatively untrained analyst to perform many MS analyses with only the minimum of expert assistance. Although a detailed knowledge of mass spectrometers is probably unnecessary for most analysts, an appreciation of the basics of operation of the vacuum system, ion source, mass analyser, and ion detection is undoubtedly of value; ensuring safe and efficient handling of the valuable instrumentation. Chapman's monograph *Practical Organic Mass Spectrometry* (3) covers most modern GC–MS techniques.

An overview of the ionization methods commonly used in MS analysis is presented below (Section 2.3), as the choice of these can greatly affect the outcome of analyses. Complementary to this is a knowledge of possible sample preparation strategies, including chemical transformations or derivatizations that can be used to enhance the structural information content of mass spectral data or improve detection limits in trace analyses. Some examples are given in Section 3 and more detailed coverage of derivatization methods is available elsewhere (4–6). In addition, the analyst needs to be aware of the capabilities of the different operational modes of the mass spectrometer that exist. For instance, the use of full mass range scanning to provide mass spectra for structure investigations versus the use of selected ion monitoring (SIM) to enhance sensitivity of detection in trace analyses. Due consideration is given to these latter points in Section 3.

2. GC–MS instrumentation

The discussion of the theory and general principles of operation of GC presented in other chapters apply equally well to combined GC–MS. *Figure 1* shows the components of a modern GC–MS instrument. Mass spectrometers operate at reduced pressure (typically $<10^{-6}$ Torr) to enable detection of the small amounts (sub-nanogram to microgram) of material that are routinely analysed. Further details concerning the theory and mode of operation of the components of modern GC–MS instruments are presented below.

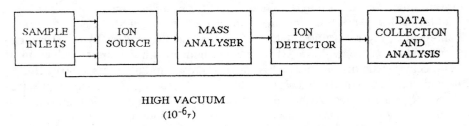

Figure 1. Components of a mass spectrometer with sample inlet from GC.

2.1 Use of packed columns

Packed columns are rarely used now in GC–MS work, as the higher carrier gas flow rates, *ca.* 50 ml/min, are incompatible with the mass spectrometer vacuum system. A separator must be used to remove the bulk of the carrier gas preferentially from the sample components eluting in the column effluent, prior to its introduction into the mass spectrometer. The Ryhage jet separator and Watson–Biemann membrane separator have been widely employed as enrichment devices. These separation devices are notoriously troublesome as they provide sites for adsorption and decomposition of higher-molecular-weight and thermally unstable compounds. As with packed GC columns, the high flow rates employed with the increasingly popular wide bore (0.53 mm i.d.) capillary columns are incompatible with high vacuum pumping systems of mass spectrometers. Separators or enrichment devices of the type referred to above are required for the use of the columns in GC–MS.

2.2 Use of capillary columns

Flexible fused silica capillary columns are now preferred for GC–MS work as they can be fed directly from the GC oven into the ion source of the mass spectrometer (7). This is the optimum arrangement as it ensures efficient transfer of analytes from the GC to the mass spectrometer. The temperature of the transfer line between the GC and MS must be maintained close to the maximum temperature used in the GC analysis, in order to avoid sample components condensing in this region. The carrier gas flows employed in capillary GC, *ca.* 2 ml/min, are readily accommodated by the vacuum pumping systems of modern mass spectrometers. Some care is required when connecting metal-clad columns directly into the sources of magnetic sector mass spectrometers, as these operate at several kilovolts potential. A method of interfacing aluminium-clad capillaries uses sodium hydroxide to dissolve the distal 5 cm of aluminium, ensuring insulation from the ion source voltages (8).

Protocol 1. Treatment[a] of aluminium-clad capillary columns for interfacing to a magnetic sector MS

1. Remove source housing back flange and ion source assembly.
2. Feed the terminal 0.5 m of aluminium-clad capillary column through the GC–MS transfer line and out into the laboratory.
3. While maintaining a normal flow of carrier gas, e.g. helium, submerse the terminal 5 cm of column in a 50%, w/v, aqueous solution of sodium hydroxide.
4. When the cladding has completely dissolved, wash the end of the column with distilled water and air dry (hair dryer).
5. Pull the column carefully back into the re-entrant jet.
6. Replace the ion source assembly and back flange.
7. Push the column into its optimum position in the ion source.

[a] The stripped 5 cm section of column ensures complete electrical isolation, as shown by the effective maintenance of all source potentials.

The open split interface is widely used to connect capillary columns to mass spectrometers (9). This mode of coupling is of particular advantage where frequent column changes are anticipated as there is no need to vent the MS vacuum system during column changes. The open-split coupling is constructed of an inlet restrictor which is fed into the MS ion source. The i.d. of the inlet restrictor is chosen such that a constant flow rate of *ca.* 2 ml/min is sampled into the MS. The end of the GC column terminates immediately adjacent to the capillary restrictor, with the junction being maintained at atmospheric pressure. A flow of helium over the column/restrictor junction prevents air being drawn into the MS. The nature of the materials used to construct the interface is worthy of consideration. Open-split interfaces constructed of Pyrex glass or flexible-fused silica overcome many of the problems associated with some commercial systems (10, 11).

As the GC column terminus is maintained at ambient pressure, no losses of chromatographic efficiency occur when using the open-split coupling. This contrasts with the direct mode of coupling, where the pressure drops between the injector and column exit in the ion volume, particularly in the case of short or wider bore columns, which means that the chromatographic system does not operate at optimum resolution.

2.3 Ion sources

Electron ionization (EI) and chemical ionization (CI) are the two most widely used ionization techniques in GC–MS. The following section describes the mechanism of operation of these two ionization techniques. Those readers requiring more detailed discussion of ionization techniques are directed to the specialist MS literature (12–14).

2.3.1 Electron ionization (EI)

EI is produced by accelerating electrons from a hot filament through a potential difference, usually of 70 eV. Organic molecules introduced into the electron beam by elution from a GC column will be ionized and fragmented. The initial product of EI in the ion source is a radical cation, resulting from the removal of an electron from the analyte molecule (*Scheme 1*). If this singly-charged species is stable it will be the highest mass ion appearing in the mass spectrum for that substance. It is referred to as the molecular ion, denoted $M^{+\cdot}$, and it provides the molecular weight (measured in atomic mass units, amu, or daltons, Da) of that substance. The electron beam usually possesses sufficient energy to induce fragmentation in organic compounds, by loss of radicals or neutral molecules (see *Scheme 1*).

Although EI produces extensive fragmentation which is of use in structural investigations of lipids, the $M^{+\cdot}$ ion is frequently weak or absent from spectra. The relative abundance of $M^{+\cdot}$ may be enhanced by decreasing the ion-source block temperature, or electron beam potential (say to 20 eV). When this latter approach is unsuccessful, chemical ionization (CI) is used to derive molecular weight information in GC–MS work.

2.3.2 Positive-ion CI

CI is used to generate mass spectra by means of ionic reactions rather than electron bombardment. The technique is termed a 'soft' ionization technique,

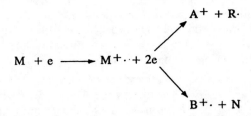

Scheme 1. Electron ionization (EI) induced by collision of an electron with a molecule to produce a molecular ion ($M^{+\cdot}$) which can give the fragment ions, A^+ (cation) or $B^{+\cdot}$ (radical cation), by loss of a radical ($R\cdot$) or a neutral molecule (N), respectively.

as the spectra usually contain essentially only molecular weight information, fragmentation being either absent, or much reduced compared to EI. Many variations of CI exist, but typically a reagent gas, e.g. methane, ammonia, or isobutane, is introduced into the ion source at a pressure of around 1 Torr. Bombardment of the gas with electrons yields a population of ions and neutral molecules. The reagent gas ions will take part in ion-molecule reactions, with sample molecules vaporizing into the source (*Scheme 2*). The most abundant reagent gas ions formed by the electron bombardment of methane are CH_5^+ and $C_2H_5^+$. Detailed descriptions of the theoretical and practical aspects of chemical ionization mass spectrometry are available elsewhere (12–14).

$$M + C_2H_5{}^+ \longrightarrow MH^+ + C_2H_4$$

Scheme 2. Reagent gas ions, produced by electron bombardment of methane, react with analyte molecules to yield positively charged quasi-molecular adduct ions.

High mass ions arising by proton transfer (see *Scheme 2*) termed pseudo or quasi-molecular ions, $[M + H]^+$, occur commonly. Other high mass ions which may be seen, correspond to molecular adduct ions, e.g. $[M + CH_5]^+$, $[M + C_2H_5]^+$, $[M + C_3H_5]^+$ etc., and arise by electrophilic addition of reagent gas ions to the sample molecules. The quasi-molecular ion and adduct ions are useful in inferring the molecular weights of sample molecules. The lack of fragment ions in CI spectra results from the low energy transfer involved in the CI process compared to EI. Furthermore, even-electron $[M + H]^+$ ions are energetically more stable than the radical $M^{+\cdot}$ ions produced by EI.

2.3.3 Negative-ion CI

Negative ions are also produced in mass spectrometers under special circumstances. The majority of organic compounds do not yield negative ions under EI conditions (involving high energy electrons, see above). However, many organic compounds will capture low energy electrons. Negative ion chemical ionization (NICI) has been used extensively in conjunction with GC/MS in trace analyses and, to a lesser extent, in structure investigations. Two major categories of negative ion formation exist.

(a) The capture of thermal electrons by sample molecules occurs with the reagent gas serving as a means of producing thermal electrons and as a source of molecules for collisional stabilization of the negative ions formed (*Scheme 3*).

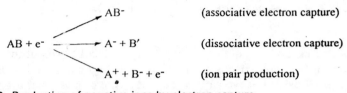

$AB + e^-$

    AB^- (associative electron capture)

    $A^- + B'$ (dissociative electron capture)

    $A^+ + B^- + e^-$ (ion pair production)

Scheme 3. Production of negative ions by electron capture.

(b) Ion-molecule reactions can occur when stable anions are formed by the reagent gas. The possibilities that exist for ion-molecule reactions between gas phase anions and sample molecules are summarized in *Scheme 4*.

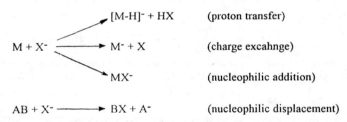

$M + X^-$

    $[M-H]^- + HX$ (proton transfer)

    $M^- + X$ (charge excahnge)

    MX^- (nucleophilic addition)

$AB + X^- \longrightarrow BX + A^-$ (nucleophilic displacement)

Scheme 4. Production of negative ions from organic molecules by ion molecule reactions with stable anions.

An example of the use of negative ion CI in conjunction with high temperature GC/MS for structure elucidation and mixture analysis is given below (Section 3.1.2).

2.4 Use of different mass analysers

A number of types of analyser are now available which will measure the mass of ions according to their mass-to-charge ratio (m/z). GC–MS instruments most commonly employ quadrupole mass filters, magnetic sector analysers, or ion-trap detectors. The following sections discuss the basic principles and advantages and disadvantages of these various mass analysers, in terms of their suitability for use in GC–MS work.

2.4.1 Magnetic sector analysers

The double-focusing analyser is the most common magnetic sector instrument configuration currently in use for the analysis of organic compounds. The usual arrangement is for an electrostatic sector to precede a magnetic sector (conventional Nier–Johnson geometry). The electrostatic sector is an energy focusing device, hence, ions produced and accelerated by the ion source are focused by the electrostatic sector according to their translational energies, irrespective of their m/z. The magnet on the other hand is a momentum analyser, where the ions are separated according to their m/z ratio. The combination of an electrostatic sector with a magnetic analyser is highly versatile, providing the capability for operation at high resolution. The latter property enables accurate mass measurements to be performed, from which the elemental compositions of ions can be determined. A number of alternative arrangements of electrostatic and magnetic sectors are available in commercially produced instruments, introducing the possibility for a wide range of scan modes (15).

2.4.2 Quadrupole mass filters

Quadrupole mass spectrometers are constructed and function in entirely different ways to magnetic sector instruments. The quadrupole analyser comprises four parallel rods of hyperbolic or circular cross-section. A radio-frequency (r_f) potential and dc voltage is connected between opposite pairs of rods. Ions are injected into the oscillating electric fields by a small accelerating voltage (10–20 V) and, under the influence of the fields, begin to undergo complex oscillations. Mass separation is achieved by varying (scanning) the voltages on the quadrupole rods, while keeping the $r_f/$dc ratio constant. At any one point in the scan only one mass can pass through the system.

Quadrupole instruments are very popular, especially in GC–MS systems, owing to their relatively simpler operation compared to magnetic sector instruments. The operation of the quadrupole ion source at ground potential greatly simplifies interfacing. Although the mass range of quadrupole instruments is substantially less than that of many magnetic sector instruments, it is sufficient for GC–MS analyses. As there are no slits in the quadrupole mass filter, the transmission of ions is very efficient. This is of special advantage in selected ion monitoring analyses where detection limits in the picogram to femtogram range are routinely attained. The faster scan rate attainable by quadrupole mass filters (and ion trap detectors: see Section 2.4.3) are preferred in quantitative GC–MS work, especially when small diameter WCOT capillary columns are used (16, 17). A number of the properties of magnetic sector and quadrupole instruments are compared in *Table 1*.

2.4.3 Ion trap (18)

The three-dimensional quadrupole ion storage trap (QUISTOR or ion trap) is a device comprising two end caps and a ring electrode. Ions produced in an

Table 1. Advantages/disadvantages of mass spectrometers using magnetic sector or quadrupole mass analysers

Magnetic sector	Quadrupole
Advantages	*Advantages*
Greater versatility	Compact
Accurate mass measurement	Easy to operate
More specific forms of selected ion monitoring	Source at earth potential makes interfacing less problematical
Metastable ion analysis	Easy to look at positive and negative ions
High mass capabilities >2000 daltons	Much less expensive than sector instruments
Disadvantages	*Disadvantages*
Highly sophisticated	Limited mass range
Relatively more expensive	Only low resolution capabilities
High source potentials present interfacing problems	Limited data types compared to magnetic sector instruments

electron impact source within the ion trap are trapped in the quadrupole field by a combination of r_f and dc potentials applied to the electrodes. Scanning of the mass range is achieved by varying the r_f and dc voltages either together or singly. Each trapped ionic species (m/z) becomes unstable, i.e. the ions develop trajectories that exceed the boundaries of the trapping field. The ions pass out of the field through perforations in the field-imposing structure and impinge on a detector such as an electron multiplier. The ion trap provides a sensitive, low-cost mass analyser for GC–MS work. The mass range and resolution capabilities are comparable to those of benchtop quadrupole GC–MS systems (18). CI can also be achieved within the ion trap using a facility for injecting ions from an external ion source (19).

2.4.4 Tandem mass spectrometers

Tandem MS refers to the linking of at least two sequential stages of mass separation. One advantage of a tandem mass spectrometer is the ability to combine processes of separation and identification in a single instrument. The use of GC with tandem MS essentially adds a further separation step with the result that very high selectivities and sensitivities can be achieved in complex mixture analysis. The high selectivity offered by the MS–MS combination means that short GC columns may be adequate, hence, shorter analysis times are required, with the result that sample throughput can be increased. Tandem MS has gained wide acceptance and numerous reviews are available containing examples of applications and detailed descriptions of instrumentation (20, 21).

The most common types of tandem MS are:

(a) four sector instruments which incorporate various arrangements of electrostatic (E) and magnetic (B) sectors, e.g. BEEB and EBEB

(b) hybrid instruments which combine electrostatic, magnetic, and quadrupole (Q) analysers, and r_f only quadrupole (q) into a single instrument, e.g. EBqQ and BEqQ

(c) triple quadrupole instruments which incorporate quadrupole mass analysers only, e.g. QqQ

Only hybrid and triple quadrupole tandem mass spectrometers have been used to any significant extent in conjunction with GC. Although tandem mass spectrometers are applied widely to enhance structure information and increase selectivity in trace analyses, it is primarily in the latter role that GC–MS–MS is used. The use of GC–MS–MS for selected reaction monitoring is discussed in detail in Section 3.2.3.

Other modes of use of tandem mass spectrometers include the recording of full product ion spectra of selected precursor ions. These product ion spectra can be used in the same way as conventional mass spectra, e.g. through interpretation or comparison of unknown spectra with those derived from authentic compounds. Precursor ion scans reveal precursor ions of different m/z that fragment to yield a common product. While neutral loss scans reveal the loss of specific neutral fragments from precursor ions. Both the precursor and neutral loss scans are used for screening mixtures for the presence of particular compound classes.

2.5 Ion detection

The small current (typically in the range 10^{-9} to 10^{-17}A) generated by ions passing through the mass analyser must be amplified and converted into a voltage that can be digitized or displayed on some form of recording device. Electron multipliers are widely used for this purpose. These devices comprise copper-beryllium dynodes which emit electrons when impacted by highly energetic ions, focused from the mass analyser. Cascade emission of electrons through a series of dynodes produces gains of the order of 10^6. The final dynode of the electron multiplier is connected to a pre-amplifier, to convert the output current to a voltage suitable for recording.

Electron multipliers are the workhorse detectors for general purpose mass spectrometers. If negative ion detection is required, then the inclusion of a conversion dynode is to be preferred. This is particularly the case on low voltage instruments such as quadrupole and ion trap mass spectrometers. Photomultipliers are also used as detectors in mass spectrometers. They are identical to the electron multiplier except that the dynodes are sealed in a glass vacuum envelope with a photocathode coated on to the internal surface of the glass in front of the first dynode. Ions striking the photocathode cause electrons to be emitted which are accelerated to the first dynode.

2.6 Data collection and interpretation

Computers are essential to the efficient operation of modern mass spectro-meters. The type of computers employed range from large, powerful computers to desk-top personal computers. The computer is usually fully integrated into modern instruments, controlling the various mass spectrometer, inlet, ioniza-tion, and scan functions. In addition, the computer provides an essential data collection and storage facility. The large volume of mass spectral data pro-duced in GC–MS analyses of complex mixtures requires substantial post-run data processing. The interpretation of this large volume of GC–MS data is aided by automated searching of mass spectral databases held in the perma-nent memory of the mass spectrometer computer. A number of database libraries of EI mass spectra are available from proprietary sources. *Figure 2* shows part of the output from a commercially available GC–MS data pro-cessing package. In addition to showing the closest matching library mass spectrum, the library search also shows a range of statistical evaluations of the search. Although potentially very useful, the results of library searches are in no way definitive. The search results must be carefully assessed on the basis of the statistical evaluation of the match and consideration of GC elution orders, where mass spectra have been obtained by combined GC–MS analyses. The possibility exists for installing mass spectral data processing software and libraries on other PCs for off-line data processing.

3. Applications

The following sections provide examples of selected applications of modern GC–MS techniques to the solution of specific problems arising in mixture and trace analysis. The examples have been selected in order to illustrate the possibilities that now exist for combining specialist GC and MS techniques, including novel derivatization methods, GC separations, ionization modes, selective detection techniques, and data processing strategies.

3.1 Mixture analysis

3.1.1 GC–MS analysis of alkylporphyrins

Alkylporphyrins occur widely in sedimentary materials, e.g. crude oils and their source rocks. Termed petroporphyrins, they often comprise highly com-plex mixtures of many tens to hundreds of closely related components per sample, and as such present a considerable analytical challenge. This applica-tion of GC–MS has been chosen to demonstrate:

(a) a novel derivatization strategy

(b) the use of high resolving flexible fused silica capillary columns and on-column injection

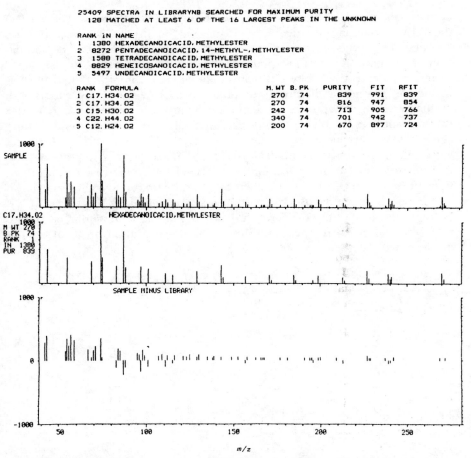

Figure 2. Example of the output from a library search package showing the most likely candidate compounds selected and the statistical evaluation of the library search.

(c) computerized post-run data processing techniques for deconvoluting complex mixtures of co-eluting components

i. Derivatization

Derivatization was required to overcome problems associated with the direct GC/MS analysis of petroporphyrins as their commonly occurring complexes of nickel(II) and vanadyl (V=O), or free base compounds obtained by demetallation. Both the metal complexes and free base alkylporphyrins were highly retained on the GC, i.e. eluted in the pseudo-Kóvats' retention index (KRI) range 5200–5500, with inferior peak symmetry and greater peak widths compared to *n*-alkanes eluting at similar retention times (22). The GC–MS approach adopted was based on the use of the bis(trimethylsiloxy)silicon(IV) derivatives originally investigated by Boylan and Calvin (23). The derivatization proceeded with the reaction of free base porphyrins (obtained by demetalla-tion of the naturally occurring metal complexes) with hexachlorodisilane to form dichloroSi(IV) porphyrins. Conversion of the dichloroSi(IV) porphyrins to the corresponding dihydroxy compounds was achieved by hydrolysis with methanolic sodium hydroxide. A significant departure from Boylan and Calvin's original method involved the preparation of the bis-(*tert*-butyldimethyl-siloxy)Si(IV) derivatives by treatment of the dihydroxysilicon(IV) porphyrin with either *tert*-butyldimethylchlorosilane/imidazole or *N*-methyl-*N*-(*tert*-butyldimethylsilyl)trifluoroacetamide (MTBSTFA). In the case of the latter, two drops of MTBSTFA was added to the dihydroxysilicon(IV) porphyrin (0.1 mg) in dry pyridine (0.3 ml) and the mixture heated for 10 h at 60°C (24). In some instances the porphyrin derivatives were vacuum sublimed before GC–MS analysis to remove impurities. The derivatization protocol is sum-marized in *Scheme 5*.

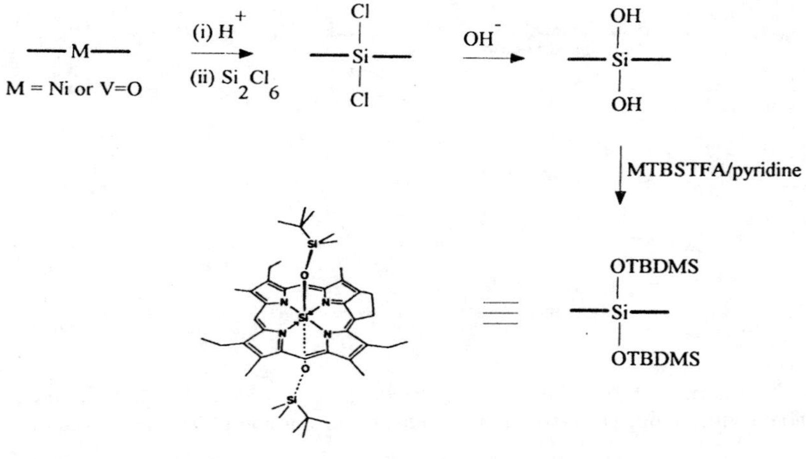

Scheme 5. Demetallaton and derivatization of alkyl petroporphyrins.

ii. GC–MS analysis

The bis-TBDMS derivatives were preferred to the corresponding TMS de-rivatives due to their enhanced chemical stability and more desirable MS properties. The bis(TBDMSO)Si(IV) derivative of alkylporphyrins displayed excellent GC behaviour on flexible fused silica capillary columns coated with cross-bonded polydimethylsiloxane stationary phases (further experimental details are given in the caption to *Figure 3*). On-column injection was pre-ferred to splitless injection in order to avoid problems of sample transfer on to the GC column, due to the relatively involatile nature of the petroporphyrin derivatives. The use of high resolution fused silica capillary columns ensured optimum resolution of the complex mixtures and allowed direct introduction of the GC column into the MS ion source (see Section 2.2). Although bis-TBDMS derivatives eluted *ca.* 380 Kóvats' retention index (KRI) units higher than the corresponding bis-TMS derivatives, the difference in temperature requirements was found to be trivial. The naturally occurring alkylporphyrins elute predominantly in the KRI range 3400–4000 (*Figure 3*). The EI mass spectra of the TBDMS derivatives yielded abundant $M^{+\cdot}$ions. The abundant $[M-R_3SiO]^+$ ion, i.e. $[M-131]^+$, produced by the bisTBDMS derivative was found to be of most use analytically in deconvoluting the complex mixtures of porphyrins that occur in geological materials.

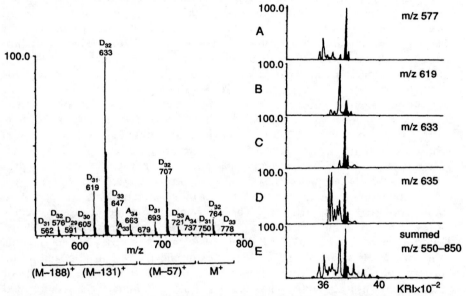

Figure 3. Selected mass chromatograms (right) and fully interpreted summed mass spectrum (left) for the shaded peak (KRI 3755–3800) of the Boscan total porphyrin chromatogram (trace E). Misassignment of mass chromatogram peaks may occur in the absence of careful examination of the mass spectra. The mass chromatograms A and D contain peaks which do not correspond to the characteristic $[M - 131]^+$ fragment ion but arise from
[13]C isotope peaks of lower mass fragment ions. A full interpretation of these data is presented in ref. 24. The retention time scale has been converted to KRI by computer interpolation from co-injected *n*-alkanes (the *n*-alkane peaks are not shown on these profiles). The analyses were performed on a 25 m × 0.31 mm i.d. cross-linked dimethyl polysiloxane coated (0.17 μm film thickness) flexible fused-silica capillary (Hewlett Packard, Ultra series) column. Helium was the carrier gas at a flow rate of 50 cm/s. The analysis was temperature-programmed, following on-column injection, ballistically from ambient temperature to 150°C, then at 3°C/min to 290°C. The GC–MS instrument comprised a Finnigan 9610 GC equipped with a SGE OCI-2 on-column injector, linked to a Finnigan 4000 quadrupole mass spectrometer. Instrument control and data aquisition and processing were performed by a Finnigan INCOS 2300 data system. (Reproduced with kind permission from ref. 24 © Elsevier Science Publishers BV.)

Richard P. Evershed

iii. Computerized data interpretation

The power of GC–MS for the analysis of complex mixtures of petroporphyrins lies in its ability to resolve the many co-eluting components through the generation of mass chromatograms. *Figure 3* shows the fully interpreted mass spectrum and selected $[M-131]^+$ mass chromatograms for the major peaks of the petroporphyrins of Boscan crude oil. The individual mass chromatograms show a multiplicity of peaks, many of which correspond to genuine structural isomers of the same molecular formula. Careful checking of the mass spectra of the individual mass chromatogram peaks was required to confirm that they were derived from genuine $[M-131]^+$ ions, and did not correspond to other isobaric fragment ions of co-eluting compounds. The conversion of the retention time (or scan number) scale to KRI values used a computer interpolation based on the co-chromatography of an n-alkane mixture. This procedure served to eliminate run-to-run variations in the GC parameters, and hence simplify comparisons between independently analysed samples. Indications were that many of the compounds present were related as homologous or pseudo-homologous series. Evidence for this came from the plots of C_n versus KRI values for the peaks in the $[M-131]^+$ mass chromatograms (24). Furthermore, by analogy with retention behaviour of authentic standards, it was found that if points on the Kóvats' plots were joined by straight lines with a gradient of *ca.* 40–50 KRI units per carbon number, then these lines represented pseudo-homologous series of structurally related petroporphyrins (25). Supporting evidence for this came from co-injection experiments with compounds of known structure. These straight-line relationships are commonly found for homologous or a pseudo-homologous series of compounds. *Figure 4* shows an example of KRI versus C_n plots for pseudo-homologous series of related alkylporphyrins bearing exocyclic alkano rings present in the Gilsonite bitumen (25). Application of these methodologies to the analysis of petroporphyrins isolated from a range of sedimentary materials has confirmed their highly complex composition. For example, computerized GC–MS analysis of the petroporphyrins of Boscan crude oil revealed at least 224 compounds, comprising five distinct structural classes (26).

3.1.2 High temperature GC–MS analysis of triacylglycerols

Natural triacylglycerol mixtures are potentially extremely complex, since the number of possible molecular species is equal to the number of fatty acids cubed. Hence, most triacylglycerol analyses only provide partial compositional information. The most common approach to the analysis of triacylglycerols is to release the free fatty acids, either chemically or enzymatically, and perform GC or GC–MS analysis after appropriate derivatization, e.g. methylation. A seemingly more attractive approach is to analyse the intact triacylglycerols. Most of the early analyses of intact triacylglycerols by MS used the direct

372

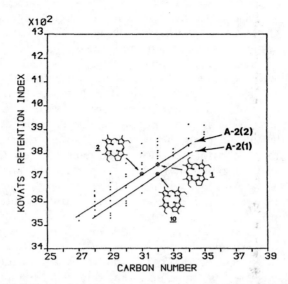

Figure 4. Plots of KRI versus carbon number for the porphyrins of Gilsonite bitumen which contain 13,15-ethano-and 13¹-methyl-13,15-ethano-rings [correspond to the components designated D_n (where n = total carbon number) in the mass spectrum displayed in *Figure 3*]. The components co-eluting with compounds of known structure are marked by arrows. (Reproduced with kind permission from ref. 25 © Elsevier Science Publishers BV).

insertion probe technique for sample introduction (27, 28) This approach can provide a certain amount of information concerning the carbon number distribution and the nature of the fatty acyl moieties present in the triacylglycerol mixtures. However, less information is obtained concerning the distribution of the various fatty acyl moieties amongst the individual triacylglycerol molecular species. The following section discusses, with examples:

(a) the use of fused silica capillary columns coated with high-temperature stable stationary phases for the GC–MS analysis of triacylglycerols

(b) the use of *different* ionization techniques to derive compositional information

(c) problems relating to the survivability of acyl lipids during high-temperature GC–MS analyses.

A notable development in the area of GC and GC–MS analysis of triacylglycerols has been the use of high-temperature stable, polarizable, immobilized stationary phases. One such stationary phase is an immobilized OV-22 (65% phenyl methyl polysiloxane) polymer, which becomes more polar with increasing temperature, and is stable up to 360°C. This stationary phase is capable of separating molecular species which exhibit only small differences

Richard P. Evershed

in polarity, e.g. triacylglycerols possessing different numbers of double-bonds. Although the effectiveness of using this stationary phase for the GC analysis of triacylglycerols has been demonstrated by several groups (29, 30), little work has been performed using GC–MS. Kuksis *et al.* have provided an elegant example of the use of a polar stationary phase of this type in the GC and GC–MS analysis of a volatile distillate of butter-oil that resembles the lower half of the molecular mass distribution of bovine milk fat (31). One problem that exists in the analysis of higher-molecular-weight triacylglycerols derives from the high level of chemical background resulting from column bleed into the mass spectrometer during high temperature GC–MS analyses employing electron ionization (EI), which can prevent satisfactory interpretation of spectra (32). A partial solution to the problem is offered by carrying out GC–MS analyses in the SIM mode, in which case the retention times of peaks in appropriate mass chromatograms are used to determine the nature of the various fatty acyl moieties in the individual molecular species separated by the polarizable column. Ions employed in this include RCO^+ and [M-OCOR]$^+$. It is well-recognized (27, 33), that at low source temperature (*ca.* 200°C) triacylglycerols containing unsaturated fatty acyl moieties yield more intense [RCO-1]$^+$ and [RCO-2]$^+$ ions rather than RCO^+. However, the use of these ions for the identification of unsaturated fatty acyl residues in high temperature GC–MS analyses employing EI has been found to be problematical (32), presumably due to thermal decomposition in the ion source. However, a high ion source temperature is essential in high temperature GC–MS analyses, in order to ensure chromatographic resolution is maintained.

An alternative strategy is based on the use of negative ion chemical ionization (NICI) (34). The spectra (*Figure 5*) obtained from triacylglycerols when using ammonia as the reagent gas at an ion source block temperature of 300°C, exhibit only abundant [RCO$_2$]$^-$, [RCO$_2$-H$_2$O]$^-$ and [RCO$_2$-H$_2$O-H]$^-$ ions, presumed to derive through an initial gas phase ammonolysis reaction (*cf.* to the ammonia NICI of steryl fatty acyl esters (35, 36)). Significantly, as the spectra are obtained at an ion source block temperature of 300°C, the technique is compatible with high temperature GC–MS. Although NICI spectra lack information concerning the carbon number of the intact triacylglycerol, they can be used very conveniently to determine the nature of the fatty acyl groups, where a complex mixture of molecular species is present. The assignment of carbon number is made by comparing the retention times of peaks in the GC–MS analyses to those of authentic compounds of known carbon number. The combination of negative ion chemical ionization with GC–MS, using capillary columns coated with high temperature stable polar stationary phase, offers a very useful technique for intact triacylglycerol analysis.

A total ion chromatogram derived from the high temperature GC–MS of butter employing NICI is shown in *Figure 6*. The chromatographic resolution is somewhat reduced compared to that obtained by GC alone due to the slow

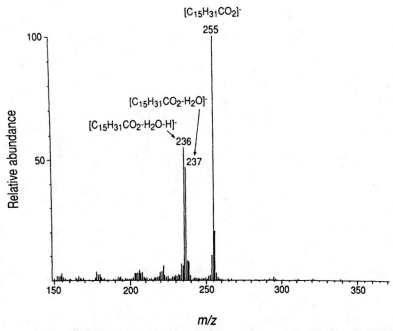

Figure 5. Ammonia NICI mass spectrum of tripalmitin recorded using high temperature GC–MS. (Reproduced with kind permission from ref. 34 1990 © J. Wiley & Sons, Ltd.)

scan cycle time (3.5 sec) of the mass spectrometer used in this study. The region of the TIC from scan 800–1050 is shown in *Figure 7*, together with the m/z 255, 281, and 283 mass chromatograms, to show the distribution of the palmitate, oleate, and stearate moieties, respectively, in the later-eluting triacylglycerol components. The clusters of peaks appearing in the mass chromatograms at each carbon number is due to resolution of individual molecular species according to the degree of unsaturation in their component fatty acyl moieties.

The mass spectra for the peaks in the TIC maximizing at scans 975 and 979 are shown in *Figure 8*. Peaks in this region correspond to triacylglycerols bearing 52 acyl carbon atoms, hence, due to the established fatty acid composition of butter, these components will comprise combinations of one C_{16} fatty acyl moiety and two C_{18} fatty acyl moieties, with the C_{18} fatty acyl moieties being either fully saturated or mono-unsaturated. Bearing this in mind, interpretation of the nature of the peaks in the TIC and mass chromatograms based on the ammonia NICI spectra is straightforward. The major RCO_2^- ions in the spectrum shown in *Figure 8a* at m/z 255, 281, and 283 derive from $C_{16:0}$, $C_{18:1}$, and $C_{18:0}$ fatty acyl moieties respectively. The presence of these fatty acyl components is corroborated by the existence of the

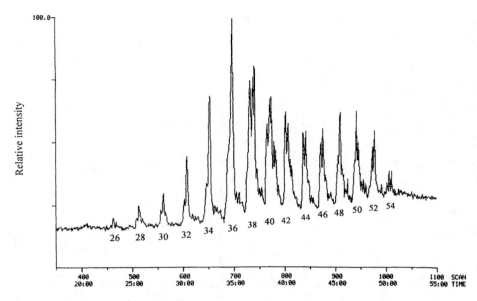

Figure 6. Total ion current (TIC) chromatogram obtained for the high temperature GC–MS analysis of butter triacylglycerols. The numbers under the peaks refer to the total number of acyl carbon atoms in each cluster. The analysis employed on-column injection into a 25 m × 0.25 mm i.d. aluminium-clad capillary column coated with methyl 65% phenyl polysiloxane (Quadrex Corporation; 0.1 μm film thickness). The GC oven temperature was held for 2 min at 50 °C before programming to 350 °C at 10 °C/min. The GC–MS transfer-line and ion source block were maintained at 360 and 300 °C, respectively. Negative ion ammonia chemical ionization was used to characterize the fatty acyl moieties associated with the individual triacylglycerols resolved by the polarizable stationary phase. The GC–MS instrument comprised a Pye Unicam 204 GC interfaced to a VG 7070H double-focusing magnetic sector mass spectrometer. The GC–MS had been modified to allow operation to 350–400 °C (35). Scan control, data acquisition, and processing were performed by a Finni an INCOS 2300 data system. N.B. The distal 5 cm of the aluminium cladding of the column must be removed (by dissolution with aqueous alkali; protocol given in Section 2.2) when such columns are used with magnetic sector mass spectrometers to prevent electrical discharging.

appropriate $[RCO_2–H_2O]^-$ and $[RCO_2–H_2O–H]^-$ ions, i.e. m/z 237 and 236 for $C_{16:0}$, m/z 263 and 262 for $C_{18:1}$, and m/z 265 and 264 for $C_{18:0}$. The mass spectrum of the latter eluting peak of the C_{52} cluster, maximizing at scan 979, is shown in *Figure 8b*. As expected, there is close similarity between this spectrum and that of the earlier eluting peak in this cluster, confirming that this too derives from a triacylglycerol bearing 52 acyl carbon atoms. A notable difference between the spectra of the two peaks is the high abundance of the RCO_2^- ion of m/z 281, and negligible abundance of m/z 283, in the spectrum of the latter eluting peak. Hence, this peak corresponds to a triacylglycerol bearing one $C_{16:0}$, and two $C_{18:1}$ fatty acyl moieties. These

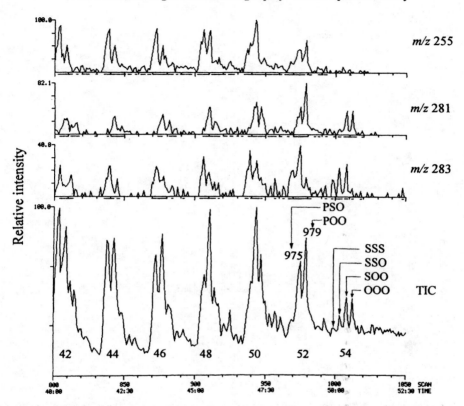

Figure 7. Expansion of the 800–1050 scan number region of the TIC shown in *Figure 6*. The mass chromatograms correspond to the *m/z* values for the RCO_2^- ions produced in ammonia NICI for $C_{16:0}$ (palmitate = P; *m/z* 255), $C_{18:1}$ (oleate = O; *m/z* 281), and $C_{18:0}$ (stearate = S; *m/z* 283). All other experimental details identical to those given in the caption to *Figure 6*.

observations are entirely consistent with the predicted elution order of these two components on the polarizable stationary phase employed, i.e. the triacylglycerol bearing one degree of unsaturation elutes before that bearing two degrees of unsaturation (see *Figure 7*). Using similar arguments, considerable progress can be made in identifying the other fully and partially resolved peaks in the TIC. The identities of the C_{54} triacylglycerols based on their NICI spectra are shown in *Figure 7* by way of a further example.

The analysis of triacylglycerols by GC–MS is a particularly demanding application of this technique, and the use of high temperature GC and GC–MS has been suggested for a number of other compound classes, e.g. wax esters, steryl esters etc. It is important to be aware that the high temperatures required for elution of some intact high molecular weight lipids (generally

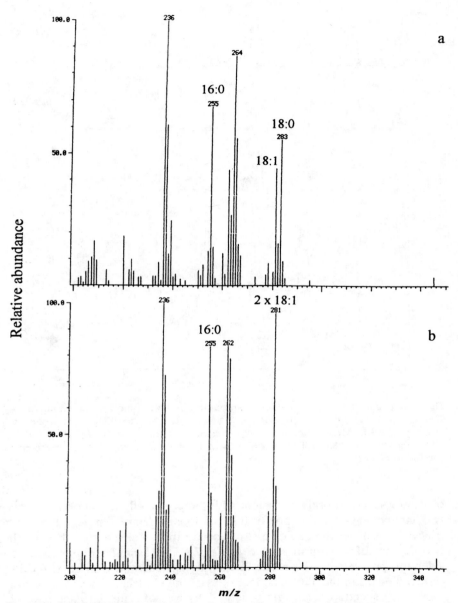

Figure 8. Ammonia NICI mass spectra for the triacylglycerol peaks maximizing at scan numbers 975 (a) and 979 (b) in *Figures 6* and *7*. See text for further descriptions of the interpretation of these spectra.

>300°C) can cause considerable losses through thermal decomposition or irreversible adsorption. Most thorough assessments of recoveries from high-temperature GC analyses have been carried out on triacylglycerols. Loss of material is dependent on the molecular weight and degree of unsaturation of a given compound (37). Hence, while good recoveries are obtained for analyses of many fats and oils, high-temperature GC and GC–MS currently appears unsuitable for the analysis of native fish oils, and the intact high-molecular-weight lipids of other marine organisms, owing to the high content of polyunsaturated fatty acids in their acyl lipids. The loss of saturated species that occurs is attributed to irreversible saturation of the stationary phase. There is negligible loss of lower molecular weight lipids, such as wax esters, diacylglycerols, and low carbon number triacylglycerols.

Losses are best assessed through the analysis of authentic compounds. Analysis of equimolar mixtures of pure analogues of the compounds to be analysed allows calibration, or correction, factors to be calculated to account for losses during the GC or GC–MS analyses. Where a given compound is not available, then only approximate correction factors can be derived from a closely related compound. A useful test for the loss of polyunsaturated components during GC or GC–MS analyses of unknowns is to perform a microscale catalytic hydrogenation, then to repeat the analysis. Major losses of polyunsaturated components are revealed by the appearance of new peaks in the chromatogram for the repeat analysis. Alternatively, chemical or enzymatic degradation, and subsequent analysis of the simpler lipid moieties released, can be used to check for components lost in high temperature GC or GC–MS analyses. One approach that has recently been employed in the high temperature GC–MS analysis of steryl fatty acyl esters bearing polyunsaturated fatty acyl moieties employed a deuterium reduction in the presence of a homogeneous catalyst (Wilkinson's catalyst) to introduce deuterium atoms specifically across double bonds (38). Selective labelling performed in this way preserves structural information concerning the carbon number and degree of unsaturation of the fatty acyl moiety. Although originally developed for use in the analysis of diacylglycerols, this technique has yet to be applied to the analysis of triacylglycerols (39).

3.1.3 Use of short columns and GC–MS and GC–MS–MS for the analysis of pesticides.

The following advantages can accrue from the use of short GC columns for some GC–MS applications.

(a) very rapid analyses are possible

(b) minimum peak broadening produces higher sample concentrations at the detector, and hence, enhanced sensitivity

(c) it is possible to analyse samples that are too volatile to be introduced by direct insertion probe

(d) it is also possible to analyse compounds that are either too involatile or thermally unstable to be amenable to GC with conventional column lengths (40–43)

Yost and co-workers (41–43) used short GC columns in combination with MS and tandem MS for the analysis of a thermally labile pesticide, aldicarb (2-methyl-2-(methylthio)propanal *O*-[(methylamino)carbonyl]oxime), and its toxic residues. One of the major drawbacks in using GC methods for the analysis of aldicarb derives from its thermal degradation to aldicarb nitrile (2-methyl-2-(methylthio)propannitrile), which can also arise via chemical degradation in the environment. Thus, unless steps are taken to remove aldicarb nitrile prior to GC analysis, the environmental degradation product can give a positive interference for aldicarb (41). Of the columns that were investigated a short (2.6 m) J & W DB5 (SE-54 equivalent) coated capillary was found to be suitable for the GC–MS analysis of aldicarb. The use of the short column eliminated problems of thermal decomposition due to the low GC oven temperature required for elution in very short analysis times. The GC–MS analysis of aldicarb is shown in *Figure 9*; the total analysis time is *ca.* 3.5 min. Further experimental details are given in the caption to *Figure 9*. The mode of injection and condition of the injection port can have a considerable influence on the recoveries achieved for any compound analysed by GC. This is particularly important in the case of more difficult compounds, such as aldicarb,

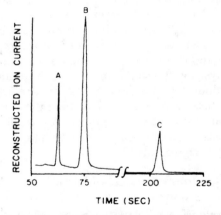

Figure 9. Reconstructed ion chromatogram for a methylene chloride extract of water spiked with aldicarb, peak C. The peak is aldicarb oxime, which is formed by hydrolysis of aldicarb. Peak B is due to ethyl benzoate which was used as an internal standard. Analysis employed a 2.6 m fused silica capillary column coated with DB5 (J & W SE-54 equivalent) bonded stationary phase. Splitless injection was used and the GC oven temperature-programmed from 40 to 100°C at 20°C/min after a hold of 1 min. Helium was the carrier gas at a column head pressure of 14 KPa. The GC/MS transfer line was held at 135°C. The instrument used was a Finnigan triple stage quadrupole mass spectrometer. (Reproduced with kind permission from ref. 4. 1984 © American Chemical Society.)

where the condition of the injection port can affect sensitivity and linearity of response. A clean silanized injection port insert was found to be necessary to obtain a linear response for 1.5 to 150 ng of aldicarb injected (41). The narrow GC peaks that are obtained when using short columns necessitates the use of a fast MS scan rate, to ensure adequate sampling of peaks, in order to achieve adequate precision in quantitative GC–MS analyses. In the analysis of adicarb oxime (2-methyl-2-(methylthio)propanal oxime), shown in *Figure 9*, a scan rate of 0.27 sec/scan was used to sample the 1.5 sec wide peak. Quadrupole mass spectrometers are especially well-suited to use in analyses requiring fast-scan rates (see Section 2.4.2).

Two other degradation products of aldicarb, aldicarb sulphoxide, and aldicarb sulphone, were found to be so thermally labile that they could not be determined under conditions used in the analysis of aldicarb. Successful were possible by use of on-column injection and a 1 m DB5-coated capillary (43). The reconstructed ion current chromatogram for the GC–MS analysis of aldicarb, aldicarb sulphone, and aldicarb sulphoxide is shown in *Figure 10*. Examination of the mass spectra of the two peaks showed the later eluting peak to be the result of the co-eluting sulphone and sulphoxide. GC–MS–MS operating in the product ion mode can be used to enhance the selectivity of the analysis shown in *Figure 10*, and, hence, compensate for the reduced chromatographic resolution achieved by the short GC column. The mode of formation of product ion mass spectra by collision-induced dissociation (CID) in a triple quadrupole mass spectrometer is shown schematically in *Figure 11*. *Figure 12* shows the product ion mass spectra for the co-eluting aldicarb sulphoxide and aldicarb sulphone, together with those of pure standards. Although GC–MS–MS does not improve the sensitivity of the analysis, it has the advantage of producing essentially interference-free spectra for the individual compounds; use of this facility can greatly ease identifications, particularly when using library searching algorithms.

3.2 Trace analysis

3.2.1 Selected ion monitoring

The advantage of using mass spectrometry in trace analyses derives from the ability to monitor pre-selected ion(s) in the spectrum of a compound. This technique, termed selected ion monitoring (SIM), leads to increases in both sensitivity, as the instrument does not then spend time scanning redundant regions of the mass spectrum, and selectivity, through the monitoring of an ion(s) (molecular or fragment) that is characteristic of the analyte of interest. Greatest sensitivity is achieved in SIM analyses when a single intense ion is monitored. In such instances picogram (10^{-12} g) detection limits are achieved routinely, with sub-picogram limits possible under favourable circumstances.

As already alluded to above, SIM is frequently used in conjunction with GC–MS to add extra selectivity to analyses, through the provision of a high

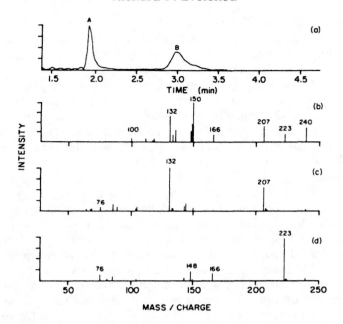

Figure 10. The reconstructed ion current (a) for the GC–MS (using positive ion methane CI) determination of aldicarb (peak A), aldicarb sulphoxide, and aldicarb sulphone (both co-eluting in peak B) separation on a 1 m × 0.25 mm i.d. DB5 (SE-54 equivalent) coated capillary column.

Product ion spectra

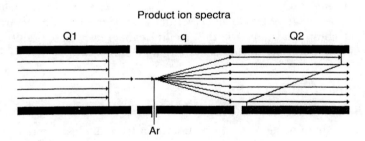

Figure 11. Mode of formation of product ions by collision-induced dissociation (CID) of a selected precursor ion in a triple quadrupole mass spectrometer.

resolution chromatographic separation step. In a trace analysis using GC–MS with SIM, confirmation of the presence of the chosen analyte would rest on the detection of the characteristic ion, as a peak in an ion chromatogram, at the expected GC retention time. The expected retention time is determined by the prior analysis of an authentic sample of the analyte of interest. Still further enhancements in analytical selectivity can be achieved by monitoring

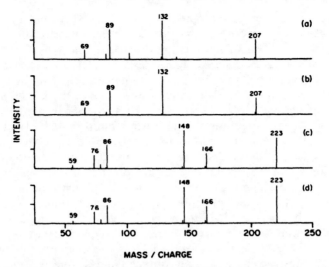

Figure 12. MS–MS product ion mass spectra for aldicarb sulphoxide [M + H]⁺, *m/z* 207, authentic (a) and the mixture (b) present in peak B in *Figure 10b*, and product ion spectra for aldicarb sulphone [M + H]⁺, *m/z* 223, the authentic compound (c) and the mixture (d) present in peak B in *Figure 10b*. (Reproduced with kind permission from ref. 43. 1986 © American Chemical Society.)

more than one ion from the mass spectrum of an analyte. The possibility then exists for confirming the presence of the analyte, by comparing the ratio of the responses of these ions determined by SIM to their abundance ratios in the full scan mass spectrum. *Figure 13* shows a number of possible hypothetical results that can be obtained from the search for a particular target analyte using GC–MS SIM for three ions.

Low resolution selected ion monitoring (SIM) is the most commonly used

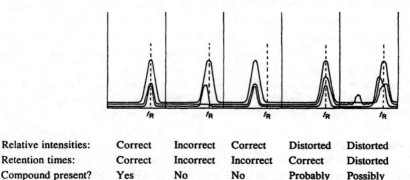

Relative intensities:	Correct	Incorrect	Correct	Distorted	Distorted
Retention times:	Correct	Incorrect	Incorrect	Correct	Distorted
Compound present?	Yes	No	No	Probably	Possibly

Figure 13. Hypothetical results from the GC/MS selected ion monitoring analysis for a targeted compound. (Reproduced from ref. 8 with kind permission. 1989 © J. Wiley & Sons, Ltd.)

technique for detecting the target analyte. Before SIM can be performed, the mass spectrum of the compound(s) of interest must be carefully inspected; the ions chosen should be of reasonable abundance and must not coincide with potential interferences, such as column bleed ions. High mass ions are preferred, as this will generally limit the chances of interferences from background ions and co-eluting impurities. The use of even mass ions is recommended, as these occur relatively infrequently in the mass spectra of organic compounds. In quantitative investigations, using isotope dilution mass spectrometry (see Section 3.2.4), it is vital to choose ions which contain the stable isotope labels. The generation of high mass ions can be greatly influenced by the choice of derivatizing agent where compounds are functionalized. When EI does not produce abundant high mass ions, CI can be used, with the added advantage that the number of potentially interfering ions will be markedly reduced. The higher scan rates used in SIM studies, compared to those employed in analyses where full scan data is acquired, ensure the frequent sampling of individual capillary GC–MS peaks necessary to achieve high quantitative precision (see Section 3.2.4 for further discussions of techniques for performing quantitative GC–MS). Quadrupole instruments are particularly effective for use in SIM work, as rapid switching between the selected masses is easily performed. *Figure 14* shows the results of a GC–MS SIM analysis for ecdysteroids (polyhydroxylated steroids which function as arthropod moulting hormones) extracted from a helminth species. The ions of m/z 567 and 561 are monitored to distinguish, respectively, between those ecdysteroids structurally related to ecdysone (lacking C-20 hydroxyl group) and 20-hydroxyecdysone (44).

3.2.2 High resolution selected ion monitoring

While *Figure 14* shows little or no evidence of chemical interferences in the mass chromatograms, co-eluting peaks produced by sample contaminants are frequently encountered in analyses of extracts of biological or environmental materials. These contaminant peaks frequently mask that of the analyte and so prevent its reliable determination. An example of this is seen in *Figure 15a* which shows the analysis for a prostaglandin in human brain tissue at mass resolution of 1000 (45). At this resolution the target analyte $PGF_{2\alpha}$ was found to elute as a shoulder on a contaminant peak, so preventing reliable confirmation of its presence, and adequate quantification. One approach to overcoming this problem relies on the use of the high mass resolution capabilities of double focusing magnetic sector mass spectrometers to monitor ions of selected elemental compositions. In the example shown in *Figure 15b*, adjusting the mass spectrometric resolution to 5000 introduces sufficient selectivity into the analysis to eliminate enough interference from the contaminant to confirm the presence of prostaglandin $F_{2\alpha}(PGF_{2\alpha})$ in the brain extract. Moreover, the resolution was sufficient to allow quantification by comparison of the peak area with that of a pentadeuteriated internal standard. The peak detected in the brain extract corresponded to *ca.* 40 pg $PGF_{2\alpha}$.

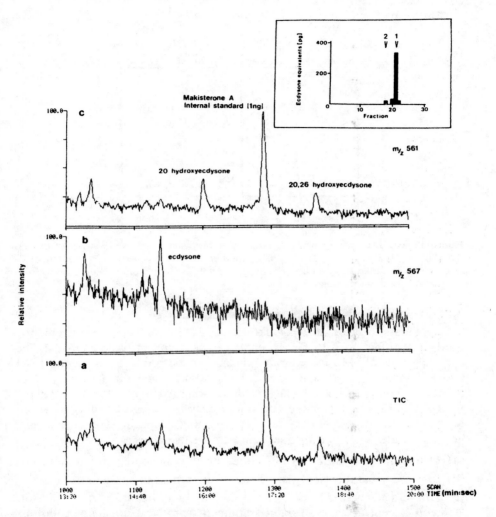

Figure 14. Partial reconstructed ion currents for (b) *m/z* 567 and (c) *m/z* 561 for the capillary GC/MS low resolution SIM analysis of the trimethylsilylated ecdysteroids released by hydrolysis of the polar conjugate fraction from the cestode, *Hymenolepis diminuta*. Analyses used a 25 m × 0.22 mm i.d. BP-1 coated (SGE; immobilized dimethyl polysiloxane; 0.1 μm film thickness) flexible fused silica capillary column. Following on-column injection at 50°C, the oven temperature was raised ballistically to 200°C, then programmed at 8°C/min to 320°C and held isothermally. Ionization was by EI at 70 eV with SIM being performed by accelerating voltage switching. The numbers 1 and 2 in the inset denote the positions of elution of 20-hydroxyecdysone and ecdysone, respectively, on reversed-phase HPLC. (Reproduced from Ref. 44 with kind permission 1987 © Elsevier Science Publishers BV.)

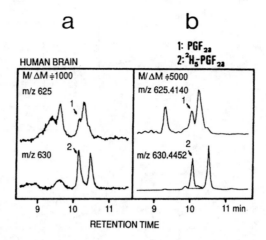

Figure 15. GC–MS SIM analyses of dimethylisopropylsilyl ether derivatives of prostaglandin $F_{2\alpha}$ (1) $PGF_{2\alpha}$ and (2) $[^2H_5]PGF_{2\alpha}$ methyl esters in an extract of human brain (arachnoid) obtained by monitoring at (a) low and (b) high resolution. (Reproduced from ref. 45 with kind permission.)

3.2.3 Use of tandem mass spectrometers for trace analysis

Tandem mass spectrometry (also referred to as MS–MS) may be conveniently used in conjunction with GC to enhance selectivity in trace analyses. The usefulness of MS–MS in generating characteristic product ion spectra from two compounds co-eluting in a single GC peak was shown in *Section 3.1.3*. A related technique, selected reaction monitoring (SRM), is the most widely used GC–MS–MS technique for trace analysis. *Figure 16* illustrates the operation of a GC–MS–MS system incorporating a triple quadrupole analyser in SRM mode. The following sequence of events occur.

(a) Compounds separated by GC elute into the ion source and are ionized.

(b) Ions are separated in Q_1 according to their mass-to-charge ratio (m/z).

(c) An ion (precursor) from Q_1 is selected and fragmented in q by collision with a neutral gas to produce fragment (product) ions, i.e. collision-induced dissociation (CID).

(d) One or more product ions are then monitored to achieve optimum sensitivity and selectivity.

The technique is analogous to conventional selected ion monitoring except that confidence in the experimental results are greatly enhanced by the selection of precursor and product ions. As discussed below, the selectivity of analyses can be improved still further by using hybrid instruments to take advantage of the high precursor ion resolution.

Gaskell *et al.* have taken advantage of the favourable properties of the TBDMS derivatives in developing highly sensitive and selective GC–MS tech-

Selected reaction monitoring

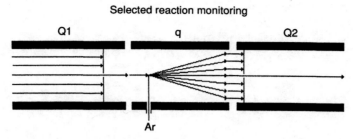

Figure 16. Mode of operation of SRM following CID of a selected precursor ion in a triple quadrupole mass spectrometer.

niques for the detection and quantification of steroids in biological matrices. An overview of the various sample preparation and MS strategies developed by this group has been reported (46). One earlier study performed by this group compared the use of high-resolution SIM and metastable peak monitoring techniques for the detection of plasma testosterone as its TBDMS ether, methyl oxime–TBDMS ether, or TBDMS oxime/TBDMS ether derivative (47). As discussed in Section 3.2.2, high-resolution SIM involves the detection of ions of a selected exact mass and, hence, of specified elemental composition, rather than all ions of a particular nominal mass (48). The EI mass spectra of the TBDMS ethers of steroids contain abundant high mass ions, e.g. $[M-57]^+$, which can serve to enhance detection limits in trace analyses. High selectivities were also achieved by the use of metastable peak monitoring. Metastable peak monitoring involves the detection of fragmentations occurring in the first field-free region of a double-focusing magnetic sector mass spectrometer. The technique is comparable to SRM performed on an MS–MS instrument, except that inferior precursor ion resolution is achieved. High selectivity is achieved in metastable ion monitoring and SRM derives from the preselection of both the precursor and product ions of specific m/z (49).

Gaskell *et al.* have combined the advantages of both high resolution SIM and metastable ion detection techniques by using a hybrid tandem double-focusing/quadrupole instrument (50). TBDMS derivatives of steroids yield favourable metastable decompositions, e.g. $[M^+] \to [M-C_4H_9]^+$. *Figure 17* shows the results of GC/MS/MS analyses of the bis-TBDMS ether of oestradiol in blood plasma, using the techniques discussed above to improve analytical specificity. At a precursor ion resolution of 5000 and SRM of the $[M-C_4H_9]^+$ product ion the detection limit for the bis-TBDMS derivative of oestradiol was 10 pg, with the chosen analyte being the only component detected.

3.2.4 Quantitative analysis

The effectiveness of MS in quantitative analysis derives from the ability to make simultaneous measurements, with equivalent specificity, of the concentrations

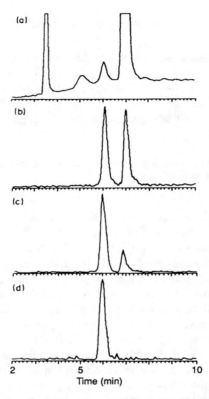

Figure 17. GC–MS analysis of oestradiol-17β in a plasma extract, as the bis-TBDMS ether: (a) low resolution (m/Δm 1000) SIM of *m/z* 500; (b) SRM of *m/z* 500→443 (precursor ion resolution 1000); (c) ɛs (b) precursor ion resolution 2500; (d) as (b) precursor ion resolution 5000. (Reproduced from ref. 50 with kind permission. 1985 © J. Wiley & Sons, Chichester.)

of the chosen analyte and added internal standard, or calibrant. Optimum precision is attained in quantitative MS when the internal standard is a stable isotope labelled (most commonly ^{2}H, ^{13}C, or ^{18}O) analogue of the analyte. This method of quantitative MS is known as isotope dilution MS. The high precision that can be attained in quantitative MS analyses has led to its adoption as a reference technique for use in assessing the performance of other less specific analytical techniques, e.g. radioimmunoassay. An overview of the methodological aspects of quantitative MS can be found elsewhere (51).

Figure 18 shows the analytical protocol recommended for use in quantitative analysis by MS. The recovery of the internal standard during the isolation procedure should be identical to that of the analyte, and it should not be differentiated from that of the analyte until the final MS detection stage. Stable isotopically labelled analogues generally fulfil this latter requirement.

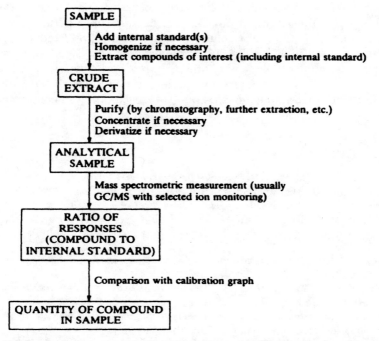

Figure 18. The analytical protocol used in quantitative mass spectrometry. (Reproduced from Ref. 52 with kind permission. 1982 © Cambridge University Press.)

However, the positioning of the isotopic label must be carefully considered in order to reduce the risk of isotopic exchange. Other factors that must be considered include the proportion of unlabelled material in the synthetic isotopically labelled internal standard, and the possibility for chromatographic separation between the analyte and the isotopically labelled internal standard. Internal standards incorporating three to four isotopic labels are considered ideal, although two labels may be used in many instances.

The amount of internal standard added should be of a comparable concentration to the analyte to be analysed. The homogenization (equilibration) step is important to ensure that the physico-chemical state of the analyte and internal standard are identical. An acceptable equilibration method would be the addition of the calibration material in a small volume (<0.02 vol. fraction of the sample) of ethanol to a plasma sample followed by several hours equilibration at 4–20 °C (51). Provided the calibration material has been well-chosen, then standard isolation and chemical derivatization procedures are employed, as in any analytical protocol; the prime concern is the maximization of the recovery of the internal standard and analyte.

GC–MS, employing isotope dilution, has been widely used in quantitative analyses due to the advantages of including a high resolution chromatographic step on-line with the mass spectrometer; this combination reduces the possibility for interference from ions of the same m/z as the analyte and internal standard. Capillary columns are generally preferred over packed columns, as their higher efficiencies produce enhanced signal-to-noise ratios which greatly improve detection limits. *Figure 15* provides an example of the use of high resolution SIM and a pentadeuteriated internal standard for the detection and quantitative analysis of a prostaglandin in human brain.

References

1. Evershed, R. P. (1987). In *Specialist periodical reports: mass spectrometry*, Vol. 9, (ed. M. E. Rose). Royal Society of Chemistry, London, pp. 196–263.
2. Evershed, R. P. (1989). In *Specialist periodical reports: mass spectrometry*, Vol. 10, (ed. M. E. Rose). Royal Society of Chemistry, London, pp. 181–221.
3. Chapman, J. R. (1985). *Practical organic mass spectrometry*. John Wiley & Sons, Chichester.
4. Knapp, D. R. (1979). *Handbook of analytical derivatisation reactions*. John Wiley & Sons, New York.
5. Blau, K. and King, G. S. (ed.) (1977). *Handbook of derivatives for chromatography*, 1st ed. Heyden & Sons Ltd, London.
6. Blau, K. and Halket, J. M. (ed.) (1993). *Handbook of derivatives for chromatography*, 2nd edn. John Wiley & Sons Ltd, Chichester.
7. Hurst, R. E., Settine, R. L., Fish, F., and Roberts, E. C. (1981). *Anal. Chem.*, **53**, 2175.
8. Evershed, R. P. and Prescott, M. C. (1989). *Biomed. Environ. Mass Spectrom.*, **18**, 503.
9. Henneberg, D., Henrichs, U., and Schomburg, G. (1975). *J. Chromatogr.*, **112**, 343.
10. Arrendale, R. F., Stevenson, R. F., and Chortyk, O. T. (1984). *Anal. Chem.*, **56**, 2997.
11. Pankow, J. F. and Isabelle, L. M. (1987). *J. High Resolut. Chromatogr. Chromatogr. Commun.*, **10**, 617.
12. Richter, W. J. and Schwarz, H. (1978). *Angew. Chem. Int. Ed. Engl.*, **17**, 424.
13. Mather, R. E. and Todd, J. F. J. (1979). *Int. J. Mass. Spectrom. Ion Phys.*, **30**, 1.
14. Harrison, A. G. (1983). *Chemical ionisation mass spectrometry*. CRC Press, Boca Raton.
15. Jennings, K. R. and Dolnikowski, G. G. (1990). *Methods Enzymol.*, **193**, 37–61.
16. Hyver, K. J. and Phillips, R. J. (1987). *J. Chromatogr.*, **399**, 33.
17. Hyver, K. J. (1988). *J. High Resolut. Chromatogr. Chromatogr. Commun.*, **11**, 69.
18. Stafford, G. C., Kelley, P. E., Syka, J. E. P., Reynolds, W. E., and Todd, J. F. J. (1984). *Int . J. Mass Spectrom. Ion Process.*, **60**, 85.
19. McLuckey, S. A., Glish, G. L., Asano, K. G., and Grant, B. C. (1988). *Anal. Chem.*, **60**, 2220.
20. Gross, M. J. (1990). *Methods Enzymol.*, **193**, 131.
21. Yost, R. A. and Boyd, R. K. (1990). *Methods Enzymol.*, **193**, 154.

22. Marriott, P. J., Gill, J. P., Evershed, R. P., Eglinton, G., and Maxwell, J. R. (1982). *Chromatographia*, **16**, 304.
23. Boylan, D. B. and Calvin, M. (1967). *J. Am. Chem. Soc.*, **89**, 5472.
24. Marriott, P. J., Gill, J. P., Evershed, R. P., Hein, C. S., and Eglinton, G. (1984). *J. Chromatogr.*, **301**, 107.
25. Gill, J. P., Evershed, R. P., and Eglinton, G. (1986) . *J. Chromatogr.*, **369**, 281.
26. Gill, J. P., Evershed, R. P., Chicarelli, M. I., Wolff, G. A., Maxwell, J. R. and Eglinton, G. (1985). *J. Chromatogr.*, **350**, 37.
27. Hites, R. A. (1975). *Methods Enzymol.*, **35**, 348.
28. Murata, T. (1977). *Anal. Chem.*, **49**, 2209.
29. Lipsky, S. R. and Duffy, M. L. (1986). *J. High Resolut. Chromatogr. Chromatogr. Commun.*, **9**, 725.
30. Kuksis, A., Myher, J. J., and Sandra, P. (1990). *J. Chromatogr.*, **500**, 427.
31. Myher, J. J., Kuksis, A., Marai, L., and Sandra, P. (1988). *J. Chromatogr.*, **452**, 93.
32. Ohishima, T., Yon, H.-S., and Koizumi, C. (1989). *Lipids*, **24**, 535.
33. Bhati, A. (1976). In *Analysis of fats and oils*, (ed. R. J. Hamilton). Elsevier Applied Science, London, pp. 207–41.
34. Evershed, R. P., Prescott, M. C., and Goad, L. J. (1990). *Rapid Commun. Mass. Spectrom.*, **4**, 345.
35. Evershed, R. P. and Goad, L. J. (1987). *Biomed. Environ. Mass Spectrom.*, **14**, 131.
36. Evershed, R. P., Prescott, M. C., Spooner, N., and Goad, L. J. (1989). *Steroids*, **53**, 285.
37. Mares, P. (1988). *Prog. Lipid Res.*, **27**, 107.
38. Evershed, R. P., Prescott, M. C., and Goad, L. J. (1992). *J. Chromatogr.*, **590**, 305.
39. Dickens, B. F., Ramesha, C. S., and Thompson, G. A. (1982). *Anal. Biochem.*, **127**, 37.
40. Riva, M. and Carisano, A. (1969). *J. Chromatogr.*, **42**, 464.
41. Trehy, M. L., Yost, R. A., and McCreary, J. J. (1984). *Anal. Chem.*, **56**, 1281.
42. Yost, R. A., Fetterolf, D. D., Hass, J. R., Harvan, D. J., Weston, A. F., Skotnicki, P. A., and Simon, N. A. (1984). *Anal. Chem.*, **56**, 223.
43. Trehy, M. L., Yost, R. A., and Dorsey, J. G. (1986). *Anal. Chem.*, **58**, 14.
44. Evershed, R. P., Mercer, J. G., and Rees, H. H. (1987). *J. Chromatogr.*, **390**, 357.
45. Ishibashi, M., Yamashita, K., Watanabe, K., and Miyazaki, H. (1986). In *Mass spectrometry in biomedical research*, (ed. S. J. Gaskell). John Wiley & Sons, Chichester, pp. 423.
46. Gaskell, S. J., Gould, V. J., and Leith, H. M. (1986). In *Mass spectrometry in biomedical research*, (ed. S. J. Gaskell). John Wiley & Sons, Chichester, pp. 347.
47. Finlay, E. M. H. and Gaskell, S. J. (1981). *Clin. Chem.*, **27**, 1165.
48. Millington, D. S. (1975). *J. Steroid Biochem.*, **6**, 239.
49. Gaskell, S. J. and Millington, D. S. (1978). *Biomed. Mass Spectrom.*, **5**, 557.
50. Gaskell, S. J., Porter, C. J., and Green, B. W. (1985). *Biomed. Mass Spectrom.*, **12**, 139.
51. Lawson, A. M., Gaskell, S. J., and Hjelm, M. (1985). *J. Clin. Chem. Clin. Biochem.*, **23**, 433.
52. Rose, M. E. and Johnstone, R. A. W. (1982). *Mass spectrometry for chemists and biochemists*. Cambridge University Press, p. 101.

A1

Combined gas chromatography–
Fourier transform infrared
spectroscopy

PETER JACKSON

1. Introduction

As exemplified in earlier chapters, there is increasing pressure on analytical scientists from lobby groups, government and legislative bodies, the courts and legal profession, forensic science, industry, and research to identify smaller amounts of material with increased certainty. As detailed earlier, combined separation science–mass spectrometry (LC–MS, GC–MS) is now considered as the major tool to meet many of these challenges. However, even with the most powerful mass spectrometers available there is often a requirement for complementary evidence to confirm identification, especially where closely related structures are possible or when specific isomeric forms are to be determined. Fourier transform mid-infrared (FTIR) spectroscopy is particularly useful in the identification of functional groups and in the finger-printing of specific isomers, and is routinely applied to identify solids, liquids, and vapour-phase samples, including, increasingly, applications in environmental monitoring and industrial process control. In conjunction with nuclear magnetic resonance (NMR), FTIR is an ideal complementary technique to MS analysis.

Recently, there have been significant advances in combined separation science and FTIR spectroscopy, such as supercritical fluid chromatography (SFC–FTIR) and liquid chromatography (LC–FTIR); an overview of the significant developments has recently been published (1). This appendix will not consider these techniques in more detail, though it is expected that as research efforts become more widely developed, commercial systems will become available in the future. At present, gas chromatography (GC) technology is most widely interfaced to FTIR instrumentation, with various commercial systems currently available. This mirrors the early situation in hyphenated separation science–MS systems, where the small samples (pg to ng), low carrier gas amounts (1–2 ml/min), and little interaction from the carrier gas (H_2, He) make GC the ideal sample inlet medium. Unfortunately, FTIR is a considerably less sensitive technique, not normally considered for trace analysis. However, systems have been developed to maximize the FTIR sensitivity and interface such a detector to a GC system. It should be noted at this point that the introduction of interferometer-based Fourier transform IR instrumentation was essential for successful interfacing to GC, allowing rapid, high sensitivity spectra to be obtained.

The most straightforward GC–FTIR interface is the 'lightpipe': a narrow tube coated internally with, typically, gold through which the GC eluent passes after chromatographic separation. The IR beam is propagated down the lightpipe, which is operated at high temperature at ambient pressure, and the components of interest are resident in the sampling tube, in the vapour phase, for a few seconds. This gives maximum signal from the eluent, but the short residence time and small sample amount means that sensitivity is limited; it is some two to three orders of magnitude less sensitive than corresponding bench-top GC–MS. Nevertheless, useful data can be generated from components in the 10 s to 100 s of ng range using typical commercial lightpipe GC–FTIR systems.

This sensitivity limitation has been addressed in two major designs. Both make use of low temperatures to store the high-temperature GC eluent after separation. The first, developed at Argonne National Laboratory (2–5) utilizes matrix isolation (MI) to trap the GC eluent. Initially developed to study unstable or reactive species, MI is an established technique, having the attractiveness that the common matrix materials, such as argon, nitrogen, or xenon are transparent to IR irradiation. In the GC–MI–FTIR technique, the eluent is mixed with excess argon after separation, forming a glassy matrix on a moving reflective surface at cryogenic temperatures. The second technique (6, 7) deposits the GC eluent directly on to a moving low-temperature IR-transparent window. In both cases, the GC eluent is stored at low temperature as a solid 'spot', allowing more time for FTIR data acquisition and signal-to-noise ratios to be improved by signal averaging. Signals can also be improved by using focused IR optics, or even FTIR microscope detection. In both cases, spectra exhibit narrower bands than in the vapour phase, and MI spectra are, in some cases, narrowed further by the elimination of inter-molecular effects, such as band broadening due to hydrogen bonding. The combination of these factors results in enhanced sensitivity, approaching that obtained by bench-top GC–MS. Both the sample storage interface designs are now commercially available in the Mattson CRYOLECT (GC–MI–FTIR) and the Bio-Rad DigiLab TRACER (GC–FTIR).[a]

---

[a] CRYOLECT is a trademark of Mattson Instruments. TRACER is a trademark of DigiLab Corp.

2. Techniques description

2.1 Lightpipe GC–FTIR

In a typical lightpipe GC–FTIR system, the GC eluent is sampled in a flow cell following chromatographic separation on column. The lightpipe is held at a temperature greater than that of the maximum programmed temperature in the GC oven. This ensures that there are no reverse temperature gradient effects and that there is little contamination of the internal surface. Lightpipe designs are varied, though a reflective inert inner coating is normally employed. In two commercial designs, the lightpipe consists of a gold-coated glass tube; the Hewlett Packard IRD[b] employs a 120 mm long and 1 mm diameter lightpipe, whereas a Bruker system uses a 200 mm long, 0.8 mm diameter system. Maximum sensitivity is obtained when the lightpipe volume is matched to the GC flow rate and peak widths.

Vapour phase spectra are recorded 'on the fly' during the course of the GC experiment, with a typical spectrometer resolution of 8–32 cm^{-1}. Due to the limited time available (related to the mean residence time of the separated GC components in the lightpipe), only a limited number of FTIR scans may be accumulated for each point on the FTIR chromatogram. Typically, between 2 and 10 scans per seconds are recorded, depending on the peak widths and sensitivity requirements. Real time processing may be employed to generate a total IR response chromatogram (usually employing the Gram–Schmidt algorithm) or, on more advanced systems featuring fast array processing, selected wavenumber traces may be displayed. These are analogous to selected ion traces in GC–MS, and may be used to improve sensitivity and selectivity.

Since lightpipe instruments operate 'on the fly' and employ a relatively clean detector, automated operation is feasible, with overnight runs possible with autosampling before GC separation.

2.2 Low-temperature matrix isolation GC–FTIR

In the Mattson CRYOLECT commercial GC–MI–FTIR system, the GC eluent (in helium carrier) is combined at a four-way open cross with an argon in helium make-up flow to give, typically, between 100:1 and 1000:1 argon-to-sample molecule ratios. The flows into the open cross are balanced to ensure that all the material eluting from the column is effectively transferred to the matrix isolation interface. The eluent is then transferred to the matrix isolation vacuum chamber via a heated transfer line (0.32 mm i.d. capillary) and heated deposition tip. The tip is held approximately 0.3 mm from the polished rim of a gold disk during deposition, and is retracted to a distance of approximately 3 cm when not depositing to avoid contaminating the sample already

---

[b] IRD is a trademark of Hewlett Packard.

laid down. The vacuum chamber pressure is maintained between 5×10^{-5} and 2×10^{-4} torr with flow rates of approximately 1 to 2 ml/min from the transfer line from the open cross. The high vacuum is necessary to prevent build up of condensed atmospheric contaminants on the cold surfaces. The argon mixture is then sprayed on to the side of the cryogenic disk, with the temperature typically held between 7 and 20 K. This is cooled by a closed cycle helium gas expander/refrigerator.

As the GC run progresses, the entire refrigerator/cryogenic disk assembly is slowly rotated using a stepper motor drive system. The entire process is completely computer controlled, so that the stepper motor position is directly related to the GC retention time, providing the necessary time synchronization to generate GC–FTIR chromatograms.

The spectrometer IR beam is focused on to the sample, reflected off the gold disk, goes back through the sample, and is then collected by a pair of parabolic mirrors. For optimum sensitivity, the sample spot size, the diameter of the focused IR beam, and the size of the detector element should be matched. In practical systems, this has been found not to be the case; in one system, (8) the detector element was 0.5×0.5 mm, the IR beam diameter was 0.37 mm, and the matrix diameter was 0.17 mm. Other workers (9) have used a matrix diameter of 0.3 mm with similar beam and detector parameters.

The chromatographic resolution obtained with the MI interface depends on the GC peak widths, peak separation, cold surface movement rate, and spot size. With suitable variation of the disk rotation rate, it is possible adequately to retain the GC separation without seriously degrading sensitivity. On the CRYOLECT, disk speed variation cannot be performed straightforwardly from software, so manual methodologies and external software programs have been developed (10), enabling disk speeds from 0.05 to 0.5 mm/sec to be achieved.

Once the sample is deposited on to the cryogenic disk, the sample can then be analysed. It should be noted that unlike GC–MS or GC–lightpipe–FTIR, spectra are not recorded 'on the fly', resulting in a time lag between GC separation and eventual spectral acquisition. On the CRYOLECT, the disk is rotated by 180° to position the matrix-isolated sample in the IR beam, and a rapid scan, low resolution 'reconstruction' is performed as the disk is slowly stepped. This is simply a rapid Gram–Schmidt absorption map of the deposited matrix; the FTIR chromatogram. The Gram–Schmidt algorithm provides a quick, convenient representation of the total absorption across the detected frequency range. The chromatogram is then displayed and, since the FTIR spectrometer is a single beam system, both peaks of interest and background positions are selected from the software with a cursor, before being ratioed to give a spectrum of the sample alone. High resolution, high sensitivity spectra can thus be obtained from the peak of interest.

The FTIR detector (mid-range MCT) is extremely sensitive to both water and carbon dioxide contaminants, either from wet samples or from air leaks.

The other major air contaminants (nitrogen and oxygen) are transparent to the FTIR experiment. Water may appear as both ice, with broad signals due to extensive hydrogen bonding, and matrix isolated molecules (including dimers, trimers, etc.), with sharp signals shifted above 3500 cm^{-1} in the FTIR spectrum, giving less interference and overlap with bands from other acidic functionalities.

The cold disk is cleaned by simply allowing the disk to warm up to room temperature. Because of the high argon to sample ratio in the matrix, the sample is effectively vapourized, even with relatively involatile components. Some disk cleaning may be required eventually, especially if very large concentrations of high boiling components are encountered.

2.3 Low-temperature solid sample deposition GC–FTIR

This interface is used in the Bio-Rad Digilab TRACER. The capillary column is connected to a deactivated fused silica transfer line which connects to the vacuum chamber of the GC–FTIR interface via a stainless steel tube. At the end of the transfer line, a fused silica deposition tip with an i.d. of 150 μm reduces the transfer line flow to approximately 1 ml/min. The deposition tip is adjusted carefully to be approximately 30 μm above the surface of a moving zinc selenide (ZnSe) window which is cooled with liquid nitrogen to 77 K. The window is separately adjusted to travel in a plane at 90° to the deposition tip. As with the CRYOLECT, the interface is located in a high vacuum chamber which is held at 10^{-5} torr to prevent condensation of atmospheric contaminants. The usual atmospheric and sample contaminants (water and carbon dioxide) give intense broadened bands when solidified on the cold window. Both the deposition tip and the transfer line are heated to the default temperature of 250°C to ensure effective transfer of the GC eluent across the interface.

The eluting GC components are sprayed onto the cold window from the transfer line and are immobilized on the cold window as solid-phase deposits. The deposited trace is approximately 100–150 μm in diameter. The IR beam is selected by means of a flip mirror and the beam is then focused on the moving window by a gold-coated aspheric mirror. This passes through the sample and the ZnSe window before being sampled by an infrared microscope consisting of two Schwartzchild objectives and a mid-range MCT detector.

As the GC run progresses, the window is continually shifted by means of an X–Y stepper motor drive. The IR beam path is adjacent to the deposition tip and the deposited eluent components are passed through the IR beam a few seconds after deposition. The cold window scan rate can be controlled from software, and in order to compensate for the general increase of the chromatographic peak width with longer retention times, a decreasing window velocity is usually programmed. This maintains the optimum concentration of sample and spot size relative to the IR beam size. Interferograms are recorded 'on the fly', generally averaging 4 scans every 2 seconds with a resolution of 8 cm^{-1}. During the run, a Gram–Schmidt reconstruction is performed to provide an immediate FTIR chromatogram. The interferograms are also stored on computer disk, allowing spectra and selected frequency (functional group) reconstructed chromatograms to be processed after the initial GC run has finished.

In addition, post-run scanning can also be performed by repositioning the window at the correct stepper motor position (retention time) under the IR beam. This allows spectra to be obtained with both higher resolution and with increased signal-to-noise ratios from components of interest. One drawback exists with the fixed deposition tip; since the IR beam is adjacent to the tip, repeated positioning of a sample within the IR beam can result in sample degradation due to contamination from components eluting from the transfer line.

The cold window has sufficient storage capacity for in excess of 10 hours GC deposition. Cleaning is effected by warming the window to room temperature in the high vacuum. Where high-boiling components are deposited, some long-term contamination may occur. In such cases, cleaning of the window must be done manually.

3. GC–FTIR sensitivity

A recent study (11) has been performed to investigate the sensitivity of the above mentioned interface types on a common sample set. A series of caffeine concentration standards was analysed on a representative lightpipe instrument, a Hewlett Packard IRD, a CRYOLECT and a TRACER. Solution strengths of 500, 100, 50, 10, 5, 1, 0.5, 0.1, 0.05, and 0.01 mg/l were prepared in toluene. The chromatographic conditions were as follows:

- GC–FTIR   IRD, TRACER, CRYOLECT
- column   IRD and CRYOLECT, CP-SIL-8CB, 25 m × 0.32 mm i.d., d$_f$ 0.25 µm
                   TRACER CP-SIL-5CB, 25 m × 0.25 mm i.d., d$_f$ 0.25 µm
- temperatures
- *i.* injection   250°C
- *ii.* program   IRD and CRYOLECT 120°C for 0 min, 10°/min to 260°C, hold for 5 min; TRACER 80°C for 2 min, 25°/min to 120°C, 10°/min to 280°C, hold for 2 min

Chromatograms and spectra were recorded on the HP IRD using both split and splitless injection. The split ratio was determined to be 38:1. In each case, 1 microlitre of sample was injected. With splitless operation, the instrument located signals from the first three samples, corresponding to 500, 100, and

50 ng of caffeine injected. With split injections, signals could only be obtained from the 500 ng sample, corresponding to 13 ng of caffeine on column. This gives an effective detection limit for caffeine of approximately 10 ng on column. On the CRYOLECT, reconstructed chromatograms and FTIR spectra were obtained following split injection, with an effective splitter ratio of 25:1. With 1 microlitre injected, caffeine spectra could be obtained from samples corresponding to between 20 ng and 40 pg of caffeine on column. At the lowest concentration, the signal-to-noise ratio for the 1660 cm^{-1} carbonyl peak was 4.5:1, with a noise level of 0.0004 A (measured at 2000 cm^{-1}). By acquiring 2048 scans, an improved signal-to-noise ratio of 18:1 was obtained. Using both split and splitless injection on the TRACER, spectra were obtained from samples down to 35 pg on column. The peak absorbance is 0.0048 A, giving a signal-to-noise ratio of 32:1 with the noise level at 0.00015 from 512 scans at 4 cm^{-1} resolution.

The sample storage interfaces thus demonstrate an increase in sensitivity over a typical lightpipe instrument of some 2–3 orders of magnitude. The TRACER optics are more closely matched to the spot size and IR beam diameter, giving an increase in sensitivity over the CRYOLECT, though the narrower bands generally found in MI–FTIR spectra offset this somewhat.

4. GC–FTIR resolution

In principle, the chromatographic resolution obtained in GC–FTIR systems is only limited by the on-column separation. In practice, however, there is a trade-off between sensitivity and chromatographic resolution for each of the interfaces considered here.

A Grob test mixture has been used (11) to test the chromatographic capabilities of a lightpipe and sample storage GC–FTIR interfaces. The samples consisted of equal concentrations of each component in dichloromethane as follows: (1) octan-2-one, (2) octan-1-ol, (3) 2,6-dimethylphenol, (4) 2,4-dimethylaniline, (5) naphthalene, (6) tridecane, and (7) tetradecane. The chromatographic conditions were as follows:

- GC–FTIR   IRD, TRACER, CRYOLECT
- column   IRD and CRYOLECT, CP-SIL-8CB, 25 m × 0.32 mm i.d., d_f 0.25 μm
  TRACER, DB-5, 30 m × 0.25 mm i.d., d_f 0.25 μm
- split ratio   IRD, 82:1; TRACER and CRYOLECT, 25:1
- amount detected   IRD 12.2 ng TRACER and CRYOLECT 2.2 ng
- temperatures
- *i.* injection   280°C
- *ii.* program   50°C for 2 min. 20°/min to 280°C, hold for 5 min.

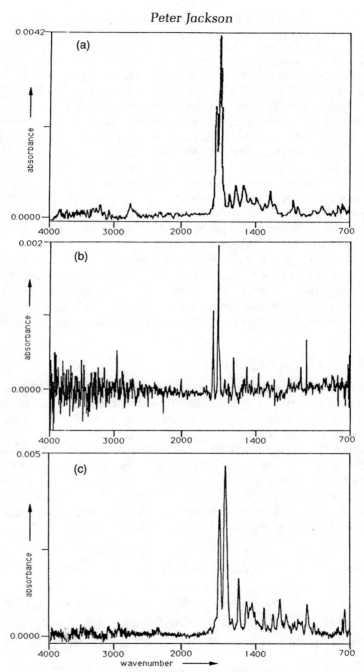

Figure 1. FTIR spectra obtained from caffeine concentration standards: (a) HP–IRD, 4 scans, 8 cm^{-1} resolution, 13 ng sample; (b) CRYOLECT, 2048 scans, 4 cm^{-1} resolution, 40 pg sample; (c) TRACER, 512 scans, 4 cm^{-1} resolution, 35 pg sample.

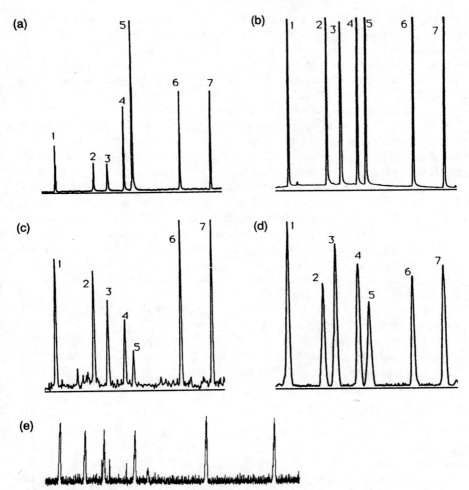

Figure 2. Chromatograms obtained from the Grob testmix: (a) GC–FID; (b) GC–MS TIC; (c) CRYOLECT GC–MI–FTIR; (d) TRACER GC–FTIR; (e) lightpipe GC–FTIR. The components are (1) octan-2-one, (2) octan-1-ol, (3) 2,6-dimethylphenol, (4) 2,4-dimethylaniline, (5) naphthalene, (6) tridecane, and (7) tetradecane.

All the interfaces used were capable of resolving components separated by GC–FID and GC–MS detection, though varying degrees of peak broadening were observed, dependent on the interface parameters used. *Figure 2* shows the lightpipe, TRACER, and CRYOLECT FTIR chromatograms as well as GC–FID and GC–MS (total ion chromatogram, TIC) chromatograms obtained from the test mix. Some variations in peak intensities between techniques are seen. This is, of course, due to the differing response factors between the respective detectors used.

If the flow rate through a lightpipe is too large or GC peaks are broad, sensitivity is degraded. On the other hand, if the flow rate is not sufficiently high and GC peaks are narrow, chromatographic resolution may be degraded, with more than one separated GC peak present in the lightpipe at any one time. The Bruker lightpipe system employs a make-up gas to compensate for low column flow rates. A detailed consideration of chromatographic requirements has been published elsewhere (12).

With both the low-temperature sample storage interfaces, the chromatographic resolution obtained is a complex relationship between the sample spot size, the disk or window velocity, the GC peak width, and separation and the IR beam width. As mentioned earlier, sensitivity must also be considered. Chromatographic resolution may be maintained by moving the cold surface more quickly, though this has the effect of spreading the sample more thinly, reducing sensitivity. On the CRYOLECT, a fixed disk velocity of 50 μm/sec is programmed. Recent work (10) has shown this to be too slow in many situations, and a computer system has been developed to allow variable disk speed programming. Other workers have used manual disk control to match disk speeds to particular chromatographic conditions. On the TRACER, the cold window may be programmed to travel at speeds appropriate to the separation; in addition, a ramp is often employed to compensate for peak broadening with increasing GC retention time.

The effects of varying the CRYOLECT cold disk rate is displayed in *Figure 3*. As the disk rate is increased from 40 to 400 μm/sec, the improvement in chromatographic resolution is apparent.

5. GC–FTIR spectra

Each interface type gives rise to spectra characteristic to the sample form and temperature. With lightpipe instruments, spectra are obtained in the vapour phase at, typically, 250 °C. Consequently, spectral bands are broadened for most molecules, with bandwidths typically of the order of 10–20 cm^{-1}. Libraries of vapour phase spectra are available, though it should be remembered that many of the components separated by GC would be solids or liquids at room temperature, and vapour phase reference spectra may not be readily available.

Spectra obtained on the TRACER solid sample storage interface are, as would be expected, characteristic of the solid state, and significant band broadening is found for hydrogen-bonded species. In addition, the sample deposition mechanism may also result in some crystallization, resulting in band splitting. The spectra are generally searchable against room-temperature FTIR databases, though the possible difference in sample form should be remembered: room temperature liquids and amorphous or waxy solids may show differences.

With matrix isolation, bands due to hydrogen-bonded functionalities may

A1: Combined gas chromatography

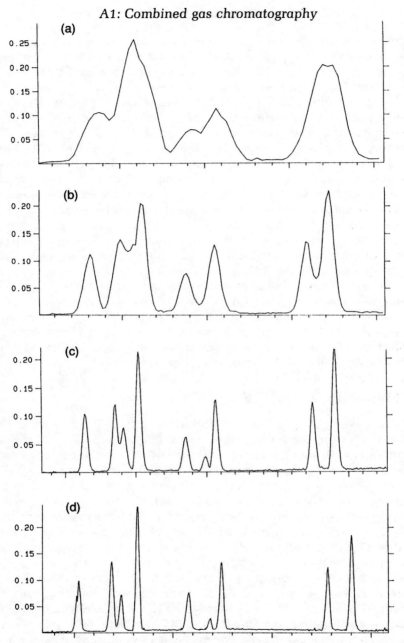

Figure 3. GC–MI–FTIR chromatograms obtained from a 9-component testmix recorded as a function of cryogenic disk rotation rate; (a) 40 μm/sec, 80 μm/sec, 200 μm/sec, and 400 μm/sec. The improvement in chromatographic resolution is apparent. The actual parameter values required to maintain optimum GC resolution may be different for each sample studied.

be dramatically narrowed, due to the elimination of intermolecular inter-actions, though some variation in band positions has been noted in experiments using different matrix gases, due to sample–matrix interactions. The above band sharpening also means that relative band peak-to-peak intensities may be significantly changed from those found in FTIR spectra recorded at room temperature. Low-temperature effects may also result in variations in band intensities. The sharp nature of MI–FTIR spectra makes library searching very selective, although the number of spectra in MI–FTIR libraries is limited (*c.* 5000). Peak-pick searching of room-temperature spectra has proven effective, though full-spectrum similarity searching is not likely to prove successful. However, as the sample concentration increases in the matrix, and sample molecules become larger, matrix effects are reduced, with spectra becoming more 'solid-like'.

As an illustration of the spectra obtained using sample storage, results from octan-2-one are shown in *Figure 4*. Features characteristic to the sample storage method are exhibited, including differences shown between spectra recorded at varying matrix:sample ratios using MI–FTIR. At the lower concentration (*Figure 4a*) the carbonyl signal is split, with components at 1715 and 1732 cm^{-1}. At the higher concentration (*Figure 4b*) only a single, broader carbonyl signal is found. The solid-phase spectrum obtained from octan-2-one on the Tracer (*Figure 4c*) shows evidence of crystallization, indicated by peak splittings (especially the CH-rock doublet at 720 cm^{-1}).

6. Quantitative GC–FTIR

There are many examples of the application of GC–FTIR analysis in the qualitative identification of mixture components. The advent of the high-sensitivity sample storage methods has led workers to investigate the quantitative aspects of GC–FTIR. In particular, the capabilities of GC–MI–FTIR have been explored. In early work, it was found that experimental precision was poor (>20% RSD). Childers and co-workers (8) have investigated the effects of the experimental parameters on the quantitative determination of semivolatile organic compounds (SVOC) in air sample extracts. The effects of the deposition tip position, disk sampling methods and experiment timing were studied using the xylene isomers and *p*-xylene-d$_{10}$ as test compounds.

Under the GC conditions used, the *m*- and *p*-xylene isomers were not resolved. This made separate analysis of the isomer concentrations impossible by both GC–FID and GC–MS detection. FTIR spectroscopy is, on the other hand, extremely isomer specific. In the FTIR spectrum of the co-eluting GC peak, separate bands are seen due to the specific *m*- and *p*-isomers. This allowed the quantitation of these separate isomers despite co-elution.

With careful experimental operation, precisions of <2% RSD were found for repeated analyses of single depositions. For multiple depositions, this increased to <4% RSD. For the three xylene isomers, samples were analysed

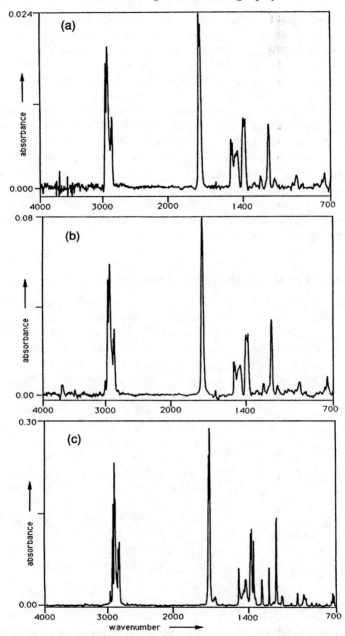

Figure 4. FTIR spectra obtained from octan-2-one: (a) CRYOLECT, 32 scans, 4 cm^{-1} resolution, 2.2. ng at a low concentration in matrix; (b) CRYOLECT, 32 scans, 4 cm^{-1} resolution, 6.6. ng at a high concentration in matrix; (c) TRACER, 128 scans, 4 cm^{-1} resolution, 2 ng sample.

between 0.87 and 86.9 ng/μl. For concentrations in excess of 52.1 ng/μl, a non-linear response was found, probably due to the increase in sample spot size or loss of matrix isolation effectiveness. For routine measurements (128 scans averaged), the estimated detection limits for the xylene isomers was 1–2 ng/μl. Spectra could be recorded at concentrations below this, though extensive signal averaging (>1000 scans) was required.

The detection limits and precision found compares well to those found by other GC detection methods. Other workers (13) have found similar precision in the detection of 2,3,7,8-TCDD in fish extracts by GC–MI–FTIR down to a limit below 0.2 ng/μl, though extensive signal averaging was required.

Despite having the potential for excellent quantitative measurement capability, the operator intensive nature, complexity of the instrumentation, and the difficulty in automating measurements makes GC–MI–FTIR unsuitable for routine quantitative analysis.

7. Multiple detector systems

The combination of GC with multiple detectors should also be considered. As mentioned at the start of this chapter, increased certainty of identification is increasingly required. If this can be achieved in a single experiment, valuable time may be saved. In addition, where more expensive techniques, such as MS–MS or sample storage FTIR is employed, a link to a cheaper more routine analysis may be required in order to simplify subsequent repeat analyses. Multiple detection can be achieved, dependent on the detectors concerned, by either series connection or by post-column splitting.

The most usual form of splitting would be to take off a small part of the GC eluent to a parallel FID from a MS or lightpipe FTIR detector. With the lightpipe, however, the non-destructive nature of the detector allows serial connection, either to a FID or a MS detector. In the most complex commercially available instrument, one CRYOLECT option offers a three-way split to a FID (20%), a MS system (40%), and a MI–FTIR interface (40%). This is achieved by use of a three-hole ferrule, with variable transfer lines. The split ratio is determined by the diameter and lengths of the transfer lines used. Sensitivity is obviously degraded by splitting of the GC eluent in this fashion, but only one experiment need be performed instead of three.

8. GC–FTIR applications

Rather than simply list the many examples of the application of GC–FTIR in analytical science, three examples have been chosen to illustrate typical application areas and to show the type of information available from derived FTIR spectra.

8.1 Industrial applications: commercial alcohols

A commercial sample of isoheptanol (a mixture of branched chain primary C_7 alcohols) was analysed as part of a technique evaluation study (11), though this sample has been analysed earlier for industrial reasons. This sample also formed part of an earlier CRYOLECT study by workers at the University of California, Riverside (14) where several specific isomeric structures were proposed.

The sample was studied as a 0.1% solution in dichloromethane. The chromatographic conditions were as follows:

- column   CP-WAX-52CB, 25 m × 0.25 mm i.d., d_f 0.2 μm
- injection mode   split
- temperatures
- *i.* injection   250°C
- *ii.* program   70°C, 0.5°/min to 120°C

Split injection was used. The chromatography required to effect good separation of the 8 components found in this commercial mixture is demanding, requiring optimum performance to resolve peaks 3 and 4, and peaks 5, 6, and 7. In the earlier CRYOLECT work on this sample, the chromatographic resolution was not adequate to resolve these components, and the operation of the cryogenic disk interface was at low speed, resulting in further degradation of the chromatographic resolution. Spectral manipulation methods (e.g. subtraction) were then required to obtain spectra from each individual component. Even so, peaks 3 and 4 were not resolved. *Figure 5* shows the chromatograms obtained (FID, MS, and FTIR) under optimized conditions. The MI–FTIR interface was operated with five times the usual disk rotation rate. Under these optimized conditions, the interface was capable of reproducing the GC chromatographic resolution, and the resolution obtained was equivalent to that obtained on a separate optimized GC–FID instrument.

Significant information can be deduced from the combined MS and FTIR data. In this case, the CRYOLECT data was also combined with ^{13}C NMR data from the mixture run as a whole. The NMR had shown only seven C_1 resonances, whereas GC separated eight components, meaning that spectral overlap must be present in the NMR data. The NMR, FTIR, and MS data were then considered in total, allowing specific isomeric structures to be assigned to each peak in the GC chromatogram. This was done by considering model NMR spectra, MS fragmentation, and FTIR splittings/CH_3 : CH_2 ratios (chain branching ratios). This identified the 8 major components in the mixture to be the following primary alcohols: (1) 2,4-dimethylpentanol, (2) 3,4-dimethylpentanol, (3) 2-ethylpentanol, (4) 2-methylhexanol, (5) 5-methylhexanol, (6) 3-methylhexanol, (7) 2-ethyl-3-methylbutanol, and (8) heptanol. Repeat analyses could then be performed using GC-FID only.

Peter Jackson

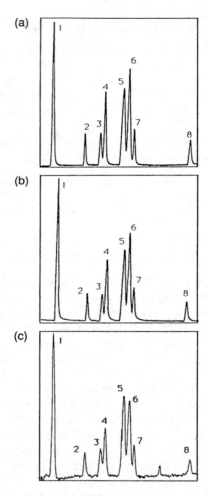

Figure 5. Chromatograms obtained from the isoheptanol sample: (a) CRYOLECT GC–FID; (b) CRYOLECT GC–MS TIC; (c) CRYOLECT GC–MI–FTIR. The components are (1) 2,4-dimethylpentanol, (2) 3,4-dimethylpentanol, (3) 2-ethylpentanol, (4) 2-methylhexanol, (5) 5-methylhexanol, (6) 3-methylhexanol, (7) 2-ethyl-3-methylbutanol, (8) heptanol.

8.2 Advanced sampling techniques

In conventional GC analysis, techniques such as thermal desorption, headspace analysis, purge-and-trap analysis, and pyrolysis are increasing the diversity of applications, and are becoming more important, especially for routine, reproducible quantitative analyses. In many respects, such sample inlet devices are unsuitable for use with GC–FTIR systems. Sample amounts may be too low for lightpipe detection, particularly with headspace analysis, and high

A1: Combined gas chromatography

water levels may lead to significant spectral distortion with the sample storage GC–FTIR interfaces. This can, of course, be avoided in GC–MS analysis by scanning at higher m/z values or by selected ion monitoring. Nevertheless, where unidentified components are found, FTIR may be the ideal complementary identification method.

Analytical pyrolysis has many applications in industry, research, geoscience, and forensic science. Two main designs are available: inductively heated (Curie point) and resistively heated. These systems are easily coupled to the split/splitless injector port of most GC systems. Pyrolysis (Py–GC) can provide detailed information about the composition, microstructure and degradation of the sample, particularly high-molecular weight polymers.

Figure 6 shows the Py–GC–FTIR results from an industrial copolymer. The pyrolysis conditions were optimized to give reproducible monomer evolution. Because high molecular weight products can cause contamination of the pyrolysis interface, the injector port and the GC column, and low molecular weight gases give little compositional information, temperature optimization is necessary for each material.

With three-dimensional cross-linked polymers, e.g. epoxy and phenolic resins, characteristic and reproducible information can be obtained. In addition, residual solvents from the original production process may also be detected and analysed. The Py–GC–FTIR analyses of natural (NR) and styrene-butadiene (SBR) rubbers are relatively simple. The NR Py-chromatogram is dominated by isoprene, whereas that of SBR has several characteristic compounds, namely butenes, benzene, toluene, and styrene. More complex materials, such as filled, cured, and finished products have also been analysed, though the results are more complex and are not widely reported due to commercial reasons.

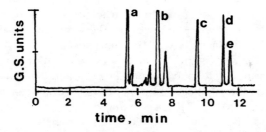

Figure 6. Py–HRGC–FTIR analysis of a MMA–BA–BMA–HEMA–MAAC copolymer (Gram–Schmidt chromatogram). (a) methacrylic acid methyl ester (MMA); (b) methacrylic acid (MAAC); (c) acrylic acid butyl ester (BA); (d) methacrylic acid butyl acid (BMA); (e) methacrylic acid hydroxyethyl ester (HEMA). Pyrolyser: PE/CDS pyroprobe 190, quartz sample tube, 600°C for 2 sec. Column: 60 m × 0.32 mm fused silica DB-1, d_f 1 μm. Temperatures: from 80°C to 180°C at 8°/min. Injection port–pyrolyser interface, light pipe, and transfer lines at 180°C. (Reproduced with kind permission from reference 12, © 1987 Alfred Huthig Publishers.)

8.3 Pesticides analysis

Many applications of the identification of pesticides have been published, including various studies on tetrachlorodibenzo-*p*-dioxins (TCDDs), (15), tetrachlorodibenzofurans (15), pesticides in groundwater (16), and hexa-chlorocyclohexane isomers (17). FTIR analysis is particularly effective at exact isomer determination. In unknown samples, the combination of GC–MS and GC–FTIR is particularly effective in the identification of pesticides.

A good illustration of the advantages of multiple detection on a single instrument is provided by the following example. A pesticides standard containing the following major components (plus minor impurities) in hexane was obtained: delta-HCH (8.3 mg/l); heptachlor (4.2 mg/l); aldrin (7.1 mg/l); telodrin (7.7 mg/l); isodrin (8.6 mg/l); alpha-endosulfan (9.6 mg/l); dieldrin (12.1 mg/l); endrin (10.2 mg/l); beta-endosulfan (7.4 mg/l), and was analysed using the following chromatographic conditions:

- column   CP-SIL-8CB, 25 m × 0.32 mm i.d., d_f 0.25 μm
- injection mode   split 10:1
- temperatures
- *i.* injection   220 °C
- *ii.* program   80 °C for 2 min, 25°/min to 120 °C, 10°/min to 280 °C, hold for 15 min.

Following separation, the column eluent was split three ways to a FID, MS, and MI–FTIR detector.

As shown in *Figure 7*, there are in fact twelve peaks detected in both the GC–FID and GC–MS traces obtained from the sample. The GC–FTIR chromatograms obtained is also shown. As can be seen, the sensitivity obtained on the FTIR instrument using the fast scan, Gram–Schmidt reconstruction approach does not allow the clear location of all twelve components found by the other detectors. However, the presence of the minor components could be deduced from the parallel detectors, allowing high-resolution, high-sensitivity spectra to be recorded at the correct retention time. In this fashion, FTIR spectra could be obtained from every peak except peak 2, which was the smallest component and had a weak response.

Figure 8 shows the MI–FTIR spectra and the MS results from peaks 7 and 10, α- and β-endosulfan. The two endosulfan isomers give rise to essentially identical EI⁺ MS data, making unambiguous identification impossible by GC–MS alone. In a typical situation, where an unknown sample is being analysed, standard materials, methodologies, and calibrated retention times may not be available. In such a case, the GC–MS data cannot provide an unambiguous solution. The GC–MS data is, though, capable of identifying the correct molecular structure, and in this case, the FTIR data can identify the exact isomer concerned by matching against library spectra. The avail-

A1: *Combined gas chromatography*

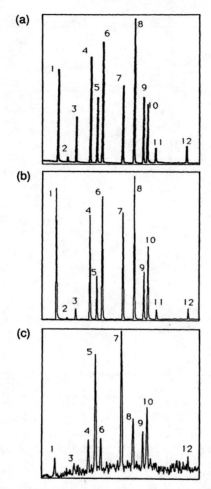

Figure 7. Chromatograms obtained from a pesticides standard sample: (a) CRYOLECT GC–FID; (b) CRYOLECT GC–MS TIC; (c) CRYOLECT GC–MI–FTIR. The components are (1) delta-HCH, (2) impurity, (3) heptachlor, (4) aldrin, (5) telodrin, (6) isodrin, (7) α-endosulfan, (8) dieldrin, (9) endrin, (10) β-endosulfan, (11) impurity, (12) impurity.

ability of the FID means that subsequent repeat analyses and quantitative work-up may be performed on a more simple GC instrument with improved routine capabilities and at lower cost.

9. Summary

GC–FTIR has been shown to be an excellent complementary method to GC–MS analysis. Whereas MS is particularly suited to molecular weight

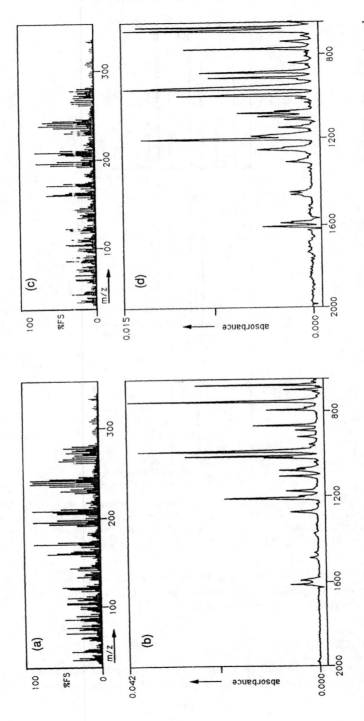

Figure 8. Spectra obtained from the two endosulfan isomers. α-endosulfan: (a) CRYOLECT MS; (b) CRYOLECT FTIR, 256 scans, 4 cm⁻¹ resolution; (c) CRYOLECT MS; (d) CRYOLECT FTIR, 256 scans, 4 cm⁻¹ resolution. β-endosulfan;

determination and evaluation of molecular structure, FTIR spectra are extremely isomer specific, and can give detailed information on functional groups.

The introduction of the low-temperature sample storage GC–FTIR interfaces extends the range of analyses present, now overlapping significantly with the detection limits achievable by bench-top GC and GC–MS. These interfaces are capable of maintaining the GC resolution and, with careful experimental technique, can give excellent quantitative reproducibility. The combination of multiple detectors on single instruments is expected to increase, allowing complementary spectral information to be obtained from separated components at the same time, delivering an extremely powerful identification tool to analysts, significantly increasing confidence in the assignment of unknown species.

Acknowledgements

The author is grateful to Dr T. Visser for the preparation of some of the figures and for permission to use some of his results, and to G. Dent, D. Carter, J. Chalmers, and D. Schofield for their contribution to some of the results described. The help of Dr P. Baugh is also acknowledged, particularly providing information on pyrolysis GC–FTIR applications.

References

1. Fujimoto, C. and Jinno, K. (1992). *Anal. Chem.*, **64**, 476A.
2. Reedy, G. T., Bourne, S., and Cunningham, P. T. (1979). *Anal. Chem.*, **51**, 1535.
3. Bourne, S., Reedy, G. T., and Cunningham, P. T. (1979). *J. Chromatogr. Sci.*, **17**, 460.
4. Reedy, G. T., Ettinger, D. G., Schneider, J. F., and Bourne, S. (1985). *Anal. Chem.*, **57**, 1602.
5. Bourne, S., Reedy, G. T., Coffey, P. J., and Mattson, D. (1984). *Am. Lab.*, **16**, 90.
6. Haeffner, A. M., Norton, K. L., Griffiths, P. R., Bourne, S., and Curbelo, R. (1988). *Anal. Chem.*, **60**, 2441.
7. Bourne, S., Haeffner, A. M., Norton, K. L., and Griffiths, P. R. (1990). *Anal. Chem.*, **62**, 2448.
8. Childers, J. W., Wilson, K. N., and Barbour, R. K. (1992). *Anal. Chem.*, **64**, 292.
9. Bourne, S. and Croasmun, W. R. (1988). *Anal. Chem.*, **60**, 2172.
10. Klawun, C., Sasaki, T. A., Wilkins, C. L., Carter, D., Dent, G., Jackson, P., and Chalmers, J. M., *Appl. Spectrosc.* (in press).
11. Jackson, P., Dent, G., Carter, D., Schofield, D. J., Chalmers, J. M., Visser, T., and Vredenbregt, M. (1993). *J. High Res. Chromatogr.* (in press).
12. Herres, W. (1987). *Capillary gas chromatography–Fourier transform infrared spectroscopy. Theory and applications.* Huthig Verlag, Heidelberg.
13. Mossoba, M. M., Niemann, R. A., and Chen, J. T. (1989). *Anal. Chem.*, **61**, 1678.

14. Baumeister, E. R., Zhang, L., and Wilkins, C. L. (1991). *J. Chromatogr. Sci.*, **29**, 331.
15. Brasch, J. W. (1987). Proc. of the 1987 EPA/APCA Symposium on Measurement of Toxic and Related Pollutants. Air Pollution Control Association, Pittsburgh.
16. Holloway, T. T., Fairless, B. J., Friedline, C. E., Kimball, H. E., Wurrey, C. J., Jonooby, L. A., and Palmer, H. G. (1989). *Appl. Spectrosc.*, **43**, 1344.
17. Visser, T. and Vredenbregt, M. J. (1990). *Vibrational Spectrosc.*, **1**, 205.

A2

Suppliers of specialist items

Advanced Separation Technology Inc., Whippany, NJ 07981, USA.

Aldrich Chemical Co. Ltd., The Old Brickyard, New Road, Gillingham SP8 4JL, UK.

Alltech Associates, Carnforth LA5 9XP, Lancashire, UK; 2051, Waukegan Road, Deerfield, IL 60015, USA.

Analytichem International Inc., 24201 Frampton Avenue, Harbor City, CA 90710, USA.

BDH Chemicals Ltd., Broom Road, Parkstone, Poole, Dorset BH12 4NN, UK.

BIO-RAD Laboratories GmbH, Kaiserwerther Strasse 207, D-4000 Dusseldorf 30, Germany.

Camlab, Nuffield Road, Cambridge CB4 1TH, UK.

Chromatography Services Ltd., PO Box 862, Deeside, Clwyd CH5 2WA, UK.

Chrompack International B.V., PO Box 8033, Kuipersweg 6, 4330 EA Middelburg, The Netherlands; 4, Indescon Court, Millharbour, London E14 9TN, UK.

Clean-Screen, Worldwide Monitoring Corporation, Morrisville, USA.

Digilab Corporation, see **Bio-Rad** above.

Finnigan MAT plc, Paradise, Hemel Hempstead HP2 4TG, Hertfordshire, UK.

Fisons plc, 41/45, Gatwick Road, Crawley, Sussex RH10 2UL, UK.

Fisons Instruments, VG Analytical Group, Floats Road, Wythenshawe, Manchester, UK; VG Mass Lab, Crewe Road, Wythenshawe, Manchester M23 9BE, UK.

Fluka, Peakdale Road, Glossop SK13 9XE, Derbyshire, UK.

Fluorochem, Wesley Street, Old Glossop SK13 9RY, Derbyshire, UK.

Hewlett-Packard Analytical Group, PO Box 10301, Palo Alto, CA 94303, USA; Cain Road, Bracknell RG12 HIN, UK.

Jones Chromatography Ltd., Colliery Road, Llanbradach, Mid-Glamorgan CF8 3QQ, UK.

J & W Scientific, 91, Blue Ravine Road, Folsom, CA 95630, USA.

Kontron AG, Bernstrasse Sud 169, CH-8010 Zurich, Switzerland.

Lancaster Synthesis, East Gate, White Lund, Morecambe LA3 3DY, Lancashire, UK.

Macherey Nagel, PO Box 307, Werkstrasse 6–8, D-5160 Dueren, Germany.

Mattson Instruments Ltd., see **Philips Scientific**.

E. Merck, Frankfurter Strasse 250, Postfach 4119, D-6100 Darmstadt, Germany.

Millipore Corporation, 80 Ashby Road, Bedford, MA 01730, USA.

Pharmacia Fine Chemicals AB, PO Box 175, S-75104 Uppsala, Sweden; Pharmacia House, Mid Summer Boulevard, Milton Keynes MK9 3HP, UK.

Phase Separations Inc., River View Plaza, 16, River Street, Norwalk, CT 06850, UK; Deeside Industrial Estate, Deeside, Clwyd CH3 2NU, UK.

Philips Scientific, York Street, Cambridge CB1 2PX, UK.

Pierce Chemical Co., PO Box 117, Rockford IL 61105, USA.

Sarstedt Ltd., 68, Boston Road, Beaumont Leys LE4 1AW, Leicestershire, UK.

SGE International Ltd., 7, Argent Place, Ringwood, Victoria 3134, Australia; SGE (UK) Ltd., Kiln Farm, Potters Lane, Milton Keynes MK11 3LA, UK.

Sigma Chemical Company, PO Box 14508, St Louis, MO 63178, USA.

Supelco Inc., Supelco Park, Bellefonte, PA 16823, USA.

Technicol, Brook Street, Stockport SK1 3HS, Cheshire, UK.

Varian Associates Inc., Instrument Group, 611, Hansen Way, Palo Alto, CA 94303, USA.

Index

Index

Index

419

Index

Index

Index

Printed in the United Kingdom
by Lightning Source UK Ltd.
506